T0213102

Automotive Product Development

Automotive Product Development
Development
A Systems Engineering Implementation

by

Vivek D. Bhise

CRC Press
Taylor & Francis Group
Boca Raton London New York

CRC Press is an imprint of the
Taylor & Francis Group, an **informa** business

CRC Press
Taylor & Francis Group
6000 Broken Sound Parkway NW, Suite 300
Boca Raton, FL 33487-2742

First issued in paperback 2019

ISBN-13: 978-1-4987-0681-0 (hbk)
ISBN-13: 978-0-367-87185-7 (pbk)

Library of Congress Cataloging-in-Publication Data

Names: Bhise, Vivek D. (Vivek Dattatray), 1944- author.
Title: Automotive product development : a systems engineering implementation
/ Vivek D. Bhise.
Description: Boca Raton : Taylor & Francis, a CRC title, part of the Taylor &
Francis imprint, a member of the Taylor & Francis Group, the academic
division of T&F Informa, plc [2017] | Includes bibliographical references
and index.
Identifiers: LCCN 2016037644| ISBN 9781498706810 (hardback : alk. paper) |
ISBN 9781315119502 (ebook : alk. paper)
Subjects: LCSH: Automobiles--Design and construction. |
Automobiles--Technological innovations. | Systems engineering.
Classification: LCC TL240 .B54 2017 | DDC 629.2068/5--dc23
LC record available at https://lccn.loc.gov/2016037644

Visit the Taylor & Francis Web site at
http://www.taylorandfrancis.com

and the CRC Press Web site at
http://www.crcpress.com

Contents

SECTION II   Tools Used in the Automotive Design Process

SECTION III Applications of Tools: Examples and Illustrations

Preface

The development of a new automotive product requires an understanding of the integration of knowledge from a number of disciplines. In this book, I have provided material that was generated and used in teaching the automotive product development process to graduate students in Automotive Engineering over many years at the University of Michigan-Dearborn.

The material provides the basic background, principles, techniques, and steps that I found to be useful in understanding the complex and coordinated activities that need to be undertaken to ensure successful development of the "right vehicle" that customers will enjoy driving. Proper implementation of the process should make the product development team members feel very proud of their accomplishments. It should enhance the reputation of the company for creating exciting new vehicles and thus, lead the company to achieve financial success beyond its imagination in terms of revenues, profits, and return on investments.

The formula for creating successful automotive products lies in the creation of a well-coordinated product development process, using the right tools and techniques, a dedicated team of highly motivated multidisciplinary professionals, and very supportive senior management.

This book is about understanding "the big picture" of how automotive products need to be developed with the sole purpose of satisfying their customers. The book resulted from my deep desire to understand how automotive products are developed, to understand the many challenges facing the auto industry, to study the methods currently used in designing automotive products, and to make our future automotive engineers realize that their main job is to satisfy the customers who use their products.

We teach our engineers to be proficient in applying specialized techniques in narrowly specialized areas such as structural analysis, vehicle dynamics, powertrain efficiency analysis, aerodynamic drag reduction, and electrical architecture design. But they need to realize that the customer buys the "whole" car, not just a collection of systems and components that they helped design, such as four wheels, a steering wheel, pedals, seats, vehicle body, lamps, wiring harnesses, and fuel tanks. All vehicle systems and their subsystems and components must "work together" to provide the "desired" feel to the customer—so that he or she is either "completely" or "very" satisfied with the vehicle.

Engineers working in the automotive industry may claim that they currently have the necessary knowledge in areas such as system design specifications, design tools, verification test procedures, test equipment, and subsequent data analysis methods. However, many cars and trucks currently satisfy only about 60%–80% of their customers; that is, the vehicles do not achieve the high scores, such as over 90%, desired by the customers and the senior management of the automobile companies. This gap between the high levels of customer satisfaction "desired" by the customers and the management and those "actually achieved" by the current automotive products in various market research surveys is largely because of failure to understand customer needs, to translate these needs into design specifications, and to confirm that the designed products are indeed the right products for the customers.

The objective of the book is to provide the necessary background for future engineering graduates and practicing engineers in the industry to ensure that they understand the automotive product development process, the issues challenging the industry, and the applications of various approaches and tools available to conduct the necessary steps in design, analysis, and evaluation to create products that will satisfy their customers.

This book is divided into three parts. The first part provides an in-depth understanding of the various phases of the product development process and the steps involved in implementing the systems engineering process. Strict and thorough implementation of the systems engineering process is a prerequisite for achieving success in any automotive product program. Otherwise, the vehicle development program may exceed its budget or time schedule, and/or the designed product may fail to meet its customer satisfaction target. The second part of the book covers many important tools and methods used in the vehicle development process. The third part provides many examples and case studies generated during the past several years of my teaching graduate courses in the Automotive Systems Engineering program at the University of Michigan-Dearborn.

The auto industry is facing fierce competition and unending pressure to reduce program timings and costs. This results in further pressure to minimize or even to eliminate many of the systems engineering tasks, and thus, endanger the successful completion of vehicle programs. The complexity of the vehicle programs is also increasing due to rapid advances in technologies, the large number of variables considered in many analyses, and our inability to measure a number of key variables, which still rely on subjective judgments. Subjective measures are used in evaluations of many vehicle attributes, such as styling, drivability, performance feel, ergonomics, interior spaciousness, and quality. It is hoped that this book will help in addressing many of the challenging issues facing the industry.

WEBSITE MATERIALS

The following files are in the Download section of this book's web page on the CRC Press website (http://www.crcpress.com/product/isbn/97814987068100).

A. Computer programs and models
   1. Automotive Product Development Chart with Present Value Calculations
   2. Program for Cost Flow by Months
   3. Program for Cost Flow by Quarters
B. Slides for Chapters 1 to 25

Acknowledgments

This book is a culmination of my education, experience, and interactions with many individuals from the automotive industry, academia, and government agencies. While it is impossible for me to thank all the individuals who influenced my career and thinking, I must acknowledge the contributions of the following individuals.

My greatest thanks go to the late Professor Thomas H. Rockwell of the Ohio State University. Tom got me interested in human factors engineering and driving research. He was my advisor and mentor during my doctoral program. I learned many skills on how to conduct research studies and analyze data, and more importantly, he introduced me to the technical committees of the Transportation Research Board and the Society of Automotive Engineers, Inc.

I would like to thank the late Lyman Forbes, Dave Turner, the late Eulie Brayboy, and Bob Himes from Ford Motor Company. Lyman Forbes, manager of the Human Factors Engineering and Ergonomics Department at the Ford Motor Company in Dearborn, Michigan, spent hours with me discussing various approaches and methods to conduct research studies on crash-avoidance research and development of motor vehicle safety standards. Dave Turner, director of the Advanced Design Studios in the Ford's Design Staff, got the Human Factors Engineering and Ergonomics department firmly anchored in the automotive design process. He also helped establish a Human Factors Group within Ford of Europe when he was the director of Ford's European Design Centre. Eulie Brayboy, chief engineer, Design Engineering in the Corporate Design, always provided support in implementing human factors inputs into the automotive design process. Bob Himes, chief engineer of the Advanced Vehicle Engineering staff, helped in incorporating ergonomics and vehicle packaging as a vehicle attribute in systems engineering implementation in the vehicle development process.

The University of Michigan-Dearborn campus provided me with unique opportunities to develop and teach various courses. Our Automotive Systems Engineering and Engineering Management programs allowed me to interact with hundreds of graduate students, who in turn implemented many of the techniques taught in our graduate programs when solving problems within many other automotive original equipment manufacturers and supplier companies. I would to thank Professors Pankaj Mallick and Armen Zakarian for giving me opportunities to develop and teach many courses in the Automotive Systems Engineering and Industrial and Manufacturing Systems Engineering programs. Roger Schulze, director of our Institute for Advanced Vehicle Systems, got me interested in working on a number of multidisciplinary programs in vehicle design. Together, we developed a number of vehicle concepts, such as a low mass vehicle, a new Model "T" concept for Ford's 100th anniversary, and a reconfigurable electric vehicle. We also developed a number of design projects by creating teams of our engineering students with students from the Product Design and Transportation Design department from the College for Creative Studies in Detroit, Michigan. I must also thank my students for working on a number of research projects—developing test setups, recruiting subjects, and collecting and analyzing data—over many years.

Over the past 40-plus years, I was also fortunate to meet, and discuss many automotive design issues with, members of many committees of the Society of Automotive Engineers, Inc., the Motor Vehicle Manufacturers Association, the Transportation Research Board, and the Human Factors and Ergonomics Society.

I would like to also thank Cindy Carelli from CRC Press—a Taylor & Francis Company—for encouragement in preparing the proposal for this book, and her production group for turning the manuscript into this book.

Finally, I want thank my wife, Rekha, for her constant encouragement and her patience while I spent many hours working on my computers, writing the manuscript and creating figures included in this book.

Vivek D. Bhise
Ann Arbor, Michigan

Author

Vivek D. Bhise is currently visiting professor/Lecturers' Employee Organization lecturer and professor in post-retirement of industrial and manufacturing systems engineering at the University of Michigan-Dearborn. He received his B.Tech. in Mechanical Engineering (1965) from the Indian Institute of Technology, Bombay, India, his M.S. in Industrial Engineering (1966) from the University of California, Berkeley, California, and a PhD in Industrial and Systems Engineering (1971) from the Ohio State University, Columbus, Ohio.

During 1973–2001, he held a number of management and research positions at the Ford Motor Company in Dearborn, Michigan. He was the manager of Consumer Ergonomics Strategy and Technology within the Corporate Quality Office, and the manager of Human Factors Engineering and Ergonomics in the Corporate Design of the Ford Motor Company, where he was responsible for the ergonomics attribute in the design of car and truck products.

Dr. Bhise is the author of recent books entitled *Ergonomics in the Automotive Design Process* (ISBN: 978-1-4398-4210-2. Boca Raton, FL: CRC Press, 2012) and *Designing Complex Products with Systems Engineering Processes and Techniques* (ISBN: 978-1-4665-0703-6. Boca Raton, FL: CRC Press, 2014.)

Dr. Bhise has taught graduate courses in Vehicle Ergonomics, Vehicle Package Engineering, Automotive Systems Engineering, Management of Product and Process Design, Work Methods and Industrial Ergonomics, Human Factors Engineering, Total Quality Management and Six Sigma, Quantitative Methods in Quality Engineering, Energy Evaluation, Risk Analysis and Optimization, Product Design and Evaluations, Safety Engineering, Computer-Aided Product Design and Manufacturing, and Statistics and Probability Theory over the past 36 years (1980–2001 as an adjunct professor, 2001–2009 as a professor, and 2009–present as a visiting professor in post-retirement) at the University of Michigan-Dearborn. He also worked on a number of research projects in human factors with the late Professor Thomas Rockwell at the Driving Research Laboratory at the Ohio State University (1968–1973).

His publications include over 100 technical papers on the design and evaluation of automotive interiors, parametric modeling of vehicle packaging, vehicle lighting systems, field of view from vehicles, and modeling of human performance in different driver/user tasks.

Dr. Bhise has also served as an expert witness on cases involving product safety, patent infringement, and highway safety.

He received the Human Factors Society's A. R. Lauer Award for Outstanding Contributions to the Understanding of Driver Behavior in 1987. He has served on a number of committees of the Society of Automotive Engineers, the Transportation Research Board of the National Academies, and the Human Factors and Ergonomics Society.

Section I

Automotive Product Development Process

1 Introduction
Automotive Product Development

INTRODUCTION

Complex Product, Many Inputs, Many Designers and Engineers

Designing and producing an automotive product is a horrendously complicated undertaking. The automotive product itself is very complex. It involves many systems: body system, powertrain system, suspension system, electrical system, climate control system, braking system, steering system, fuel system, and so on. All the systems must work together under all possible combinations of road, traffic, and weather conditions to satisfy drivers and users with varied characteristics, capabilities, and limitations. The automotive product development (PD) process requires many resources over several years and includes many intricate, coordinated, and costly design, evaluation, production, and assembly processes. The complex automotive product must also meet hundreds of requirements to satisfy customers, applicable government regulations, and the goals and needs of company management.

Developing a new automotive product requires the efficient execution of a number of processes, and the implementation of systems engineering is essential to coordinate varied technical and company management needs. The proper implementation of systems engineering ensures that the right product is developed within the planned timing schedule while avoiding costly budget overruns. To understand the complexity in the PD process, we will begin this chapter with a clear explanation of processes, systems, and systems engineering and then proceed with the details of the automotive PD process.

BASIC DEFINITIONS OF PROCESS, SYSTEM, AND SYSTEMS ENGINEERING

Process

A process is where the "work gets done." A process generally consists of a series of steps, tasks, or operations that are performed by people (i.e., human operators) and/or machines (e.g., robots, computers, or automated equipment) using a number of inputs (e.g., information, raw materials, energy sources). People may also use one or more tools (e.g., hand tools, power tools, or software applications) in performing any of the tasks. The process can be studied and also defined by following a component (e.g., a part, an assembly, a transaction, a tracking paper, a drawing, a computer-aided design [CAD] model), or a person (e.g., one who moves from a workstation to other workstations and performs one or more tasks at each workstation) through a series of steps or tasks. The beginning and ending points of each process must be clearly defined. The purpose of the process, that is, the reason for the creation of the process, and its function, that is, what work is performed in the process, must be also clearly defined and documented.

To create (i.e., to design and produce) a product (e.g., a vehicle), many processes are required (e.g., the customer needs determination process, the vehicle concept development process, the detailed engineering process, the systems verification process, the production tools development process, and the vehicle assembly process).

SYSTEM

A system consists of a set of components (or elements) that work together to perform one or more functions. The components of a system generally consist of people, hardware (e.g., parts, tools, machines, computers, and facilities), or software (i.e., codes, instructions, programs, databases) and the environment within which it operates. The system also requires operating procedures (or methods) and organization policies (e.g., documents with goals, requirements, and rules) to implement its processes and get its work done. The system also works under a specified range of environmental and situational conditions (e.g., temperature and humidity conditions, vibrations, magnetic fields, power/traffic flow patterns). The system must be clearly defined in terms of its purpose, functions, and performance capability (i.e., abilities to perform or produce output at specified level in a specified operating environment).

Some definitions of a system are

1. A system is a set of functional elements organized to satisfy specified objectives. The elements include hardware, software, people, facilities, and data.
2. A system is a set of interrelated components working together toward some common objective(s) or purpose(s) (Blanchard and Fabrycky, 2011).
3. A system is a set of different elements so connected or related as to perform a unique function not performable by the elements alone (Rechtin, 1991).
4. A system is a set of objects with relationships between the objects and between their attributes (Hall, 1962).

The set of components has the following properties (Blanchard and Fabrycky, 2011):

1. Each component has an effect on the whole system.
2. Each component depends on other components.
3. The components cannot be divided into independent subsystems.

SYSTEMS ENGINEERING (SE)

Systems engineering (SE) is a multidisciplinary engineering decision-making process involved in designing and using systems and products throughout their life cycle. The implementation of SE is very beneficial, as without it, the likelihood of creating the "right system or product" that the customers really want (in terms of its attributes, such as performance, safety, styling, and comfort) within the targeted timings and costs can be substantially reduced (see INCOSE [2006], NASA [2007], and Kmarani and Azimi [2011] for more information on SE).

Systems Approach

The word "systems" in "systems engineering" is used to cover the following aspects of different systems in an automotive product:

1. An automobile product is a system containing a number of other systems (e.g., body system, powertrain system, chassis system, and electrical system).
2. Thus, the design of the whole automobile will involve designing all the systems within the automobile such that the systems work together (i.e., the systems are interfaced or connected with other systems, and each system performs its respective functions) to create a fully functional vehicle and meet customer needs.
3. Professionals from many different disciplines (e.g., industrial design, mechanical engineering, electrical engineering, physics, manufacturing engineering, product planning, finance, and business and marketing) are required to design (i.e., to make decisions related to the design of) all the systems in the vehicle.
4. The vehicle has many different attributes (i.e., characteristics that its customers expect, such as performance, fuel economy, safety, comfort, styling, and package). Simultaneous inputs from professionals from many disciplines and specialists with deep knowledge about each of the vehicle systems are required to make decisions about proper consideration of levels of all the attributes and trade-offs between the attributes in designing all the systems within the vehicle.
5. The automotive product is a component of other, larger systems (e.g., one or more vehicle platforms [which may be shared with other vehicle models], the highway transportation system, the petroleum consumption and fuel distribution system, the financial system, and so forth).
6. The automobile works within different environmental and situational conditions (e.g., driving on a winding road at night in a thunderstorm).
7. All phases of the life cycle, from conceptualization of a new automotive product to its discontinuation (i.e., its disposal, scrappage, recycling, replacement, plant dismantling or retooling), must be considered during its design.

Thus, the systems approach comprises simultaneous consideration of many systems, many attributes, trade-offs between the attributes, life cycle, disciplines, other systems, and working environments in solving problems (i.e., decision making). The systems approach is thus a primary and necessary part of SE.

Multidisciplinary Approach

SE is a multidisciplinary approach, that is, it obtains inputs from people from many different disciplines working together and considering many design and operational issues and trade-offs between different issues, to enable the realization of a successful product or a system. It is important to realize here that even when one discipline, such as electrical engineering, has the primary responsibility for designing an electrical system, other disciplines can raise a number of issues related to the design and operation of the system and thus assist in the design of the system by simultaneous consideration of multiple views and issues.

SE involves both technical and management activities from the early conceptual stage of a product (or a system) to the end of the life cycle of the product (i.e., when the product is removed from service and disposed of). The management activities help ensure that all requirements and design considerations are taken into account along with the key goals of meeting the product performance, developmental schedule, and budget of the product program.

Customer Focused

SE begins with an understanding of customer needs and development of an acceptable concept of the product (or system). It focuses on defining customer needs and required functionality early in the development cycle, documenting requirements, and proceeding with the design synthesis and system (product) validation while considering the problem as a whole (INCOSE, 2006).

The objective of SE is to ensure that the product (or the system) is designed, built, and operated so that it accomplishes its purpose of satisfying customers in the most cost-effective way possible by considering performance, safety, costs, schedule, and risks.

Basic Characteristics of SE

The basic characteristics of the SE approach are

1. *Multidisciplinary*: SE is an activity that knows no disciplinary bounds. It involves a collection of disciplines throughout the design and development process. It involves professionals from different disciplines working together (simultaneously and preferably co-located under one roof), constantly communicating, reviewing the design issues, and helping each other on all aspects of the product. The types of disciplines to be included depend on the type and characteristics of the product and the scope of the product program.

   For example, SE application for developing an automotive product will require personnel from many disciplines, such as engineers (including many specializations within engineering, e.g., mechanical, materials, electrical, computer and information science, chemical, manufacturing, industrial, human factors, quality, and SE), scientists (e.g., in physics, chemistry, and the life sciences) for research related to the design and production of new technological features of the vehicle, industrial designers (who define the sensory form and craftsmanship characteristics of the vehicle, i.e., the look, feel, and sound of the interior and exterior of the vehicle, such as the styling and appearance of surfaces of the vehicle, the touch feel of the surface and material characteristics, the sounds of operating equipment, and the smell of materials), market researchers (who define the customers, market segment, customer needs, market price, and sales volumes), management (e.g., program and project management personnel, including product planners, accountants, controllers, and managers), plant personnel involved in manufacturing and assembly, distributors, dealers, and even insurers to ensure that costs associated with fixing a vehicle damaged in an accident can be reduced and covered by the insurer.

It is important to get inputs from all the disciplines that affect or are affected by the characteristics and uses of the vehicle at the early stages of the PD. This ensures that their needs and concerns, and trade-offs between different multidisciplinary issues, are considered and resolved early, and costly changes or redesigns in the later phases are avoided.

2. *Customer Focused*: SE places continuous focus on the customers; that is, the product design should not deviate from satisfying the needs of the customers. The customers should be identified and involved in defining the vehicle specifications and designing the vehicle, and in subsequent evaluations, to ensure that the vehicle being designed will meet their needs. The customer needs are translated into vehicle attributes, and attribute requirements are developed to ensure that each vehicle attribute is managed (i.e., reviewed, verified, and validated) during the life cycle of the vehicle program. The vehicle attribute requirements process is described in Chapter 2.

3. *Product-Level Requirements First*: SE places concentrated effort on initial definition of the requirements at the overall product (i.e., the "whole" vehicle) level. For example, at the product level, the requirements for an automotive product will be based on all the basic attributes (derived from the needs of its external and internal customers) of the vehicle, such as safety, fuel economy, drivability (ability to maneuver, accelerate, and decelerate, and cornering or turning), seating comfort, thermal comfort, body-style, styling, costs, size, and weight.

It is important to realize that the customer buys the vehicle for his/her use as a "whole" product, not as a mere collection of the many components that form the product. (Note that an automotive product typically contains about 6,000–10,000 components.) Thus, the requirements for the systems, subsystems, and components of the product should be derived only after the product-level requirements are clearly understood and defined. This issue of cascading of the product-level attribute requirements to the system and lower-level entities is covered in Chapters 2 and 9.

4. *Product Life-Cycle Considerations*: SE includes considerations of the entire life cycle of the product being designed—through all stages from "Concept Development to Disposal of the Product" (from lust to dust). Thus, it is the applications of all relevant scientific and engineering disciplines in all the phases of the product, such as concept development; designing, manufacturing, testing and evaluation; uses under all possible operating conditions;

service and maintenance; and disposal or retirement from service, that the product encounters throughout its life cycle.

5. *Top-Down Orientation*: SE takes a "top-down" approach, which first views the product (or the entire system) as a whole and then sequentially breaks down (or decomposes) the product into its lower levels, such as systems, subsystems, sub-subsystems and components. Thus, the lower-level systems are designed to meet the requirements of the higher-level systems. (Note that if a manufacturer decides to use a carryover [i.e., existing] component or system in a new product, the top-down approach will need to be modified. This issue is covered in Chapter 2.)

6. *Technical and Management*: SE is both a technical and management process. It involves making all the technical decisions related to the product during its life cycle as well as management of all the tasks to be completed in a timely manner to implement the SE process and apply the necessary techniques.

7. *Technical Process*: The technical process of the SE is the analytic effort necessary to transform the operational needs of the customers into a design of the product (or system) with proper size, configuration, and capacity (e.g., performance level). It creates a documentation of the product requirements and drives the entire technical effort to evolve and verify an integrated and life cycle–balanced set of solutions involving the users and the product in its usage situations.

8. *Management Process*: The management process of the SE involves assessing costs and risks, providing needed resources, integrating the engineering specialties and design groups, maintaining configuration control, and continuously auditing the effort to ensure that cost, schedule, and technical performance objectives are satisfied to meet the original operational need of the product and the product program.

9. *Product and Organization-Specific Orientation*: The details of the SE implementation (such as steps, methods, procedures, team structure, tasks, and responsibilities) depend on the program objectives, the product being produced (i.e., its characteristics), and the organization (company) producing it (i.e., different companies generally have somewhat different processes, timings, organizational responsibilities, and brand-specific requirements).

PRODUCT DEVELOPMENT

The majority of PD programs do not involve designing a product from "scratch" (i.e., a totally new product) or a product of a type that did not exist before. The process of designing a product is therefore typically called the *product development process* in most industries (including the automotive industry) rather than the *product design process*. However, the terms *product development* and *product design* are interchangeable and are used in the same context in many industries. (After the product has been designed, the process of producing the product [i.e., manufacturing various systems and assembling the systems to create the whole product] is generally called the *production process* [see Figure 1.1].)

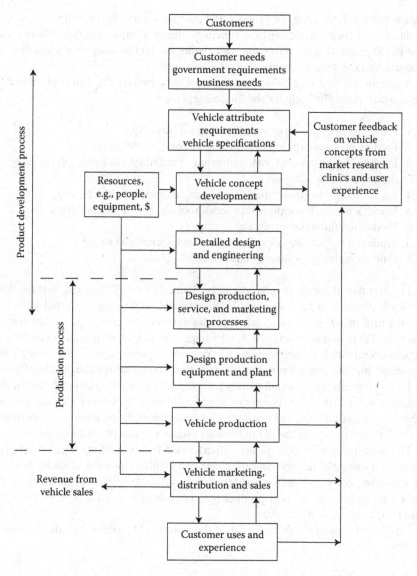

FIGURE 1.1 Flow diagram of automotive product development and production processes.

PROCESSES AND PHASES IN PRODUCT DEVELOPMENT

It is important to realize that any work is generally performed by using one or more processes. A process usually involves inputs (e.g., raw materials, energy), equipment (one or more workstations with tools, machines, robots, or computers), and human operators that are configured in a sequence of steps (operations or tasks) to produce a specified output. Designing a product is also performed by using a process (defined earlier as the PD process). The PD process, depending on the complexity of the product, can involve many processes within and outside the organization (e.g., suppliers) responsible for developing the product. PD processes vary due to differences in the products (i.e., their characteristics, functions, features, and demand volume), the type of PD program (e.g., refreshing an existing product or designing a totally new product), and the design organization (or company).

A generic process of product creation and use involves the entire product life cycle, which generally includes the following phases:

1. Pre-concept or pre-program (pre-program planning)
2. Product concept exploration (alternative concepts development)
3. Product definition and risk reduction (feasibility analyses, preliminary design, and risk analysis)
4. Engineering design (detailed engineering design including testing)
5. Manufacturing development (process, tooling, and plant development)
6. Production (manufacturing and assembly)
7. Product distribution, sales, marketing, and operational support
8. Product updating or discontinuation and disposal

The first five of the above phases can be defined as the PD process, and the fifth and sixth phases can be considered as the production process. It should be noted that the fifth phase of manufacturing development can be considered as the transition from PD to manufacturing. It is very important to include product manufacturing considerations (e.g., applications of "design for manufacture" and "design for assembly" methodologies) very early during the product design (i.e., during Phases 1 to 4, by implementing simultaneous [concurrent] engineering) to ensure that the transition in the fifth phase (involving designing of manufacturing processes and the creation of required tools and equipment in the manufacturing plants) occurs seamlessly without changes in the PD in the later phases to meet production needs.

The work in each of these phases is performed by undertaking specialized processes. For example, the pre-concept phase can involve a process of understanding the customer, corporate needs, and regulatory requirements to decide on the type and characteristics of the new product (i.e., product specification) and preparing a plan for the subsequent activities.

Ulrich and Eppinger (2015) described the generic PD process with the following phases:

1. Planning
2. Concept development
3. System-level design
4. Detail design
5. Testing and refinement
6. Production ramp-up

It should be noted that Ulrich and Eppinger (2015), in their fourth "detail design" phase, included detailed component design (e.g., part geometry, material selection, and specification of tolerances), definition of production processes, tooling design, and beginning of tooling procurement. The fifth phase involves all product verification tests (i.e., performance, reliability, and durability) and refinements of assembly processes, including training of the production workforce. The production ramp-up phase involves the evaluation (validation tests) of early production outputs and the beginning of full operation of the production system.

AUTOMOTIVE PRODUCT AS A SYSTEM

An automotive product is considered as a system that involves a number of lower- (or second-) level systems: the body system, the chassis system, the powertrain system, the fuel system, the electrical system, the climate control system, the braking system, and so on. Each of the systems within the automotive product can be further decomposed into subsystems, sub-subsystem, sub-sub-subsystems, and so on, till the lowest-level components are identified. For example, the body system includes the body frame subsystem, the body panels subsystem, the closure subsystem (which includes the hood sub-subsystem, the doors sub-subsystem, and the trunk or liftgate sub-subsystem), the exterior lamps subsystem, the seats subsystem, the instrument panel subsystem, the interior trim components subsystem, and so forth.

Table 1.1 illustrates the major systems, subsystems, and sub-subsystems or components within a typical automotive product. The definitions and contents of the various vehicle systems illustrated in this table can vary somewhat between different vehicle makes and models. Further, the implementation of different technologies used in performing different vehicle functions can have a major effect on the design of any vehicle system. In fact, one of the challenges facing vehicle engineering groups is how to divide the entire vehicle into different systems, subsystems, sub-subsystems, and so on and to assign design responsibilities to various engineering teams. This issue of division or decomposition of an automotive product for management of various PD activities and their interfaces is covered in Chapters 7, 8, and 12 and Appendix I.

The key tasks of systems designers are to ensure that each system performs its functions and that the systems, through their interfaces with other systems, work harmoniously to meet the customer needs of the whole product. Thus, the task of designing the vehicle requires a lot of understanding of systems and coordination between systems, their functions, and trade-offs between vehicle attributes to come up with a balanced vehicle design, This issue is covered in more detail in Chapters 2 and 8.

AUTOMOTIVE PRODUCT DEVELOPMENT PROCESS

What is Automotive Product Development?

The automotive PD process involves the designing and engineering of a future automotive product. The automotive product (i.e., a vehicle) can be a car or a truck or a variant such as a station wagon, a sports utility vehicle (SUV), or a van. The manufacturing and assembly operations are generally assigned to different groups. However, selected representatives from manufacturing and assembly operations must actively participate in the teamwork during the PD process.

TABLE 1.1
Major Systems and Their Subsystems in a Typical Automotive Product

Vehicle System	Subsystems of the System	Sub-Subsystems or Components of the Subsystem
Body system	Body-in-white	Body frame, cross members, body panels, front and rear fascia/bumpers
	Closures system	Doors (door frame, exterior panels, hinges, latches, inside trim panel power window mechanisms, door handles, window and mirror controls), hood and trunk-lid (or liftgate)
	Seat system	Driver's seat, front passenger seat, and rear seat(s)
	Instrument panel	Instrument panel fascia, instrument cluster, switches, glove box, brackets (for other components such as climate controls, entertainment and navigation controls and displays, passenger airbag) and trim components
	Exterior lamps	Front lighting system (headlamps and front signal lamps), rear signal system (tail lamps, stop lamps, turn signal lamps, back-up lamps, license plate lamps, rear reflectors), and side marker and clearance lamps
	Glass system	Windshield, backlite, side window glasses (also called *glazing surfaces*)
	Rear vision system	Inside mirror and outside mirrors, camera systems, and rear and side target sensing systems
Chassis system	Underbody frame work	Front subframe, rear subframe (cradle), cross members for mounting other chassis systems such as steering system and brake system
	Suspension system	Front and rear suspensions (includes arms, links, knuckles, joints, springs, shock absorbers)
	Steering system	Steering linkages, steering column, steering wheel and stalk controls
	Braking system	Brake disks/drums, brake pads and actuators, master cylinder, and pedal linkages
	Wheels and tires	Wheels and tires
Powertrain system	Engine	Engine block and cylinder heads, power conversion system (pistons, connecting rods, crank shaft, bearings), intake and exhaust system, fuel supply system, engine electrical and control system, cooling system, and lubrication system
	Transmission	Transmission casing, gears and shafts, clutches, valves and linkages, sensors, lubrication and oil cooling system
	Shafts and joints	Drive shaft, universal joints, convel joints and bearings
	Final drive and axles	Differential casing, shafts, gears, and bearings

(Continued)

TABLE 1.1 (CONTINUED)
Major Systems and Their Subsystems in a Typical Automotive Product

Vehicle System	Subsystems of the System	Sub-Subsystems or Components of the Subsystem
Fuel system	Fuel tank	Tank, fuel system module (fuel pump, pressure valve, fuel filter, fuel level sensor), carbon canister, filler pipe and fuel cap
	Fuel lines	Fuel lines, hoses, and connectors
Electrical system	Battery	Battery
	Alternator	Casing, rotor, and stator
	Wiring harnesses	Wiring harnesses, connectors, and clips
	Power controls	Switches, sensors, relays, electronic control units, fuse box and fuses
Climate control system	Heater	Heat exchanger, blower, air ducts, valves, and hoses
	Air conditioner	Heat exchanger, compressor, valves, tubing, hoses, and refrigerant
	Climate controls	Controls and displays (for setting temperature, fan speed, and mode)
Safety and security system	Air bag system	Air bag units, sensors and actuators, wiring, electronic control units
	Seat belt system	Seat belts, belt anchors, belt buckles, belt movement control mechanisms, sensors, and wiring
	Wiping and defroster systems	Windshield wipers, wiper motors, wiper control system, defroster system, and defroster control system
	Security lighting and locking systems	Exterior courtesy lamps, door locks, locking mechanisms, theft protection system, wiring and control units
	Driver assistance systems	Collision avoidance systems such as automatic braking, lane-departure warning system, driver alertness system, and adaptive cruise control system
Driver interface and infotainment system	Primary and secondary vehicle controls and displays	Driver controls and displays, wiring, and connectors
	Audio system	Audio controls and displays, audio chassis and circuit board, antenna, wiring, USB port
	Navigation system	Microprocessor, display, wiring, antenna, map database, and data ports
	CD/DVD player	CD/DVD player chassis and mechanism, microprocessor, wiring, USB port

Automotive products are generally produced in large quantities (about 10,000 to 700,000 vehicles per year per model and at rate typically of about 40–70 vehicles/hour on an assembly line), shipped to dealers in many locations, and sold to customers to meet their transportation needs. The vehicles must be safe, efficient, economical, dependable, "fun to drive and use," and "pleasing" to the customers. The vehicles also must have necessary characteristics such as performance (i.e., operating capabilities), styling/appearance (form), quality (customer satisfaction), and craftsmanship (perception of being well made). The customers must "enjoy owning the vehicles"—that is, the vehicles must have all the necessary attributes and the right features to meet their lifestyles.

FLOW DIAGRAM OF AUTOMOTIVE PRODUCT DEVELOPMENT

The vehicle development process generally begins with understanding customer needs and ends with the customers providing their feedback after using the vehicle. Figure 1.1 shows the major phases in the vehicle development process along with the production, marketing, sales, and vehicle usage phases. Based on an understanding of customer needs, government requirements, and the business needs of the company, a design team consisting of members from different disciplines (e.g., industrial designers, product architects, engineers, manufacturing personnel, product planners, and market researchers) generally develops attribute requirements at the vehicle level and creates the vehicle specifications. The information is used by the team to develop one or more vehicle concepts (in the form of sketches, drawings, CAD models, mockups, or bucks). The vehicle concepts are iteratively improved by using customer feedback and suggestions by different team members and are market researched to determine whether a leading concept can be selected for the detailed design and engineering work. Based on the selected product design, manufacturing processes and suppliers are selected. The production equipment and plants are designed and built or modified for manufacturing and assembly. Marketing, sales, and distribution plans are developed. The early production parts and systems are assembled into prototype vehicles. All entities, from components to major vehicle systems, are tested to verify that they meet their respective requirements. The assembled systems are installed into vehicle bodies, and prototype vehicles are created. These prototype vehicles are further tested to verify and validate vehicle-level requirements. Final approval to produce the vehicle is given by senior management, and the vehicle is "launched" (i.e., production begins). The produced vehicles are shipped to the dealerships for sale. As the purchased vehicles are used by the customers, feedback from the customer experience (i.e., data from field operating performance, customer likes/dislikes, vehicle repairs, and warranty work) are continuously collected and provided for improving existing products and designing future products.

To support the entire vehicle development process, resources (e.g., dollars, people, equipment, and facilities) are needed. Budgets and schedules are created to manage the entire PD process. The organization begins to make money from revenues generated from the vehicle sales. The program management and financial analysis issues are covered in Chapters 12 and 19.

TIMING CHART OF AUTOMOTIVE PRODUCT DEVELOPMENT

Figure 1.2 provides a timing chart illustrating various activities during major phases of an automotive PD program. The length and location of the horizontal bars indicate duration and beginning and ending times of each activity within each program phase.

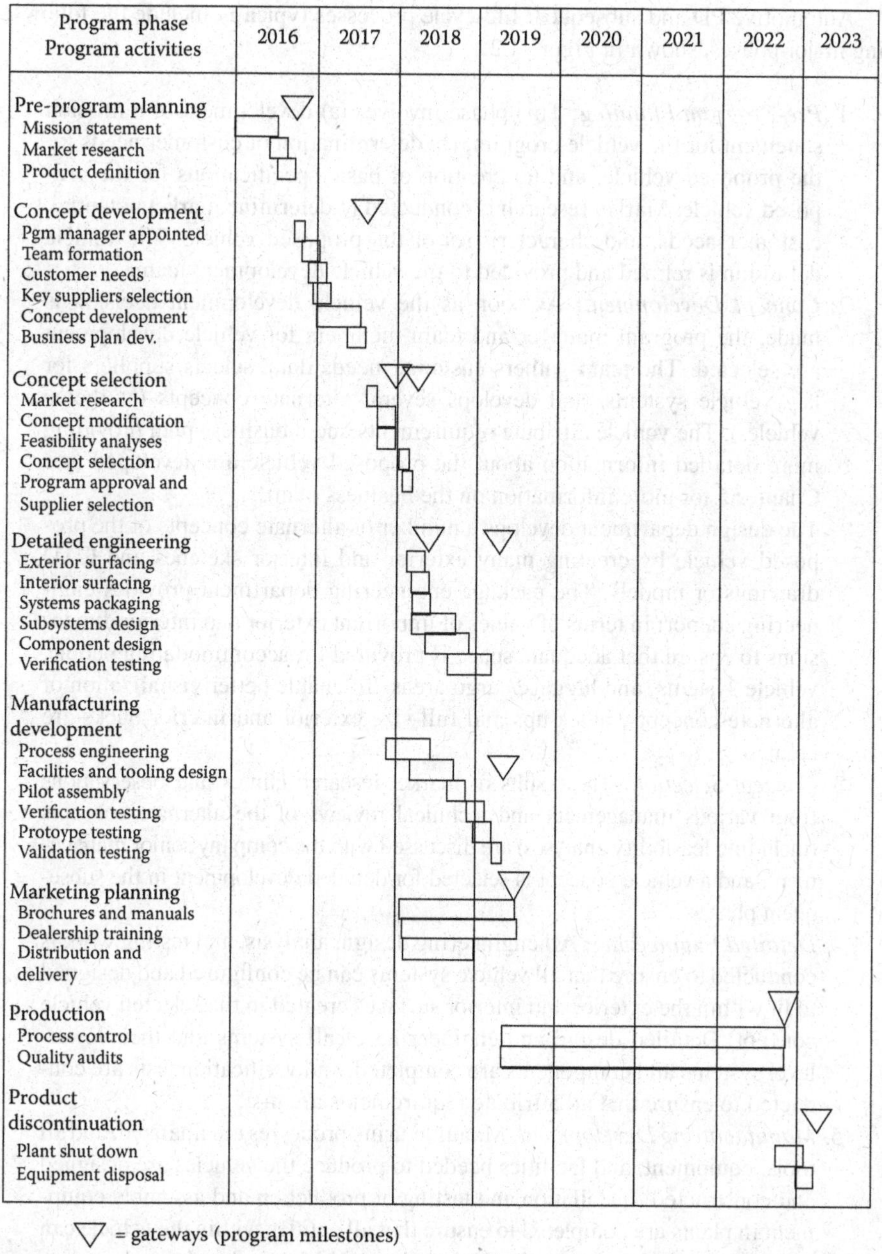

Program phase / Program activities	2016	2017	2018	2019	2020	2021	2022	2023
Pre-program planning								
Mission statement								
Market research								
Product definition								
Concept development								
Pgm manager appointed								
Team formation								
Customer needs								
Key suppliers selection								
Concept development								
Business plan dev.								
Concept selection								
Market research								
Concept modification								
Feasibility analyses								
Concept selection								
Program approval and								
Supplier selection								
Detailed engineering								
Exterior surfacing								
Interior surfacing								
Systems packaging								
Subsystems design								
Component design								
Verification testing								
Manufacturing development								
Process engineering								
Facilities and tooling design								
Pilot assembly								
Verification testing								
Protoype testing								
Validation testing								
Marketing planning								
Brochures and manuals								
Dealership training								
Distribution and delivery								
Production								
Process control								
Quality audits								
Product discontinuation								
Plant shut down								
Equipment disposal								

▽ = gateways (program milestones)

FIGURE 1.2 Timing chart of a vehicle program.

Automotive PD and subsequent life-cycle processes typically include the following major phases, shown in Figure 1.2:

1. *Pre-Program Planning*: This phase involves (a) development of a mission statement for the vehicle program, (b) determination of customer needs for the proposed vehicle, and (c) creation of basic specifications for the proposed vehicle. Market research is conducted to determine market potential, customer needs, and characteristics of the proposed vehicle. The vehicle definition is refined and provided to the vehicle development team.

2. *Concept Development*: As soon as the vehicle development decision is made, the program manager and team members for vehicle development are selected. The team gathers customer needs data, selects suppliers for key vehicle systems, and develops several alternate concepts (or theme vehicles). The vehicle attribute requirements and a business plan providing more detailed information about the proposed vehicle are developed (see Chapter 5 for more information on the business plan).

   The design department develops a number of alternate concepts of the proposed vehicle by creating many exterior and interior sketches and CAD drawings or models. The package engineering department provides engineering support in terms of values of important exterior and interior dimensions to ensure that adequate space is provided for accommodating people, vehicle systems, and luggage/cargo areas. To enable better visualization of alternate concepts, mock-ups and full-size exterior and interior bucks are created.

3. *Concept Selection*: The results of market research clinics and observations from various management and technical reviews of the alternate concepts (including feasibility analyses) are discussed with the company senior management, and a vehicle concept is selected for detailed development in the subsequent phases.

4. *Detailed Engineering*: All engineering design, analysis, and testing work is conducted to ensure that all vehicle systems can be configured and designed to fit within the exterior and interior surfaces created in the selected vehicle concept. Detailed design and engineering of all systems and their lower-level systems and components are completed, and verification tests are conducted to ensure that all attribute requirements are met.

5. *Manufacturing Development*: Manufacturing processes are finalized, and all tools, equipment, and facilities needed to produce the vehicles are designed and constructed. Installation and testing of production and assembly equipment in plants are completed to ensure that all entities within the vehicle can be manufactured and assembled to produce vehicles at the planned production rate and high quality (e.g., meeting all manufacturing tolerances and fit and finish requirements). Early prototype/production vehicles are used for validation testing to ensure that the right product was produced.

6. *Marketing Planning*: Marketing plans are created, and dealerships are provided with the necessary information and training for sales, marketing, maintenance, and repair work of the vehicles.

7. *Production*: Early production vehicles are tested to verify and validate that the vehicles meet all the attribute requirements. Customer and management reviews are completed. Plant equipment calibrations and production output quality are monitored during production. The plant output is adjusted on an ongoing basis to match the vehicle demand through dealer orders and sales forecasts.

8. *Product Discontinuation*: Plant is shut down to discontinue production and retooled for the next vehicle model. Obsolete and unneeded equipment is removed and disposed of.

Preparation of vehicle and systems development timing plans is a very important activity in managing vehicle programs. A proper amount of time must be allocated to accomplish the hundreds of tasks performed by various design and engineering departments. The tasks must be carefully analyzed and selected to ensure that they are needed, and the time required for each of the tasks should be estimated by experienced and specialized professionals from each activity. The product planning department generally takes the time estimates from all key design and engineering activities and creates an overall program timing chart, such as the one shown in Figure 1.2.

UNDERSTANDING CUSTOMER NEEDS

The SE work begins with the definition of the vehicle to be developed. The vehicle definition should include a description of its type (body-style), size (overall dimensions), and market segment (i.e., the market location and customer characteristics). The description should be as detailed and specific as possible, as it will be used by all the team members (designers and engineers) involved in the vehicle development process.

For the vehicle to be successful in the market, the vehicle definition should be based on the needs of its customers. This means that its prospective customers should be identified, and their demographic and ergonomic characteristics and needs for specific vehicle characteristics and features must be determined and used during the vehicle development process. The description of the customer needs should be comprehensive and complete, in the sense that all aspects of the vehicle covered by all the attributes of the vehicle must be obtained. The customer needs should be focused on the vehicle as a whole and not on its lower-level entities. Chapter 3 provides more information on how customer needs and other needs arising from government requirements and corporate business needs are obtained and used in the PD process.

PROGRAM SCOPE, TIMINGS, AND CHALLENGES

SCOPE OF VEHICLE DEVELOPMENT PROGRAMS

An automotive PD program is initiated to modify and improve an existing vehicle design or to replace it with a totally new vehicle. The modifications or changes can range from minor refreshments to an existing vehicle to replacing the existing vehicle with a completely new vehicle design. Vehicle development programs can thus be classified as follows:

1. *Minor Refreshment Program*: Small changes in vehicle exterior (e.g., changes in exterior colors, wheels, rear lamps, grill and headlamps, interior colors, interior materials, and/or graphics in displays)
2. *Program with Medium Changes*: Changes in appearance of some body panels and functionality of some vehicle systems (e.g., restyling shapes of hood, fenders, lamps, instrument panels, and performance improvements in selected systems or subsystems)
3. *Program with Major Changes*: New powertrain, changes in vehicle body and chassis, adding variations in vehicle body-styles (e.g., adding a coupe and/or a station wagon to an existing sedan)
4. *Totally New Design*: Replacing an existing vehicle with a completely new vehicle, which usually involves a new vehicle exterior (body), new powertrains and chassis, and a new interior (instrument panel, door trim panels, consoles, and seats)

The scope of the vehicle program has a direct effect on the number of tasks, timings, and costs associated with the program.

PROGRAM TIMINGS

An automotive PD program generally extends over 12 to 48 months, depending on the scope of the program and how the beginning and end points of the program are defined. A large PD program may involve developing a totally new vehicle platform, a new powertrain, and one or more product variations, for example, similar body-style but different exterior panels and interior components for different corporate brands (e.g., Chevrolet, Buick, and Cadillac; Toyota and Lexus; Ford and Lincoln), or adding more body-styles or variants (e.g., sedan, coupe, hatchback, station wagon, and SUV). A large vehicle program may thus extend over several years. A small program may involve merely refreshing an existing vehicle with minor changes to vehicle exterior, such as changes in front fascia, grill, wheel covers, exterior colors, headlamps and tail lamps, and other minor changes to the interior, such as changes in audio components, graphics, and interior materials and colors. A small vehicle program may take from a few months to about 18 months to complete its vehicle development activities.

No two vehicle programs (in terms of tasks to be performed), even within the same automotive company, are alike (because of differences in people working in various program activities, constraints related to time and budget, changes in customer needs, technology-related changes, etc.). Thus, vehicle programs can be very different between different vehicle manufacturers in terms of differences in design tasks, phases, timings, test procedures, organization, and management style.

A major vehicle program can cost upwards of a billion dollars over several years and involve about 600–1200 professionals from different disciplines; many design and engineering computer systems with specialized software; hardware fabrication shops, laboratories, and test facilities with specialized equipment, tooling, and fix- tures; design and building shops; and modifications to manufacturing and assembly plants.

Depending on the program size, the vehicle development program timing plan can range from a few months to several years. The timings of the vehicle program are estimated from a list of all the tasks that need to be accomplished and the time and resources needed to complete the tasks. The costs for each of the tasks are estimated and added to come up with the estimates of total time needed, program timings, and program costs. These cost-related issues are cov- ered in Chapter 19.

The success of an automotive company primarily depends on development of the "right" products that its customers truly want. Thus, PD is probably the most important process in an automotive company. The objective of the PD process is to develop one or more products that will be purchased by customers to meet their transportation needs. A successful product not only increases revenues and profits but raises the company's reputation and status, that is, how it is perceived in terms of its image, brand value, and prestige.

Important Considerations in Managing Vehicle Programs

Vehicle programs are influenced by the priorities of various customer needs and approaches used by the company management in developing the vehicle. Important considerations in managing the vehicle programs are

1. *Implement Co-Located Product Design Teams*: Co-location involves mov- ing the offices, design studios, and test facilities of all key team members into one building. The co-location facilitates more frequent interaction between team members. It also eliminates transportation time, as team meetings are held in the same building.
2. *Enable Constant Communication*: More opportunities for communications (formal planned meetings and informal discussions) between team mem- bers allow quicker identification and resolution of problems.
3. *Ensure Availability of Latest Vehicle Design, Program Status, and Reference Materials*: Online access and availability of latest data on vehi- cle design, program changes, and reference information from common data bases (e.g., benchmarking data, design standards, test procedures, and gov- ernment requirements) to all team members reduces delays in obtaining information on the latest changes and thus reduces rework or duplication of effort.
4. *Adopt Simultaneous/Concurrent Engineering Methods*: Simultaneous development involves performing many tasks within overlapping time intervals (i.e., reducing sequential scheduling of tasks). Concurrent engi- neering does not only reduce overall program time; it also reduces major rework and improves quality by communicating on issues being resolved using concurrent inputs from many disciplines.

5. *Minimize Number of Design Changes after Program Definition*: Any design change made after the product specification has been approved generally results in more changes (in all entities affected by the changes) and rework. This is especially true because automotive products are complex (i.e., they involve many systems, subsystems, and components that have many interfaces).

6. *Use Computer-Aided Methods to Reduce Costs of Physical Model Building and Testing*: Computer-aided methods do not only reduce time (by use of functions such as copy, paste, mirror, and extrude); they also reduce errors in data transfers and facilitate conducting many design iterations to optimize the design.

7. *Use Carryover Parts*: If existing components can be used (i.e., reused) in developing a new product, this can reduce design, engineering, and manufacturing time and costs. The carryover components, however, reduce design flexibility and the possibilities of incorporating innovative design ideas. The carryover content can range from reuse of a few selected components or systems from an existing vehicle model to use of an existing vehicle platform (i.e., a collection of a large number of systems and large body and chassis parts that determine the characteristics of major tools and fixtures used in manufacturing and assembly plants).

8. *Use "Book-Shelved" Technologies*: A book-shelved entity (i.e., a component or a system) is one that has already been studied, researched, and developed and is ready to be incorporated in a future complex (automotive) product. This eliminates time required to design and develop the new entity.

9. *Incorporate Design Reviews throughout the Program*: Design reviews facilitate additional critical reviews and analyses by experts and managers from different disciplines and departments, which may not have been directly involved during the earlier design work. The design reviews thus help in identifying and fixing problems in the vehicle design and related processes early.

10. *Define and Follow Gateways*: Gateways are important points (or events) in the program timeline. Gateways are also called *milestones* in some organizations. The gateways indicate when certain key events are projected to occur. They are used to guide and coordinate all activities in PD to ensure that the vehicle program progresses according to the pre-developed timing plan. They are usually tied to events such as completion of certain activities (e.g., completion of concept development, engineering steps, management reviews and approvals). Some important gateways are presented in Table 2.1. The definitions and number of gateways vary widely between different programs of different auto manufacturers. The definitions and timings of gateways are usually developed by the program planning departments with constant communication between all major areas (e.g., design, engineering, manufacturing, finance, and marketing). Gateways for each major activity, such as design, engineering, and manufacturing, will include additional lower-level gateways to coordinate their more detailed activities

with the overall program timings. Chapter 2 presents the gateways used in a vehicle development program in relation to the SE process used in the program.

SOME FREQUENTLY ASKED QUESTIONS DURING VEHICLE DEVELOPMENT

The team members involved in an automotive PD program face many questions. A few commonly asked questions are

1. Are we designing the right vehicle? (Does the vehicle have the characteristics and features that its customers truly desire? Would the vehicle sell well?)
2. Can an actual vehicle be created with the same characteristics as shown in the vehicle concept? Would such a vehicle concept be feasible, considering engineering and manufacturing challenges and tasks?
3. Can this vehicle compete well with its toughest competitors when it is introduced, many months from now?
4. Can we build the vehicle with the required level of quality and within the planned price range?
5. Do we have the capabilities, plant capacity, and resources to build such a vehicle?
6. Can we meet the program timings and stay within the budgeted resources?

DECISION MAKING DURING PRODUCT DEVELOPMENT

It should be noted that many decisions are made during each step of the PD process. Some examples of questions related to decisions involved in PD are: What type of product to make? What should be its dimensions? What type of power source would be planned for the vehicle? What should be the capacity of the power source? What types of materials should be used for each component? What types of joining or assembly methods would be used? What should be the height of the seat from the vehicle floor and the ground? What fields of view would the driver need to drive the vehicle safely? In which assembly plant would the vehicle be produced?

Making the right decisions at the right time during the PD is very critical to meet the timings of the vehicle program. Early decisions usually involve the selection of characteristics related to the basic type and configuration of the vehicle (e.g., sedan vs. SUV, front-wheel drive vs. rear-wheel drive). If any of the key parameters of the vehicle configuration, such as the type of powertrain or the wheelbase, are changed in the later phases of the vehicle program, then many other design decisions and parameters that are dependent on the key parameters will also change. The changes generally require redesign of many systems, and they can be very time consuming and costly, especially when the changes are made during the later phases of the program. Thus, all important disciplines need to be involved during the early decision making to avoid late changes.

DISCIPLINES INVOLVED IN AUTOMOTIVE PRODUCT DEVELOPMENT

Development and production of an automotive product requires professionals from many disciplines. In addition, professionals with work experience in past vehicle programs can provide a lot of knowledge during the resolution of a number of issues. The professionals from specialized disciplines (e.g., mechanical engineering, structural engineering, vehicle dynamics, aerodynamics, and electronics) needed in different functional areas are

1. Product planning (mechanical engineers, market research specialists, business management specialists, economists, operations researchers, financial planners)
2. Market research (market research specialists, business management specialists, economists, operations researchers, financial planners)
3. Industrial design (studio designers [interior designers and exterior designers], studio engineers, CAD modelers, graphic artists, color and trim specialists, craftsmanship specialists, clay modelers, computer-aided surfacing modelers, buck builders)
4. Body engineering (mechanical engineers, package engineers, CAD modelers, computer systems engineers, structural engineers, safety engineers, materials engineers, aerodynamics engineers, lighting design engineers, electrical engineers)
5. Powertrain engineering (mechanical engineers, CAD modelers, electrical and electronics engineers, chemical engineers, environmental and emissions engineers, materials engineers, fuel systems engineers, aerodynamics engineers)
6. Chassis engineering (mechanical engineers, suspension engineers, CAD modelers, vehicle dynamics engineers, brake engineers, tire engineers, electrical and electronics engineers)
7. Electrical systems engineering (electrical engineers, electronics engineers, computer systems engineers, telematics specialists, mechanical design engineers, audio engineers, display technologists)
8. Human factors engineering and ergonomics (industrial engineers, engineering psychologists, ergonomists, human factors engineers, mechanical engineers)
9. Climate control engineering (mechanical engineers, thermodynamics engineers, aerodynamics engineers, electrical and electronics engineers)
10. Manufacturing, production, and assembly engineering (mechanical engineers, manufacturing process engineers, materials engineers, metallurgists, numerical control specialists/programmers, industrial engineers, plant engineers, tool designers, tool engineers, ergonomists, industrial hygienists, safety engineers)

SELECTING THE PROGRAM LEADER

Selecting the leader for the vehicle program is probably the most important decision faced by the senior company management. The vehicle development process involves making many decisions related to the characteristics of the vehicle being

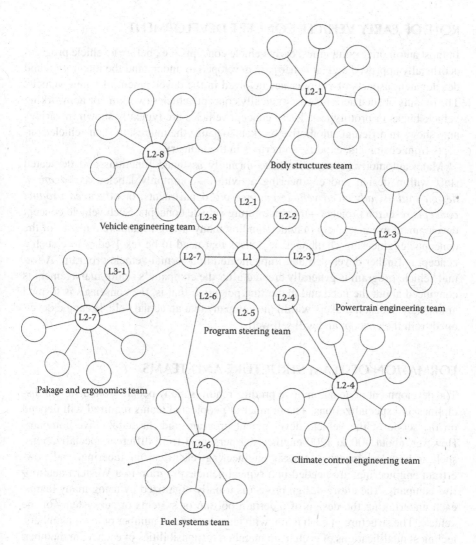

FIGURE 1.3   Illustration of linked team structure (only partial team structure is shown).

designed. The program leader (or program manager) must oversee the vehicle development activities and make all key decisions. The program leader should be a big-picture thinker and must have the skills to perform many roles, functioning as an integrator, a decision maker, a time and cost controller, a team builder, a coach, a motivator, and a communicator.

Womack et al. (1990) have compared the leadership issues in Western auto companies with Toyota and found that the reduced PD cycles and better quality in Toyota vehicles in the 1980s were due to implementation of the *shusha* concept. The shusha (or chief program engineer) is given the complete authority to make all decisions on the vehicle and its program management. Additional information on the program management tasks are provided in Chapter 12.

ROLE OF EARLY VEHICLE CONCEPT DEVELOPMENT

In most automotive companies, early vehicle concepts (i.e., before a vehicle program is officially approved and launched) are developed to understand the integration and development aspects of many issues involved in the development of a new vehicle. The outputs of such activities are typically concept vehicles (working or nonworking vehicle bucks or prototypes). These concept vehicles are typically shown in various auto shows in different automotive markets to gauge the interest in such vehicle concepts from customers, experts, and critics in the industry.

Many automotive companies have formally assigned functions and dedicated staff within design and engineering activities—commonly labeled as *advanced design studios*, *advanced vehicle engineering* departments, or *advanced product concepts* research projects—to create future vehicle concepts. Such vehicle concept development exercises help in understanding many strengths and weaknesses of the concepts, engineering challenges, and risks that need to be resolved before such a concept is further developed and implemented in a formal vehicle program. A formal vehicle program is generally created after the company's senior management is convinced about the need and marketing potential; that is, the consensus is formed among key decision makers within the company that an actual vehicle can be developed from the concept and will sell well.

FORMATION OF TEAM STRUCTURE AND TEAMS

The development of an automotive product requires many people from different disciplines and specializations. The number of people and teams required will depend on the scope of the vehicle development program and the automotive company. However, about 400 to 1200 engineering personnel from different specializations, such as body engineering, chassis engineering, electrical engineering, and powertrain engineering, are needed in a typical vehicle program in a Western automotive company. The entire design project is usually organized by using many teams, each undertaking the design of a certain portion or systems or subsystems of the vehicle. The structure of each team, with team leader and number of team members, technical qualifications of each team member, responsibilities of each team member, progress reporting, and problem resolution and communication methods, is strictly enforced to ensure that all vehicle systems and interfaces between the systems can be designed to meet all identified engineering requirements.

The highest-level team in a vehicle program is typically headed by the vehicle program manager, and the membership of the team consists of high-level managers of major activities and chief engineers of major engineering offices. In some auto companies, this is called the *vehicle program steering team*. The organizational structure of the vehicle program steering team, with the top level (Level 1) and next level (Level 2), is illustrated in Figure 1.3.

Vehicle program steering team:

L1 = Vehicle program manager (Level 1)
L2-0 = Program management manager (Level 2)

L2-1 = Body engineering chief engineer (Level 2)
L2-2 = Chassis engineering chief engineer (Level 2)
L2-3 = Powertrain chief engineer (Level 2)
L2-4 = Climate control chief engineer (Level 2)
L2-5 = Electrical engineering chief engineer (Level 2)
L2-6 = Fuel system chief engineer (Level 2)
L2-7 = Package and ergonomics engineering chief engineer (Level 2)
L2-8 = Vehicle engineering chief engineer (Level 2)
L2-9 = Manufacturing engineering chief engineer (Level 2)
L2-10 = Chief designer (Level 2)
L2-11 = Vehicle attribute engineering chief engineer (Level 2)

The next-level teams, headed by each Level 2 chief engineer with membership of
Level 3 managers, can be illustrated as follows:
Body engineering team:

L2-1 = Body engineering chief engineer (Level 2)
L21-1 = Body structural engineering manager (Level 3)
L21-2 = Body closures engineering manager (Level 3)
L21-3 = Body safety systems engineering manager (Level 3)
L21-4 = Body electrical engineering manager (Level 3)
L21-5 = Body lighting engineering manager (Level 4)
L21-6 = Instrument panel engineering manager (Level 3)
L21-7 = Seating systems engineering manager (Level 3)
L21-8 = Body trim components engineering manager (Level 3)

Vehicle attribute engineering team:

L2-11 = Vehicle attribute engineering chief engineer (Level 2)
L211-1 = Vehicle dynamics engineering manager (Level 3)
L211-2 = Aerodynamics engineering manager (Level 3)
L211-3 = Thermal management engineering manager (Level 3)
L211-4 = Noise, vibrations, and harshness engineering manager (Level 3)
L211-5 = Craftsmanship engineering manager (Level 3)
L211-6 = Weight engineering manager (Level 3)
L211-7 = Vehicle cost management manager (Level 3)

Similarly, the next-level teams headed by each of the Level 3 managers with
membership of Level 4 supervisors are

L21-2 = Body closures engineering manager (Level 3)
L212-1 = Hood engineering supervisor (Level 4)
L212-2 = Front doors engineering supervisor (Level 4)
L212-3 = Rear doors engineering supervisor (Level 4)
L212-4 = Trunk/liftgate engineering supervisor (Level 4)
L21-5 = Body lighting engineering manager (Level 3)

L215-1 = Front lamps engineering supervisor (Level 4)
L215-2 = Rear lamps engineering supervisor (Level 4)
L215-3 = Side marker and courtesy lamps supervisor (Level 4)

Depending on the issues being covered in any meeting of any of the above teams, other team members and specialists are invited to help resolve the issues.

TREATING SUPPLIERS AS PARTNERS

It is important to realize that depending on the automotive company, about 35–75% of the content of the automotive products is produced and supplied by supplier companies. Thus, the quality of the vehicle depends on the quality of the entities supplied by the suppliers and how these entities interface and work together with entities supplied by different suppliers and produced by the automotive company. Many of the suppliers are selected early, and their personnel are asked to participate in the PD process (as team members in different teams related to their supplied entities) and are given the tasks of designing the entities that they will produce. Thus, the suppliers should be treated as partners during the entire PD, production, and automotive assembly processes.

It is therefore very important to select the right set of suppliers. Supplier selection criteria typically include (a) expertise in SE and specialized disciplines needed to develop the entities, (b) production capability in terms of required levels of quantities with specified quality and price, (c) demonstrated flexibility in quickly incorporating engineering changes during early design stages, (d) dedication and responsiveness in meeting key product requirements (e.g., high fuel economy), (e) ability to incorporate innovative methods and technologies, and (f) ability to support globally (on products marketed in many countries).

OTHER INTERNAL AND EXTERNAL FACTORS AFFECTING VEHICLE PROGRAMS

Automotive PD programs are affected by many factors. The program management needs to be constantly on the lookout to determine whether these factors will affect various attributes of the vehicle, program timing, and costs. Major factors related to issues both internal and external to the automotive company that can affect the vehicle programs are listed in the following subsection.

INTERNAL FACTORS

1. Constant change due to the iterative nature of the PD process
2. Company's senior management directives and decisions related to the program (e.g., budgets, cycle plans, preferences for certain vehicle features)
3. Balancing costs, manpower, and timings across all vehicle programs within the company
4. Availability of manpower with required qualifications and expertise
5. Ability to select suppliers and integrate their involvement in the vehicle program teams

6. Ability to outsource design and production work and manage the supply chain
7. Ability to maintain confidentiality of information related to product plans and designs
8. Program management (organization, communication, and control)
9. Commonality and shared entities: platforms, systems, and components
10. Ability to meet quality characteristics of the product, including variety of expected features and delights

EXTERNAL FACTORS

1. Political changes and other situations (e.g., adverse weather) in the vehicle-producing country
2. Economic conditions, such as employment levels, tax, interest, and inflation rates
3. Changes in government regulations affecting the product
4. Availability of energy and materials sources related to the vehicle performance needs and prices
5. Global factors such as political and economic conditions affecting other countries and markets related to the product
6. State of competitors and their product plans (e.g., new products introduced by the competitors)
7. Trends and changes in vehicle design and related technologies
8. Supplier abilities to meet quality, cost, and timing targets

IMPORTANCE, ADVANTAGES, AND DISADVANTAGES OF SYSTEMS ENGINEERING

IMPORTANCE OF SYSTEMS ENGINEERING

SE, with assistance from the other engineering disciplines, establishes the vehicle configuration, allocates functions and requirements to all vehicle systems and their lower-level entities, establishes measures of effectiveness for ranking alternative concepts/designs, and integrates the design with all specialty disciplines. SE is, thus, a "glue" that bonds together all the vehicle systems and the disciplines required to create a vehicle that the customers want.

SE is responsible for verifying that the developed vehicle (with all its systems) meets all the important requirements defined in the vehicle attributes and systems specifications. SE also plans for all necessary analyses that need to be conducted and ensures that design reviews are conducted to meet program timings. Thus, products developed with the application of SE principles, processes, and techniques will benefit from the following:

1. The right products will be developed, because the SE will make sure that
   (a) the customer needs are obtained and translated into requirements,
   (b) the requirements are used by multidisciplinary teams for PD, (c) the best

product configurations are selected through iterative and recursive refinements, (d) all product entities are verified to ensure compliance with their requirements, and finally, (e) the whole product is validated using customers and pre-selected test procedures. Thus, the customers will like the products and will be very satisfied.

2. PD time can be reduced by avoiding costly delays.
3. Costly redesign and rework problems will be reduced.
4. The product will remain on the market for a longer time.

ADVANTAGES AND DISADVANTAGES OF THE SYSTEMS ENGINEERING PROCESS

The major advantages of the implementation of the SE process in the development of a complex product program are

1. It will help in reducing costs and time overruns.
2. It will help in creating products that the users want (i.e., it ensures customer satisfaction).

The disadvantages of the incorporation of SE functions in a PD program are

1. It adds people (systems engineers) to the payroll and thus increases the costs of the program.
2. It creates an additional documentation burden with the SE management plan.
3. It creates more work for the team members in communicating with the SE personnel and following the activities incorporated in the SE management plan (see Chapter 12).

CONCLUDING REMARKS

Undertaking a vehicle development program is very challenging due to the complexity of managing many tasks performed by many professionals from many disciplines to design all the vehicle systems and making sure that all the vehicle specifications and requirements are met. Vehicle programs are also affected by a number of unforeseen and uncontrollable internal and external factors. The competition between many vehicle manufacturers is also very fierce, and vehicle development teams are pressured to reduce development times and budgets under fast-paced technological changes. The subsequent chapters present the concepts, methods, and processes used to meet the design challenges.

REFERENCES

Bhise, V. D. 2014. *Designing Complex Products with Systems Engineering Processes and Techniques*. Boca Raton, FL: CRC Press. ISBN:978-1-4665-0703-6.

Blanchard, B. S. and W. J. Fabrycky. 2011. *Systems Engineering and Analysis*. 5th Edition. Upper Saddle River, NJ: Prentice Hall PTR.

Hall, A. D. 1962. *A Methodology for Systems Engineering*. New York, NY: D. Van Nostrand.

International Council of Systems Engineering (INCOSE). 2006. Systems engineering handbook. Website: http://disi.unal.edu.co/dacursci/sistemasycomputacion/docs/SystemsEng/SEHandbookv3_2006.pdf (Accessed: June 6, 2016)

Kmarani, A. K. and M. Azimi (Eds). 2011. *Systems Engineering Tools and Methods*. Boca Raton, FL: CRC Press.

National Aeronautics and Space Administration (NASA). 2007. Systems engineering handbook. Report SP-2007-6105, Rev 1. Website: http://ntrs.nasa.gov/archive/nasa/casi.ntrs.nasa.gov/20080008301.pdf (Accessed: June 6, 2016)

Rechtin, E. 1991. *Systems Architecting, Creating and Building Complex Systems*. Englewood Cliffs, NJ: Prentice Hall.

Ulrich, K. T. and S. D. Eppinger. 2015. Product Design and Development. 6th Edition. New York, NY: Irwin McGraw-Hill.

Womack, J. P., D. T. Jones, and D. Roos. 1990. *The Machine that Changed the World*. New York: Free Press.

2 Steps and Iterations Involved in Automotive Product Development

INTRODUCTION

Systems engineering implementation is an iterative process. The iterations are necessary because many decisions that are made during the vehicle development process require consideration of alternative configurations of systems and system characteristics. The type of technologies used in the operation of each of the systems also affects their characteristics and configurations and trade-offs between vehicle attributes. Many trade-offs between vehicle attributes, such as performance versus costs (e.g., acceleration capabilities of the vehicle vs. powertrain costs), vehicle weight versus performance, energy consumption versus performance, and performance versus packaging space, need to be carefully considered to ensure that the systems meet their attribute requirements, work together, and fit within the vehicle envelope. Further, many of the design issues are dependent on the importance of each of the vehicle systems and its features to customers. And many unexplored combinations of system characteristics require extensive analyses and evaluations (e.g., testing) to determine which of the design alternatives would be feasible and most economical and would best meet customer needs.

Systems engineering implementation also involves simultaneous consideration of inputs from professionals from many disciplines. Simultaneous (or concurrent) engineering requires constant communication between professionals from all disciplines to ensure that requirements for all vehicle attributes and trade-offs between the attributes are considered. The communications between professionals occur in many informal and formal information exchanges and design review meetings. The product visualization in the design reviews is facilitated through reviews of drawings, computer-aided design (CAD) models, and physical models (e.g., mock-ups, bucks, prototypes). Physical properties or three-dimensional CAD models with fly-through views (i.e., camera views from different locations or paths) are particularly useful in visualizing the space available to package all affected systems within the vehicle space when studying configurations, interfaces, interferences, and clearances between different vehicle systems (see Chapter 13 for more details).

For example, powertrain packaging involves understanding the spaces required to package the engine, transmission, suspension system, steering system, wheels and tires, shafts, final drive, and braking system within the vehicle body and chassis systems. The vehicle body system is configured to accommodate the needs of the occupants and requirements for vehicle attributes such as styling, aerodynamics, fuel economy, comfort, and safety. There are many trade-offs (e.g., occupant space vs. powertrain space, powertrain space vs. acceleration performance) that need to be carefully evaluated to come up with a balanced vehicle design. The problem is further complicated when a number of options, such as body-styles and different combinations of engines, transmissions, wheels, and optional features, are offered in the same vehicle program.

This chapter provides a basic understanding of the systems engineering process and its iterative nature.

SYSTEMS ENGINEERING PROCESS AND MODELS

THE PROCESS BEGINS WITH UNDERSTANDING CUSTOMER AND BUSINESS NEEDS AND GOVERNMENT REQUIREMENTS

The vehicle development process begins with a thorough understanding of customer needs, business needs, and government requirements. An automotive company has many types of customer. Most of us think of the customers as those who actually purchase and use the vehicles. They are generally referred to as the external customers; that is, they are outside the organization of the auto company. Their needs must be satisfied; otherwise, they may purchase their next vehicle from another manufacturer. Service personnel who repair and maintain the vehicles are also external customers, and their needs must be considered. Shareholders and investors are also external customers, whose needs must be satisfied to ensure that they supply the capital to finance the product programs in return for dividends and/or interest payments from the company and capital appreciation. There are also internal customers, who are primarily company employees who perform their work by receiving information, hardware (e.g., tools), software, and in-process work from other employees. For the employees to work together cohesively, it is important that the needs of these internal customers are met.

The auto company also has its own business need to grow its revenues and profits by satisfying its internal and external customers. It needs to ensure that right products are designed and introduced in the market at the right time and that its products compete well with other products from its competitors. Thus, benchmarking of existing vehicles, both competitors' vehicles and the company's own products, must be performed to understand how different vehicles are designed and manufactured using available technologies (see Chapter 4 for more detail on benchmarking).

A thorough understanding of government requirements that must be met during the lifetime of the vehicle being developed is crucial, because if the vehicle fails to meet any of the requirements, it may be subject to costly recalls, penalties, fines, and repair liabilities. The National Highway Traffic Safety Administration's Federal Motor Vehicle Safety Standards (NHTSA, 2015) and the Environmental Protection Agency's fuel economy and emissions requirements (EPA and NHTSA, 2012) on greenhouse gases are major requirements that must be met (see Chapter 3 for more details). In addition, the product liability climate requires vehicle manufacturers to make sure that the products sold are free from design and manufacturing defects that could cause injuries to vehicle users and others (see Bhise, 2014).

SYSTEMS ENGINEERING PROCESS

Figure 2.1 presents a flow diagram illustrating the systems engineering process during vehicle development. The top of the diagram shows the three important needs described above. These needs are generally translated into vehicle attribute requirements to ensure that the vehicle possesses the attributes that its customers expect. In addition, a number of new vehicle features that may surprise and delight the customers are also considered (Bhise, 2012, 2014). The information gathered during

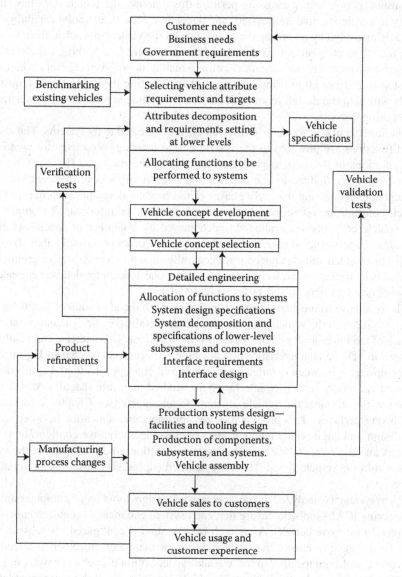

FIGURE 2.1 Flow diagram illustrating the systems engineering process in vehicle development.

benchmarking of a number of competitive vehicles aids in this process. The vehicle attribute requirements are cascaded (i.e., allocated) to lower-level subattributes of each attribute and to the vehicle systems (see Chapter 9 for more details). This process also produces more detailed specifications of the vehicle being developed (see Chapter 7).

This information is used to determine a list of vehicle functions and requirements for each of the functions. The functional requirements specify what the vehicle must be able to achieve to perform its functions. For example, one of the key functions is to transport people and/or cargo. To perform this function, the vehicle must have the ability to accelerate, maintain speed, and decelerate. The acceleration capability is typically measured by recording the time taken by the vehicle to reach a given speed (e.g., from 0 to 60 mph in 6 s). Similarly, the deceleration or braking capability is measured by recording the distance within which a moving vehicle can come to a full stop (e.g., from 60 to 0 mph within 120 ft). Meeting these functional requirements will help the design team to decide on the characteristics of the powertrain and braking systems.

The functional requirements of the vehicle are allocated to its systems. The allocated functions are provided to the system design teams as objectives for creating configurations of the vehicle systems. The systems are packaged (located) within the vehicle space defined by the exterior and interior surfaces of different vehicle concepts created during the early phases of the product development process. The vehicle concepts are also concurrently refined as additional information is obtained. The vehicle concepts are evaluated and reviewed by a number of specialists and management personnel to narrow them down to a few concepts (usually about two to four). The selected vehicle concepts are generally shown to representative groups of customers in market research clinics to help select one concept for detailed engineering work (see Chapters 10 and 11 for more details).

The challenge to the engineering team begins with the allocation of functions to various systems in the vehicle and deciding on the details (i.e., design configuration) of each of the lower-level systems of each vehicle system. This exercise can result in changes in vehicle characteristics and attribute requirements (shown in Figure 2.1 by the up-arrow between detailed engineering and concept selection). Connections between various systems (i.e., interfaces) are studied to ensure that all systems can work together to meet the vehicle attribute requirements (see Chapter 8 for more details on interfaces). This process requires constant communication between various design and engineering teams to discuss possible alternative configurations of systems and trade-offs between various system functions and packaging of the systems within the vehicle space. Thus, the detailed engineering design is an iterative process.

Each system is analyzed using specialized design tools (e.g., computer-aided engineering [CAE] tools) to ensure that each system can meet its requirements (see Chapter 16 for more details). After each basic system is configured, its subsystems must be designed to ensure that all the subsystems can work together as intended and can be packaged together in the available space within the vehicle envelope. The subsystem designs are further decomposed into lower levels till the component level is reached. Each subsystem also needs to be analyzed to ensure that it can perform its intended functions and that it fits within the space allocated for its parent system. Similarly, components are designed to ensure that they can work together and fit within the allocated space for the subsystem.

As the detailed engineering work progresses, the manufacturing process engineers work simultaneously to determine the tools and facilities required to produce the vehicle. The vehicle engineering and manufacturing engineers work together and make the necessary design changes to meet both the vehicle attribute requirements and the manufacturing requirements. This interface is shown in Figure 2.1 by both directional arrows between the "detailed engineering" and "production systems design" boxes. (Note that Figure 2.1 differs from Figure 1.1 because it includes more detail on systems engineering tasks such as attribute requirements development, cascading and allocation of vehicle functions to systems, and concept selection.)

Early production parts (prototype components) undergo verification tests to ensure that all applicable component-level requirements are met. The components are then assembled to form subsystems, and the subsystems are tested to verify that they meet their respective subsystem-level requirements. This process of assembly and verification is continued to form higher-level systems till the whole assembled vehicle is available. The early versions of such assembled vehicles (called *prototype vehicles*) are also subjected to a number of verification tests to ensure that all key vehicle-level requirements are met. The verification testing process is generally a part of the detailed engineering work. Once the early production vehicles are available, they are evaluated by a number of customers and company personnel to run validation tests (or drive evaluations) to determine their acceptability. The validation testing is described in Chapter 14.

After management approval of the vehicle validation, the production of the vehicles formally begins, and the produced vehicles are transported to the dealers. The feedback from the customers who purchase and use the vehicles is continuously monitored throughout the vehicle life cycle to ensure high levels of customer satisfaction. Any "manufacturing process changes" and "product refinements" resulting from the customer feedback (called *customer experience*) are shown in the boxes on the lower left-hand side of Figure 2.1.

SYSTEMS ENGINEERING "V" MODEL

The systems engineering "V" model presents all important steps in the product life cycle. The model is presented in Figure 2.2. The model is known as the systems engineering "V" model because the steps are arranged in a "V" shape, with succeeding steps shown below or above the preceding steps (see Blanchard and Fabrycky [2011] for more details). The model is described in the next section in the context of development of a new automotive product.

The model shows basic steps of the entire vehicle program on a horizontal time axis, which represents time (t) in months before Job#1. In the automotive industry, "Job#1" is defined as the event when the first vehicle is shipped out of the assembly plant for sale. The vehicle program generally begins many months prior to Job#1.The beginning time of the program depends on the scope and complexity of the program

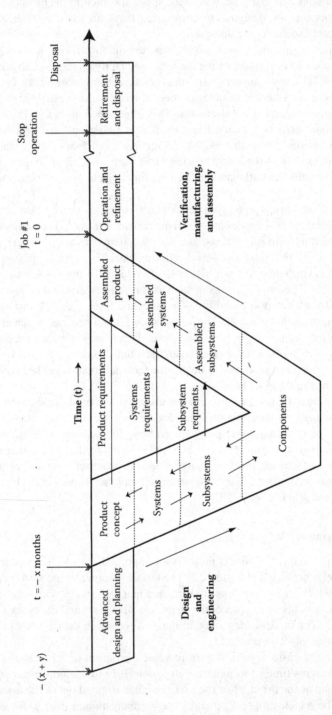

FIGURE 2.2   Systems engineering model showing the life-cycle activities of a new vehicle from the advance planning activities to the disposal of its production facilities.

(i.e., the changes in the new vehicle as compared with the outgoing product) and the state of management's approval to begin the vehicle development process.

Left Side of the "V": Design and Engineering

In the early stages prior to the official start of the vehicle program, an advanced design and product planning activity (which usually involves an advanced vehicle planning department or a special vehicle planning team) determines the vehicle characteristics and its preliminary architecture (e.g., vehicle type [body-style], size and type of powertrain, locations of the drive wheels [front-wheel drive/rear-wheel drive/all-wheel drive]), its performance characteristics, the intended market (i.e., countries where the vehicle will be sold), and so forth. It also provides a list of reference and competitors' vehicles (used for benchmarking) that the new vehicle may replace or compete with. A small group of engineers and designers (usually about 10–15) from the advanced design group are selected and asked to generate a few early vehicle concepts to understand the design and engineering challenges. A business plan, including the projected sales volumes, the planned life of the new vehicle, the vehicle program timing plan, the facilities and tooling plan, the manpower plan, and the financial plan (including estimates of costs, capital needed, revenue stream, and projected profits), is developed and presented to the senior management along with all other vehicle programs planned by the company (to illustrate how the proposed program will fit within the overall corporate product plan and business strategy). Chapter 5 presents additional details on business plan development.

The vehicle program typically begins officially after the approval of the business plan by the company management. This program approval event is considered to occur at x months prior to Job#1, as shown in Figure 2.2. The figure also shows that the advanced design and planning activity begins at (x + y) months prior to Job#1. (Depending on the scope of the activity, the value of y can range from about 3 to 12 months.)

At minus x months, the chief vehicle program manager is selected, and each functional group (such as design [styling], body engineering, chassis engineering, powertrain engineering, electrical engineering, aerodynamics engineering, packaging and ergonomics/human factors engineering, manufacturing engineering) within the product development and other related activities is asked to provide personnel to support the vehicle development work. The personnel are grouped into teams, and the teams are organized to design and engineer the vehicle and its systems and subsystems.

The first major phase after the team formation is to create an overall vehicle concept (labeled "Product concept" in Figure 2.2). During this phase, the industrial designers and the package engineers work with different teams to create the vehicle concept, which involves (a) creating early drawings or computer-assisted design (CAD) models of the proposed vehicle, (b) creating computer-generated 3-D lifelike images and/or videos of the vehicle (fully rendered with color, shading, reflections, and textural effects), and (c) physical mock-ups (foam-core, clay, wooden, or fiberglass bucks to represent the exterior and interior surfaces of the vehicle). The images and/or models of the proposed vehicle are shown to prospective customers in market research clinics and to the management. Their feedback is used to further refine the product concept. The market research clinics are described in Chapter 21.

As the vehicle concept is being developed, each engineering team decides on how each of the vehicle systems can be configured to fit within the vehicle space (defined by the exterior envelope and interior surfaces of the vehicle concept) and how the various systems can be interfaced with other systems to work together to meet all the functional, ergonomic, quality, safety, and other requirements of the product. This step is shown as the "Systems" step in Figure 2.2. Any problems discovered during this phase may require iterating the process back (to the previous phase) to refine or modify the product concept. (This feedback represents the up-arrow from "Systems" to "Product concept" shown in Figure 2.2.)

As the systems are being designed, the next phases involve a more detailed design of the lower-level entities, that is, design of subsystems of each system and components within each of the subsystems. These subsequent steps, straddled in time to the right, are shown as "Subsystems" and "Components." The steps described in this section, forming the left half of the "V," represent the time and activities involved in "Design and engineering." The up-arrows in the left half of the "V" indicate the iterative nature of the systems engineering loops shown in Figure 2.1.

Right Side of the "V": Verification, Manufacturing, and Assembly

The right half of the "V," moving from the bottom to the top, involves manufacturing components (or lower-level entities) and testing to verify that they meet their functional characteristics and requirements (developed during the left half of the "V"). The components are assembled to form subsystems, which are tested to ensure that they meet their functional requirements. Similarly, the subsystems are assembled into systems and tested; and finally, the systems are assembled to create the whole vehicle. At each of the steps, the corresponding assemblies are tested to ensure that they meet the requirements considered during their respective design steps (i.e., the assemblies are verified). These requirements are shown as the horizontal arrows from the left side to the right side of the "V" in Figure 2.2. The right side of the "V" is therefore labeled "Verification, manufacturing, and assembly." It should be noted that down-arrows between various assembly steps in the right half of "V" are not shown in Figure 2.2. The down-arrows would indicate failures in the verification steps. When failures occur, the information is transmitted to the respective design team for incorporation of design changes to avoid repetition of such failures.

The engineers and technical experts assigned to various teams in the vehicle program work through all these steps and continuously evaluate the vehicle design to verify that the vehicle users can be accommodated and they will be able to use the vehicle under all foreseeable usage situations. Early production vehicles developed just before the Job#1 are usually used for additional whole-vehicle evaluations for product validation purposes (see Chapter 14 on validation testing).

Right Side of the Diagram: Operation and Disposal

After Job#1, the vehicles are produced and transported to the dealerships and sold to the customers. The model in Figure 2.2 also shows a time period called "Operation and refinement." During this time period, the produced vehicles are purchased, used, and maintained by the customers and serviced by the dealers (or other repair shops). The vehicle may also be refined with some changes (i.e., revised with minor changes during the existing model cycle, or updated as a refreshed new model every few years) during its operational time. When the vehicle becomes old and outdated, it is pulled from the market. This marks the end of the production of the vehicle. At that point, the assembly plant and its equipment are recycled or retooled for the next vehicle (as a next model year product or a totally new product), or the plant is closed. As the products reach the end of their useful life, the products are sent to the scrapyards, where many of the components may be disassembled. The disassembled components are either recycled for extraction of the materials or sent to the junkyards.

The website of this book contains an Excel spreadsheet illustration of a "V" model of an automotive product development process. It also shows a Gantt chart with various activities and monthly cost estimates of the activities (see Chapter 19).

SYSTEMS ENGINEERING MODEL WITH FIVE TYPES OF LOOP

Figure 2.3 illustrates a variation of a commonly used systems engineering model with five types of loop. The five loops in the models are (1) requirements loop, (2) design loop, (3) control loops, (4) verification loop, and (5) validation loop. The loops illustrate the iterative nature of the systems engineering process, beginning with the customer needs, business needs, and regulatory requirements, which are translated into vehicle attribute targets and vehicle attribute requirements. The vehicle attribute requirements are cascaded into requirements of the lower-level entities (i.e., vehicle systems, subsystems, and sub-subsystems down to the component level) along with the function analysis and function allocation. The generated requirements and functional allocations are iterated and synthesized into possible and feasible product configurations till a balanced vehicle design is achieved.

A balanced vehicle design is a configuration of the vehicle that is found to be acceptable by taking into account all the attribute requirements and trade-offs between attributes. The vehicle configuration includes agreed-on allocations of vehicle functions to its systems and assignment of spaces to the systems (i.e., packaging of the systems) within the vehicle space by achieving the required interfaces between the systems.

MANAGEMENT OF THE SYSTEMS ENGINEERING PROCESS

Management of the systems engineering process can be better understood, first, by studying the processes and techniques involved in performing the following core facilitating functions:

1. Defining and locating gateways in the program timing schedule
2. Managing by vehicle attributes
3. Target setting at the vehicle level
4. Decomposing the vehicle into manageable lower-level entities
5. Defining the relationship between vehicle attributes and vehicle systems
6. Interfacing between vehicle systems

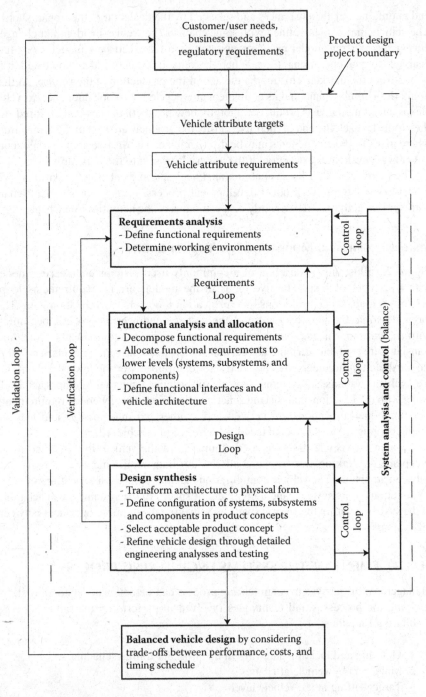

FIGURE 2.3   Systems engineering process with five types of loop.

7. Setting requirements (requirements analysis)
8. Conducting evaluations, verifications, and validations

The systems engineering management plan (SEMP) described in Chapter 12 provides documentation on how all these functions are coordinated and performed by various teams in the vehicle development process.

DEFINING AND LOCATING GATEWAYS IN VEHICLE PROGRAM TIMINGS

To manage the vehicle development program and its process, all major program activities are mapped on a timing chart such as a Gantt chart (see Figure 1.2) and a gateways timing chart (see Figure 2.4). When certain key tasks are completed, the

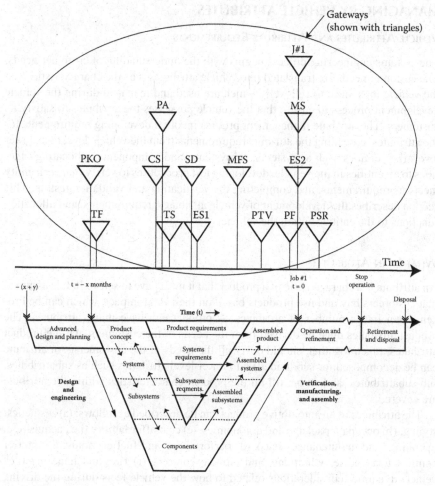

FIGURE 2.4   Gateways timing chart in the vehicle development program.

vehicle design and program progress is reviewed by technical experts and management personnel. The reviews are conducted to ensure that all the tasks planned during the previous steps were completed satisfactorily, and necessary design changes are made to solve problems identified during the prior reviews. These task completion events and management reviews are defined as *gateways*. The management approvals at the gateways signify that the vehicle program will proceed to the next tasks or program phase. The gateways are placed on the timing charts. The gateways are usually identified using a triangular symbol with acronyms defining each of the tasks. Figure 2.4 provides an example of gateways placed on the timeline of the "V" model of the vehicle development program shown in Figure 2.2. Table 2.1 provides the definitions of various gateways used in a typical vehicle development program. Each auto company generally has a unique set of definitions and identification labels for its gateways.

MANAGING BY VEHICLE ATTRIBUTES

VEHICLE ATTRIBUTES AND ATTRIBUTE REQUIREMENTS

The systems engineering process begins with the understanding of customer needs. The customer needs are translated into vehicle attributes (i.e., the characteristics that the vehicle must have to sell well), which are used and managed during the vehicle development process to ensure that the vehicle possesses the attributes to satisfy its customers. The attribute management process involves developing requirements for the attributes, cascading the attribute requirements from the vehicle level to its lower levels (i.e., entities such as systems, subsystems, and components), allocating functions to all entities in the vehicle, developing test procedures to verify that each entity meets its requirements, and completing the verification and validation testing. This section describes the development of vehicle attributes, requirements, and allocation functions to the entities within the vehicle.

WHAT IS AN ATTRIBUTE?

An attribute is a characteristic of a product that it must have to sell well. It is assumed that customers buy and use products based on their total impact, which can be broken down into a number of attributes. A product can have many attributes. The product attributes must be derived from the needs of the customers. All the product attributes, taken together, should cover all the needs of the customers. An attribute can be decomposed (or subdivided) into lower-level attributes such as subattributes, sub-subattributes, and so on, till all the product characteristics within the attribute are covered.

The attributes of an automotive product can be described as follows: (a) aesthetics/styling, (b) occupant package and ergonomics, (c) cost/affordability (i.e., acquisition, operating, and maintenance costs), (d) performance and fuel economy, (e) interior comfort (e.g., noise, vibrations, and climate control), (f) ride and handling (i.e., vehicle dynamics considerations related to how the vehicle feels during the driving

TABLE 2.1
Definitions and Labels for Gateways

Description of Phase	Description of Step	Gateway Description	Label
Pre-program work	The objective of this gateway is to define the program for developing one or more vehicles (base vehicle and its variants).	Program definition	PD
Vehicle definition, target setting, concepts development, and systems design	The time at which the proposal to develop the new vehicle is formally accepted by the senior management and the program is defined. The program is kicked off with selection of program leader and team leaders.	Program kick-off	PKO
	Teams formed and begin work on gathering customer data and trends in vehicle design and technology. Benchmarking of selected competitive vehicles begins.	Team formation	TF
	Targets set on functional specifications at vehicle and systems levels. Design and engineering teams create several vehicle concepts.	Targets set	TS
Vehicle concepts evaluation and concept selection	Vehicle concepts selected by management are prepared for internal and external market research.	Concepts reviewed	CR
	Management meets to review market research results and program team recommendations, and selects a concept for further development.	Concept selection	CS
System design engineering work	Engineering teams begin system-level design work on each major vehicle system. Functional aspects of each system and interfaces between systems are constantly reviewed to study engineering feasibility.	Engineering launch	EL
Systems approval and engineering sign-offs	Manager responsible for each major vehicle system (jth system) reviews system design with other systems managers and obtains approval of the system-level design.	System design approval	SD(j)
	Leaders in each engineering activity sign off on the overall vehicle design, stating the current vehicle and systems design is ready for further detailed design work.	1st engineering sign-off	ES1

(Continued)

TABLE 2.1 (CONTINUED)
Definitions and Labels for Gateways

Description of Phase	Description of Step	Gateway Description	Label
Detailed engineering, verification testing, tooling and facilities design, and manufacturing readiness reviews	Vehicle program is approved by management, and funds are released for detailed system design and integration work.	Program approval	PA
	Verification tests are completed. Tests are conducted at component, subsystems, and systems level. Test results are used to incorporate changes in hardware and software and retested as needed.	Verification tests (ith entity)	VT(i)
	Prototype vehicles are assembled for vehicle-level verification tests.	Prototype test vehicles	PTV
	Marketing and field support (dealerships, part sales and service) personnel are provided with necessary technical information, hardware, and service support tools.	Marketing and field support plan launch	MFS
	Manufacturing and assembly plants retooled and conduct tests to ensure vehicle build capabilities and begin building early production vehicles for training and validation tests.	Production Readiness	PR
Production prototype vehicles building, validation testing, and sign-offs	Final production prototype vehicles are made available for engineering and validation tests and reviews by experts, management, and customers.	Prototypes final (final prototypes)	PF
	All engineering leaders sign off on the functioning, reliability, and durability of the production vehicles.	2nd engineering sign-off	ES2
	Manufacturing and plant managers sign-off on the functionality and build quality of the production vehicles.	Final manufacturing sign-off	MS
Vehicle production, sales, and service	Management approves release of the production vehicles for sales. The time at which the first production vehicle rolls out of the assembly plant is called "Job#1."	Job#1	J#1
	Customer feedback, sales, warranty, and costs data are reviewed periodically to determine future changes to manufacturing process or products.	Program status reviews (jth review)	PSR(j)

maneuvers), (g) thermal and aerodynamics, (h) weight, (i) safety, (j) security, (k) emissions (harmful gases/pollutants generated by the vehicle during its operation), (l) information and entertainment (i.e., providing needed information and entertainment to the vehicle occupants), (m) customer life-cycle experience (i.e., overall experience of the customer during vehicle usage in the customer's life stages or changes), and (n) product and process compatibility.

Table 2.2 presents the attributes and their subattributes of an automotive product. It should be noted that the definition of the attributes is not standardized, and thus,

TABLE 2.2
Attributes of an Automotive Product

Serial No.	Vehicle Attribute	Subattributes
1	Package	Occupant seating package, entry and exit, luggage/cargo package, fields of view, powertrain package, suspensions and tire package, other mechanical and electrical package
2	Ergonomics	Locations and layouts of controls and displays, hand and foot reach, visibility and legibility, posture comfort, and operability
3	Safety	Front impact, side impact, rear impact, rollover and roof crush, air bags and seat belts, sensors and ECMs, other safety features (visibility, active safety)
4	Styling and appearance	Exterior—shape, proportions, stance, and so on; interior—configuration, materials, color, texture, and so on
5	Thermal and aerodynamics	Aerodynamics, thermal management, water management
6	Performance and drivability	Performance feel, fuel economy, long-range capabilities, drivability, manual shifting, trailer towing
7	Vehicle dynamics	Ride, steering and handling, and braking
8	Noise, vibrations, and harshness (NVH)	Road NVH, powertrain NVH, wind noise, electrical and mechanical systems NVH, brake NVH, squeaks and rattles, passerby noise
9	Interior climate comfort	Heater performance, air-conditioning performance, water ingestion
10	Weight	Body system weight, chassis system weight, powertrain weight, electrical system weight, fuel system weight
11	Security	Vehicle theft, contents/component theft, personal security
12	Emissions	Tailpipe emissions, vapor emissions, on-board diagnostics
13	Communication and entertainment	Internet connectivity, within-vehicle coactivity, vehicle to infrastructure communication, vehicle to vehicle communication, audio reception
14	Costs	Cost to the customer, cost to the company
15	Customer life cycle	Purchase and service experience, operating experience, life stage changes, system upgradability, disposal and recyclability
16	Product and process complexity	Commonality, reusability, carryover, product variations, plant complexity, tooling and plant life-cycle changes

Note: ECM, electronic control module.

the list and definitions of the attributes and their subattributes can vary between and within different vehicle programs and organizations of an auto manufacturer.

Quality is another vehicle attribute, which is generally combined with other attributes such as styling, package, safety, and comfort. Similarly, durability is an attribute that can be combined with other attributes such as performance, comfort, styling and appearance, and safety. For example, durability can be incorporated into part of many attribute requirements stating that the applicable requirements should be met over the life of the vehicle (e.g., 15 years or 150,000 miles of driving).

To help the reader in understanding the areas covered by product attributes, another example is a laptop computer. The attributes of this product are (a) physical size, (b) weight, (c) display size, (d) ergonomics (i.e., ease in using the keyboard, touch pad, audio, display, and disk drive), (e) processor capabilities (e.g., capacity and speed of processing data), (f) data storage capacity, (g) battery capacity (e.g., hours of operation between recharges), (h) input/output ports, (i) wireless connectivity, (j) aesthetics (i.e., styling/appearance), (k) durability, and (l) life-cycle costs (i.e., costs incurred by the customer during the life cycle of the computer). This list of attributes should cover all the customer needs that the laptop design team should consider during the entire life cycle of the laptop computer.

ATTRIBUTE REQUIREMENTS

The requirements specified to achieve the product attributes can be defined as the *attribute requirements*. To manage the attributes, each attribute can be further divided into subattributes, sub-subattributes, and so on, at different levels. All attribute requirements should be cascaded (i.e., assigned) to various vehicle systems during the functional analysis and allocation.

Additional information on requirements and guidelines for developing "good" requirements are provided in a later section of this chapter.

ATTRIBUTE MANAGEMENT

In some automotive companies, the customer requirements are specified in terms of attribute requirements. An attribute manager is assigned to each attribute, and the responsibility of each attribute manager is to ensure that the requirements of his or her attribute are allocated to a proper set of systems and lower-level entities, and evaluated constantly. The attribute management responsibility is generally assigned to independent core engineering functions that are different from the line engineering activities responsible for designing and developing various systems of the product. For example, a manager assigned to the "comfort and convenience" attribute will have to review the entire design of the vehicle being developed and analyze each vehicle system to determine whether the system will have an effect on occupant comfort and convenience. The vehicle systems having an effect on the attribute will then be analyzed and evaluated in detail to ensure that all requirements of the attribute and trade-offs between the attribute and other related attributes or subattributes, such as ergonomics of driver interfaces, seat comfort, entry/exit ease, luggage loading convenience, engine service ease, thermal comfort, and so forth, are considered.

Importance of Attributes

Instead of determining product requirements directly from customer needs, as shown in Figure 2.2, some organizations have found that defining the product attributes from the customer needs and managing each of the product attributes (i.e., managing all product requirements that deliver or contribute to a given product attribute) is a better approach—especially for the development of complex automotive products. It should be noted that customer needs obtained from interviewing customers may not be complete; that is, they may not take into account many engineering considerations. Therefore, the attribute requirements should cover both the customer needs and the engineering needs. For a vehicle to possess a certain attribute, its attribute requirements are cascaded (or assigned or allocated) from the vehicle level down to lower levels in the system hierarchy (i.e., from the product to its systems to subsystems and components).

The management of product requirements based on attributes has three major advantages:

1. The attribute requirements help everyone in the product development process to understand the traceability of the requirements to certain product attributes (i.e., any requirement can be traced back to one or more product attributes).
2. People specialized in an attribute can be made responsible for ensuring that the product is being designed to meet the attribute requirements (so that the product will possess the attribute).
3. The presence of attribute requirements ensures that all the product attributes are studied (i.e., tracked and evaluated) at every product development phase, and compliance with the attribute requirements is reviewed at all major milestones in the product program.

Thus, the attributes management approach ensures that the customer needs are not overlooked (i.e., the customers will be satisfied) during the design and subsequent phases of the product life cycle. This topic is further discussed in Chapter 9.

VEHICLE-LEVEL TARGET SETTING

Before the vehicle is defined, it is essential to agree on an overall corporate goal (i.e., product planning and marketing strategy) in developing the vehicle. Some examples of the goal are (a) to create a vehicle to replace an aging existing vehicle with a similar market positioning, (b) to create a vehicle that will be the best in its class, and/or (c) to create a vehicle that will appeal to customers in a different market segment (e.g., a vehicle designed for the U.S. market to be also sold in China). Thus, the process of definition of the vehicle to be developed will depend on the strategy adopted by the automotive company in target setting.

TARGET SETTING AND MEASURES

Target setting needs to be undertaken minimally at the vehicle level (i.e., for the selected vehicle as a whole) and at its attribute and subattribute levels. The measures used to define the targets should be objective (i.e., the targets can be measured by using one or more physical instruments) as well as subjective (i.e., based on information or ratings provided by customers or experts).

Some examples of the vehicle-level measures that can be used are

1. Best-in-class vehicle (i.e., best overall score among all the selected vehicles in its class)
2. Percentage of customers satisfied (e.g., at least 80% of customers should state that they will be "very" or "completely" satisfied with the vehicle)
3. Percentage of design guidelines and/or requirements met (e.g., at least 95% of the vehicle-level attribute requirements met)
4. Minimum, maximum, or range of values of evaluation measure achieved in a pre-defined test (e.g., fuel economy in terms of minimum miles per gallon, maximum aerodynamic drag coefficient)
5. Percentage of customers preferring your target vehicle to other selected competitors' vehicles in a predefined evaluation test (e.g., at least 15% improvement on the selected competitor's 60–0 mph stopping distance)
6. Number (or percentage) of target vehicles sold in a specified market segment (e.g., at least 20% market share in the entry-luxury mid-size passenger car segment)
7. Rank of target vehicle in a specified evaluation test against other selected competitors' vehicles (e.g., must be within the top three of the selected competitors)
8. Percentage of selected criteria met by the target vehicle
9. Average or weighted score of the target vehicle in a selected set of evaluation tests with multiple evaluation criteria
10. Vehicle based on a specified platform and sharing at least 55% of the platform (components and systems of other vehicles developed from the platform)

Some Examples of Attribute-level Measures

A new four-door sedan vehicle, when compared with its reference vehicle (e.g., current model or its leading competitor), should meet certain levels of selected attributes of the vehicle. The attributes of the vehicle are characteristics that the vehicle must have for it to sell well. For example, fuel economy, overall vehicle weight, powertrain type, ride and handling, and interior space are a few attributes of a vehicle. Some examples of vehicle targets based on the attributes are

1. The new vehicle shall provide at least 15% better fuel economy as compared with the previous vehicle model.
2. The overall weight of the vehicle shall be reduced by 6% (or 300 lb) as compared with the previous vehicle model.
3. The powertrain of the new vehicle shall have options to provide a turbo-boost engine and an eight-speed transmission.
4. The new vehicle shall have safety features comparable to those of the reference vehicle.

5. The handling, ride, and braking performance of the vehicle shall be at least 10% better than those of the reference vehicle.
6. The vehicle shall be equipped with speech recognition and text to speech capability to allow communication of e-mail messages safely.
7. The interior of the vehicle shall be more spacious and shall provide at least 15 mm increase in driver's legroom, headroom and shoulder room as compared with the reference vehicle.
8. The exterior of the vehicle shall be perceived to be more aerodynamic than that of the reference vehicle.
9. The climate control shall allow automatic temperature control and setting of independent separate controls for the driver, the front passenger, and the rear passengers.
10. The interior noise level at 70 mph shall be at least 2 dBA lower than for the reference vehicle.
11. The vehicle shall have the following optional security features: peripheral courtesy lighting, panic auditory alarm, theft control alarm, heart rate monitoring, and emergency messaging system with airbag activation.

To decide on the levels of each of the attributes, the measures (i.e., variables) that will be used for measurement of each of the targets and their measurement procedures must be determined and accepted by the company management.

DECOMPOSITION OF A VEHICLE INTO MANAGEABLE LOWER-LEVEL ENTITIES

MANAGING A COMPLEX PRODUCT

The need for systems engineering arose with the increase in complexity of products. Increased product complexity, in turn, increases the number of interactions (i.e., relationships) between many components and also increases the challenges in designing for high levels of reliability. This complexity, due to the higher number of entities in the product, results in larger organizations requiring management of many teams involving professionals from many disciplines, needing to meet many requirements, and needing to make a multitude of decisions. (Note: designing very simple products can be accomplished by a very small number of professionals.)

Therefore, a complex product should be divided (or decomposed) into a number of manageable entities. This decomposition is a useful step in managing larger systems in the product development process. The product can be decomposed into many systems, systems into subsystems, and subsystems into components. Some products can be divided into many hierarchical levels; that is, systems can be divided into many levels of subsystems, such as subsystem, sub-subsystem, and sub-subsystem. Product design responsibilities can also be divided and assigned to individuals within various groups or teams. The number of levels of divisions depends on many factors, such as past design experience and problems encountered in developing similar products, the ability of the design team to deal with many design issues simultaneously, the level of the design and engineering details that need to be analyzed and evaluated, stringency in meeting requirements, and the program schedule.

DECOMPOSITION TREE

Figure 2.5 presents a product decomposition tree. This is a tree diagram (an upside-down tree) showing the top-down progressive decomposition of a product (P) into its systems (S1 to S5), each system into its subsystems (e.g., SS11, SS12, SS21, ..., SS53), and each subsystem into its components (C111, C112, ..., C533).

Figure 2.6 presents an example decomposition tree of an automotive audio system. The audio system is shown as a Level 3 (L3) system, as it is developed within the vehicle electrical and electronics system [which is a Level 2 (L2) system]. The vehicle is considered to be the Level 1 (L1) system. The subsystems of the audio system are audio source, sound system, wiring and connectors, and controls and displays (which are Level 4 systems). Each of the subsystems is further decomposed into its sub-subsystems (shown here as L5-level systems). The L5-level systems can be further decomposed into several lower-level systems, down to the individual component level or to a level at which it is purchased as a supplier-provided assembled unit (which requires no further decomposition from the vehicle manufacturer's viewpoint). Examples of Level 5 components are AM receiver, CD player, wiring harness, and touch screen.

The decomposition of the product into its lower-level entities also requires a careful cascading of attribute requirements from the product level to lower-level entities. It should be noted that each lower-level entity exists to serve at least one or more functions necessary for the product to meet its requirements. It is also important to understand and keep track of the functions of each entity within each system, because the design team involved in designing each system must make sure that the system performs its functions. This topic is covered in greater detail in Chapter 7.

RELATIONSHIP BETWEEN VEHICLE ATTRIBUTES AND VEHICLE SYSTEMS

It is important to realize that to meet the requirements of a given attribute, one or more systems are needed. Thus, each system exists to provide one or more vehicle attributes. The relationships between the vehicle attributes and the vehicle systems can be shown using a matrix diagram. Table 2.3 provides a relationship matrix between the vehicle attributes and the vehicle systems. The vehicle attributes are listed as rows of the matrix, and the major vehicle systems are represented in the columns of the matrix. The strength of the relationship between each attribute and each system is shown in the cell defined by the intersection of the row and the corresponding column. The number in the cell shows the strength of the relationship: 9 = strong relationship, 5 = medium relationship, 1 = weak relationship, and 0 (or blank) = no relationship.

The relationship numbers provided in Table 2.3 show that the following attributes have a relationship (or are affected by) all vehicle systems: package, weight, costs,

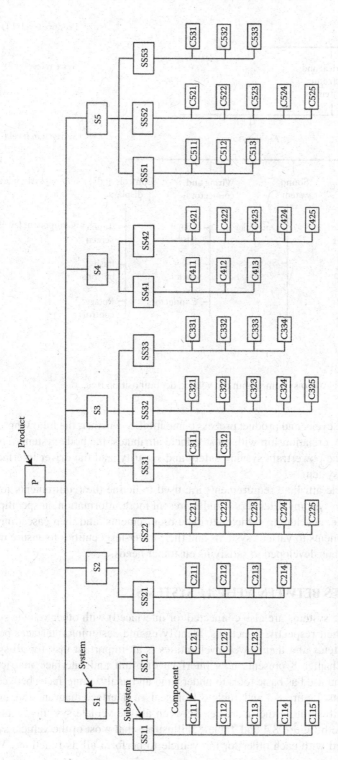

FIGURE 2.5 Illustration of top-down decomposition of a product into systems, subsystems, and components.

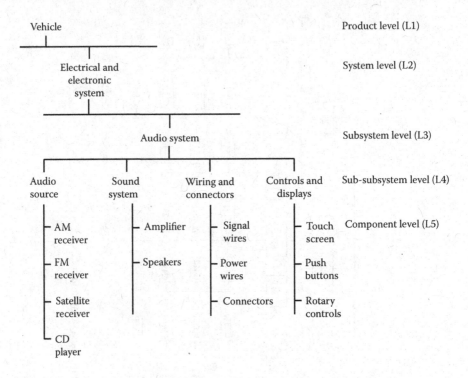

FIGURE 2.6   Audio system within the vehicle decomposition tree.

customer life cycle, and product process compatibility. Further, the following vehicle systems have a relationship with most vehicle attributes: the body system, the chassis system, the powertrain system, safety and security, and the driver interface and information system.

The vehicle attribute requirements are used to define the requirements for each of the vehicle systems. Chapters 7 and 9 present more information on specifications of the vehicle, developing vehicle attribute requirements, and then cascading attribute requirements to various systems and their lower-level entities to ensure that the whole vehicle is developed to satisfy its customer needs.

INTERFACES BETWEEN VEHICLE SYSTEMS

Many vehicle systems are also connected (or interfaced) with other vehicle systems to perform their respective functions. Identifying and designing interfaces between different systems and their lower-level entities is an important task for all systems designers. Chapter 8 presents how interface diagram and interface matrices are developed and used as basic tools to understand and analyze interfaces between different systems. Figure 8.4 and Table 8.1 present an interface diagram and an interface matrix illustrating the interfaces between all the vehicle systems included in Table 2.3. Both Figure 8.4 and Table 8.1 illustrate that most of the vehicle systems are interfaced with each other for the vehicle to perform all its functions. Vehicle

TABLE 2.3
Relationship Matrix between Vehicle Attributes and Vehicle Systems

Legend:
- Serial No.
- Vehicle Attribute
- Subattributes
- Vehicle Systems and Subsystems (in Parentheses)
- Importance Rating: 10 = Very Important, 1 = Not At All Important

Serial No.	Vehicle Attribute	Importance Rating	Body System (Body frame, Panels, Doors, Hood, Trunk/Liftgate, Fascias, Bumpers, Running Boards, Steps, Handles, Lamps, Instrument Panel, Seats, Trim Parts)	Chassis System (Wheels, Tires, Suspensions, Brakes, Steering System)	Powertrain System (Engine, Transmission, Final Drive, Cooling System)	Fuel System (Fuel Tank, Fuel Pipes, Fuel Pump, Filter, Pressure Relief Valve)	Electrical System (Alternator, Battery, Wiring Harness, Power Regulator, Fuse/Relays/Switches)	Climate Control System (Heat Exchanges, Blower, Air-conditioning System, Hoses and Tubes, Air Ducts)	Safety and Security System (Air Bags, Seat Belts, Active Safety Systems, Sensors, ECMs, Wiring Harnesses, Headlamps, Signaling Lamps, Flood Lamps)	Driver Interface and Infotainment System (Controls and Displays, Audio System, Navigation Systems)
1	Package	10	9	9	9	3	3	3	1	1
2	Ergonomics	8	3	1	9		1	3	9	9
3	Safety	10	9	9	9	9	3		9	3
4	Styling and appearance	8	9	1	1				3	1
5	Thermal and aerodynamics	7	9	3	3				1	
6	Performance and drivability	8	9	9	9		3			
7	Vehicle dynamics	8	9	9	3			1	3	
8	Noise, vibrations, and harshness (NVH)	6	5	9	9			9		
9	Interior climate comfort	8	9		3		5	5		1
10	Weight	6		9	9	9	5			
11	Security	5					5		9	9
12	Emissions	8			9	5				
13	Communication and entertainment	5					3		1	9
14	Costs	8	9	9	9	3	3	3		1
15	Customer life cycle	3	1	1	1	1	1		1	1
16	Product and process complexity	2	3	3	9	3	3	3	9	9
			712	550	602	259	203	202	312	261
			23.0	17.7	19.4	8.4	6.5	6.5	10.1	8.4
										3101

Note: Strength of relationships shown in Table 2.3 is defined as 9 = strong, 3 = medium, and 1 = weak. The subattributes of the attributes are described in Table 2.2.

systems designers need to first develop requirements for each of the interfaces to ensure that interfacing systems can work together. The interface requirements should then be used during the design of the interfacing vehicle systems. Chapter 8 provides more information on interface design considerations.

SETTING AND ANALYZING REQUIREMENTS

The requirements analysis is critical to the success of the vehicle. The requirements should be developed from customer needs, business needs, and regulatory requirements and assigned to product attributes. The requirements should be clearly documented, easy to interpret, actionable, measurable, testable, and traceable. The requirements should be analyzed to ensure that the vehicle designed to meet the requirements will produce a high level of customer satisfaction. This analysis should also verify that either the existing requirements are appropriate or new requirements (which are more appropriate for the mission/operation of the product) need to be developed.

The requirement analysis should also include (a) development of measures suitable for ranking alternative designs in a consistent and objective manner, and (b) evaluations of the impact of environmental factors and operational characteristics of the vehicle systems on the performance of the product and minimum acceptable functional requirements. These measures and evaluations should also consider the impact of the design on costs and schedule. Each requirement should be periodically examined for validity, consistency, desirability, and attainability. (See Chapters 3 and 7 for more information on requirements.)

The following subsections present basic information on requirements, reasons for their specification, and characteristics of good requirements.

What Is a Requirement?

A requirement defines one or more product characteristics and their accomplishment levels needed to achieve a specific objective (e.g., a function to be performed, performance level to be achieved, or maximum weight and/or size limits on the product) under a given set of conditions. Requirements are developed to meet customer needs, government regulations, and corporate needs (e.g., brand-specific features). The requirements are created to achieve certain attribute characteristics, functions, or performance of the product.

Why "Specify" Requirements?

Clearly stated requirements provide the information and direction needed to begin the product design process. The information provides (a) clear visibility across different teams (responsible for different systems within the vehicle) into how and why the requirements are allocated, thus helping to understand cross-functional interactions between all systems within the product, (b) clear responsibilities assigned to the design teams to meet the requirements, (c) early assurance that all top-level requirements are fully satisfied by the product, with traceability to where they are satisfied,

(d) checks to prevent unintentional addition of features and costs (i.e., avoids "gold plating"), (e) checks to avoid unwelcome surprises in later phases of the product development, (f) quick assessment of the impact of any changes made to the requirements, and (g) procedures for early and thorough verification and validation of the product design in meeting the requirements.

How Are Requirements Developed?

Most requirements are not entirely developed during the early stages of a product program. In fact, developing requirements "from scratch" requires a lot of data gathering, analysis, testing, and evaluations—and thus, requirements development is a very time-consuming and expensive process.

Most requirements are adopted from previously developed and proven requirements available from various sources such as (a) standards (e.g., internal company standards or external standards developed by international, national, or local government agencies, professional societies, and trade associations), (b) design guidelines developed by the manufacturer and its suppliers, (c) test and evaluation procedures and practices within product development organizations, (d) experiences (past failures and successes), customer feedback, lessons learned and insights gained from previous programs for similar products. The requirements should also be continuously evaluated to ensure that they are not outdated due to advances in technologies, design trends, new materials, and changes in customer needs and government regulations.

Implementation of new technologies and features into new models of previously developed products (e.g., the development of an electric vehicle as compared with vehicles with traditional internal combustion engines) will require considerable additional work in understanding issues such as how the product will be used by the customers, customer concerns, problems during operation of the product during its life cycle, and development of new technologies to get them ready for implementation. Further trade-offs between different attributes also need to be considered during the requirements development process. The requirements development process thus requires inputs and reviews from experts from different disciplines.

Characteristics of a Good Requirement

Many characteristics are to be considered in determining whether a requirement is "good" (i.e., useful, nonconfusing, and implementable). Thus, the considerations in developing a "good" requirement are

1. The requirement must state "The product shall" (i.e., shall do, shall perform, shall operate, shall provide, shall weigh, and so forth) followed by a description of what must be done.
2. The requirement should be unambiguous, clearly stated, and complete. It should be worded to minimize confusion and differences in its interpretation between different individuals. To ensure that a requirement is complete, it should provide contextual details, such as situation, environment,

operating conditions, time durations, urgency/priorities, and characteristics of its users, under which the product is expected to function.

3. The requirement should use consistent terminology to refer to the product and its lower-level entities.
4. The requirement should clearly state its applicability (i.e., when, where, types of system or hierarchical system level where it is applicable and where it cannot be applied).
5. The requirement should be verifiable by a clearly defined test, test equipment, test procedure, and/or independent analysis.
6. The requirement should be feasible (i.e., it should be possible to create the system or product without extraordinarily large development time and costs).
7. The requirement should be consistent and traceable with other requirements above and below it in the system hierarchy.
8. Each requirement should be independent of other requirements. This characteristic will help in controlling and reducing variability in the product parameters and hence, product performance.
9. Each requirement should be concise; that is, it should be stated with minimum information content.

    (Note that considerations 8 and 9 meet the two basic axioms [Axiom 1: Independence Axiom—maintain the independence of functional requirements, and Axiom 2: Information Axiom—minimize the information content in the design] of the axiomatic theory considered in product design [Kai and El-Haik, 2003]).

The National Aeronautics and Space Administration (NASA) Systems Engineering Handbook (NASA, 2007) also provides more information in its Appendix C on "How to Write a Good Requirement" and provides a "Requirements Validation Checklist."

EVALUATIONS, VERIFICATION, AND VALIDATION TESTS

Every entity at every level, from component level to whole-vehicle level, must undergo evaluations during specified design and production stages to ensure that it meets its respective requirements. The requirements are generally specified in the system design specifications or standards. The evaluations typically involve making a number of observations and/or measurements on selected samples of each entity. The measurements can involve simulations, bench tests, laboratory tests, field tests, and so forth depending on the entity and its stage in the product development cycle. The collected data are reviewed by experts and analyzed using applicable analysis tools (e.g., CAE tools, statistical tests) from applicable disciplines. The output values of parameters specified in the requirements are compared with their acceptance levels.

More detailed descriptions of different types of evaluations are covered in Chapters 20 and 21.

CONCLUDING REMARKS

Automotive products are complex, because they have many attributes and many systems. The vehicle development system involves many steps performed by many individuals from many disciplines in many iterations to determine configurations that meet many requirements through many trade-offs. This chapter provided basic information on the following topics to help in understanding their role in developing a new vehicle: (a) flow of product development process, (b) systems engineering "V" model, (c) systems engineering process, (d) product decomposition, (e) vehicle systems and interfaces between the systems, (f) managing by vehicle attributes and attribute requirements, and (g) relationships between vehicle attributes and vehicle systems. The succeeding chapters of this book provide additional material on these topics and present methods to implement the systems engineering process in the vehicle development process.

REFERENCES

Bhise, V. D. 2012. *Ergonomics in the Automotive Design Process*. Boca Raton, FL: CRC Press. ISBN 978-1-4398-4210-2.

Bhise, V. D. 2014. *Designing Complex Products with Systems Engineering Processes and Techniques*. Boca Raton, FL: CRC Press. ISBN: 978-1-4665-0703-6.

Blanchard, B. S. and W. J. Fabrycky. 2011. *Systems Engineering and Analysis*. Upper Saddle River, NJ: Prentice Hall.

EPA and NHTSA (2012). 2017 and Later Model Year Light-Duty Vehicle Greenhouse Gas Emissions and Corporate Average Fuel Economy Standards. *Federal Register*, Vol. 77, No. 199, October 15, 2012, Pages 62623–63200. Environmental Protection Agency, 40 CFR Parts 85, 86, and 600. Department of Transportation National Highway Traffic Safety Administration, 49 CFR Parts 523, 531, 533, 536, and 537. [EPA–HQ–OAR–2010–0799; FRL–9706–5; NHTSA–2010–0131]. RIN 2060–AQ54; RIN 2127–AK79.

Kai, Y. and B. El-Haik. 2003. *Design for Six Sigma: A Roadmap for Product Development*. New York: McGraw-Hill.

NASA. 2007. *NASA Systems Engineering Handbook*. Report no. NASA/SP-2007-6105 Rev1. NASA Headquarters, Washington, D.C. 20546. Website: http://ntrs.nasa.gov/archive/nasa/casi.ntrs.nasa.gov/20080008301_2008008500.pdf (Accessed: May 10, 2016)

NHTSA. (2015). *Federal Motor Vehicle Safety Standards and Regulations*. U.S. Department of Transportation. Website: www.nhtsa.gov/cars/rules/import/FMVSS/ (accessed: June 14, 2015)

3 Customer Needs, Business Needs, and Government Requirements

INTRODUCTION

Customers purchase vehicles to meet their needs. It is, therefore, very important for vehicle designers and engineers to identify the customers and understand their needs, so that vehicles can be designed to satisfy the customers. Often, the customers may not be able to tell what characteristics or features they would like in their vehicles. This is because (a) the customers themselves may not be familiar with all the features, (b) they may not be in a position to understand new features, or (c) they may not be aware of future trends in vehicle designs. But the challenge to the vehicle development team is to thoroughly understand (a) the needs of the customers, (b) design and technological trends, (c) future government requirements, and (d) economic considerations.

Using this information, the design team can define and develop possible vehicle concepts for the new vehicle. The design concepts and/or concept vehicles could then be shown to the customers, and their responses to the concepts or concept vehicles would be measured and summarized. The collected data would be used to determine the acceptability of each of the concepts and also to understand the customer needs. These are typically obtained through customers' responses, such as reactions, opinions, verbatim comments, ratings, and gestures recorded during the concept evaluations. The collected data are translated into technical terms as vehicle attribute requirements that can be used by designers and engineers to determine the specification of the proposed vehicle.

The process of translation of customer needs must be complete, that is, it must provide the specification of all major functions and characteristics of the proposed vehicle. Therefore, many automobile companies use the attributes requirements process introduced in the previous chapter. The attributes collectively define all the characteristics of the vehicle.

Vehicle development teams must know the attribute requirements of the vehicle they are asked to design. In addition to meeting the customer needs, the vehicle must meet all the applicable government requirements that will be enforced during its life cycle. Otherwise, the government agencies will not permit the auto manufacturer to sell the vehicles. The new vehicle program must also make business sense, that is, it must meet the business needs of the company so that the company will stay in business.

Clearly defined requirements from these three areas (customer needs, business needs, and government requirements) constitute inputs to the new vehicle program. This chapter covers the basic considerations and issues related to these three types of needs.

INPUTS TO THE AUTOMOTIVE DEVELOPMENT PROCESS

CUSTOMER NEEDS

The customer needs are primarily obtained by market researchers (or other team members, such as product planners and ergonomics engineers) by interviewing potential customers and asking them about what they would like or dislike in their new vehicle. This interview method works better if some item such as a concept vehicle, computer-generated images of the vehicle, or hardware with some vehicle features is shown to the customers to help them in visualizing the vehicle and evoke responses during the interviews. The customers' reaction to properties can also be observed. In some cases, the customers could be shown several products (concept vehicles along with other existing vehicles, e.g., the manufacturer's current outgoing model and its leading competitors' vehicles) and/or some product features and asked to provide their ratings, preferences, and comments. In addition, questions can be asked to obtain the reasons for their preferences or nonpreferences (i.e., what made them like or dislike certain characteristics or features of each vehicle).

Since customers have many needs, it is also essential to obtain importance ratings and reasons for the importance of each of the needs. The needs can be then classified into categories such as "must have," "nice to have," or "no interest in having." The importance ratings can be obtained by using a 10-point importance scale (e.g., 10 = absolutely important and 1 = not at all important). The importance ratings of a set of features can be also obtained from paired comparisons of the features by using techniques such as Thurstone's method of paired comparisons and the analytical hierarchy techniques covered in Chapter 21.

One of the major challenges in interviewing customers is determining whether the customer has the ability or knowledge to understand the issues related to any given item (e.g., a vehicle feature) that he or she is being asked to judge or rate during the interview. The customer may not be knowledgeable, or may not have seen or used the feature that is referred to in the question. This is especially a problem when the customer is asked to rate items or features related to future trends in design or technologies (see Chapter 6 for more details).

The data gathered on customer needs are used to develop vehicle attribute requirements (see Chapters 2 and 9). The customer needs data can also be used in applications of the quality function deployment (QFD) technique, which helps in translating the customer needs into engineering specifications (see Chapter 18).

LIST OF CUSTOMER NEEDS

As soon as the vehicle concept development team is formed, its first major task is to develop a complete list of customer needs for the vehicle. The preliminary lists of customer needs developed by the design teams for three vehicle development projects are presented below.

Mid-Size Sports Utility Vehicle (SUV)

The design team gathered data on recent models of the following three mid-size SUVs: Ford Escape, Honda CRV, and Toyota RAV-4. After interviewing a dozen customers who currently own one of these vehicles, the team prepared the following list of customer needs for a 2021 model year (MY) mid-size SUV:

1. It should be able to carry five passengers comfortably.
2. It should have lighter curb weight.
3. It should be easy to navigate through narrow city roads compared with full-sized SUVs.
4. Its ride height should be larger than that for C-platform hatchbacks.
5. It should have a powerful engine for observably undiminished performance under full load with five passengers and luggage.
6. The engine should have a more intelligent management of power.
7. If equipped with a hybrid electric powertrain, the price should not be too high.
8. It should have an all-wheel-drive option for winter driving and maintain good road grip in all conditions.
9. It should have reasonable fuel economy for daily commuting.
10. With a full gas tank, it should have a range of 385 miles in the city or 520 miles on the highway.
11. It should have significantly more cargo space than C-platform hatchbacks and subcompact crossovers.
12. It should have a 60/40 split rear seat that folds down flat.
13. It should have protective liners for the cargo area.
14. It should have additional storage space (e.g., pockets/cupboards, underneath trunk floor).
15. It should have a remote-operated powered lift gate.
16. It should have a roof rack.
17. It should have a trailer hitch option.
18. It should have a center stack media center with intuitive menu organization and with voice-activated commands capable of recognizing common terms or phrases.
19. It should have a media center that is more integrated with the customer's smartphone with a Bluetooth connection.
20. The seats should be comfortable, with ample shoulder, hip, leg, and head room.
21. The seats should be easy to adjust, and the vehicle should have optional heated and cooled seats.

22. The vehicle should have an optional heated steering wheel.
23. The vehicle should provide 110 V outlets, which should be accessible from the front and rear seats.
24. The vehicle should offer the following optional features: remote engine start button, adaptive cruise control, active lane-keeping system, parking sensors, rear view camera, blind spot detection capability, and active park assist for parallel and perpendicular parking.

Heavy-Duty Pickup Truck

Heavy-duty pickup trucks in the United States are dominated by Ford F-350, RAM 3500, and Chevy Silverado 3500. The customers for heavy pickup trucks are generally male, blue-collar, skilled tradesmen, or outdoorsy types. The target customer can also be described as a "Do-It-Yourself" type, and someone who tows and hauls regularly. The customer needs list for the 2021 MY heavy-duty pickup included the following important variables:

1. Towing capacity
2. Payload
3. Cargo bed (bed liner, size, capacity, wall height)
4. Interior space/comfort
5. Technology (global positioning system [GPS], Wi-Fi, cameras, etc.)
6. Safety features
7. Off-road capability
8. Fuel economy
9. Access (ride height, running boards, step gate, box step)
10. Entertainment and convenience
11. Audio system (Satellite/HD radio)

Primary Vehicle Controls

In another vehicle program, the customers were interviewed and asked to discuss their expectations of primary controls (i.e., steering wheel, pedals, gear shifter, and stalks) in a new 2021 MY mid-size luxury four-door sedan. The customers described their needs as follows:

1. Adjustable steering wheel position (tilt and telescopic)
2. Steering wheel surface not affected by external temperature (nonconductive material)
3. Steering wheel surface nice to touch
4. Steering wheel surface offering a good grip (rim cross section and compressive material)
5. Low steering effort at low speeds
6. Accurate steering (no slop)
7. Good steering wheel feedback
8. Appropriately positioned pedals (placed at "about right" location)
9. Linear throttle pedal
10. Linear brake pedal

11. Nice-feeling gear-shifter knob surface
12. Hand-operated paddles for shifting gears
13. Expensive-feeling stalks
14. Exciting liquid crystal display (LCD) instrument cluster with the shifter position display

BUSINESS NEEDS

The primary objectives of a vehicle program generally are to develop one or more vehicles that their customers will buy, and to make money (i.e., generate net cash flow of revenues minus costs) for the auto company from selling the vehicles. The approval of a new vehicle program and the acquisition of resources to undertake the program involve complex decisions. This is because the company management is faced with many alternatives and has to share limited resources within many vehicle programs and projects. Some examples out of many alternatives are (a) deciding whether to produce a new car or a new truck or to allocate resources to other existing vehicle programs, (b) defining the characteristics of the vehicle to develop (many possible combinations of characteristics related to the vehicle's size, body-style, level of luxury, and so forth need to be carefully considered), and (c) deciding the level of platform sharing with other vehicle models and brands. Examples of issues related to limited resources are (a) sharing of available resources, such as overall product development budget, production and assembly capacities (available plants), (b) costs of building new specialized fixtures and processing equipment, and (c) available manpower to undertake new vehicle programs (e.g., number of product development engineers available within the company). Other external factors that need consideration include the economic situation (e.g., customers' ability to purchase new vehicles), the state of market shares of different products made by the auto manufacturer, the product plans of major competitors, and the availability of suppliers and their capabilities.

The vehicle program team generally prepares a business plan to facilitate management's decision to approve the program. The business plan provides a detailed description of the proposed vehicle and other information such as competitors of the proposed vehicle, program timings, estimates of costs and revenues over the life cycle of the program, and risks in undertaking the program. More information on the business plan and related decision-making issues is provided in Chapters 5 and 17.

GOVERNMENT REQUIREMENTS

The federal, state, and local government requirements for vehicle characteristics, such as minimum and maximum limits on vehicle dimensions, weight, fuel consumption, greenhouse gas emissions, and safety systems and features (e.g., vehicle lighting, braking, and crashworthiness), must be met, that is, they are mandatory. These requirements are covered in a later section of this chapter (see "Government Requirements in Safety, Emissions and Fuel Economy") and in Chapter 9 under the safety and security attribute and its subattributes.

OBTAINING CUSTOMER INPUTS

Customer needs can be obtained by the use of a number of methods. Methods of observation, communication, and experimentation can be used to gather data on customer needs.

The following list describes the approaches commonly used to obtain customer needs:

1. Ask customers to describe their needs and summarize inputs obtained from a large number of representative customers.
2. Ask customers to rate the importance of customer needs developed from a preselected list.
3. Show a vehicle (or a vehicle concept), ask them to use the vehicle if possible, and then ask them to describe what they liked or disliked and rate the importance of customer needs developed from a preselected list.
4. The customer may be asked to participate in an experiment involving several vehicles (or vehicle designs). The experiment may involve the customer performing several tasks using each vehicle. The data collected during the performance of each task can be analyzed to determine the tasks and vehicle designs in which the customers had the best performance (e.g., using measures such as task completion time and errors committed). In addition, after the completion of each task, or after the entire experiment, a number of questions can be asked to determine the customers' likes, dislikes, and preferences when performing different tasks with different vehicle designs.

The three basic methods are described in the following section. Many combinations of these three basic methods can be used to obtain customer needs.

OBSERVATION METHODS

In observation methods, information is gathered by direct or indirect observation of users (i.e., customers as drivers or passengers) in different vehicle usage situations to determine how different vehicle features are employed by the users. One or more observers can directly watch, or video cameras can be set up and their recordings played back at a later time for observation and analysis. The customers can be observed in their natural vehicle usage situations, or they can be asked to perform certain tasks, such as to take a trip on a predetermined route, and during driving they can be asked to operate certain features, such as to tune in to a radio station with a specified frequency or to follow the directions provided by the onboard navigation system. Their actions, their errors, problems encountered during use, and the duration of the use (or tasks) can be observed (measured).

The observers must be trained to identify and classify different types of predetermined states (e.g., events, problems encountered, or error committed by the user) of the product (i.e., the vehicle or its selected vehicle feature) and the behavior of its users during the observation period. The observers can also record details such as the durations of different types of event, the number of attempts made to perform an operation, the number and sequence of controls used, the number of glances made, and so forth.

The information gathered through observations of the users during selected uses in selected vehicles can provide an understanding of user behavior, which can be translated into frequency of usage of various features, and ease or difficulties experienced by the users during such usage. The customers in general will use the features that they understand and can easily use as compared with features that they cannot understand and are difficult to use. Thus, user observations provide insights into what they like (i.e., want or want more of) or dislike (i.e., hate, do not want, or need to improve) in a vehicle.

COMMUNICATION METHODS

Communication methods involve asking the users or the customers to provide information about their impressions and experiences with a vehicle or vehicle feature. The most common technique involves a personal interview in which an interviewer asks each participant/customer a series of questions. The questions can be asked prior to, during, or after the vehicle usage. The participant/user can be asked questions that will require them to (a) describe the product (vehicle) or impressions about the product and its attributes (e.g., styling, package and ergonomics, comfort), (b) describe the problems experienced while using the product (e.g., could not locate or view a critical item such as windshield defrost button), (c) categorize the product or its performance using a nominal scale (e.g., acceptable or unacceptable, comfortable or uncomfortable, looks tough vs. flimsy), (d) rate the product on one or more scales describing its characteristics and/or overall impressions (e.g., ease in maneuvering, ride comfort), (e) compare the product with other competitors' products presented in pairs based on a given attribute (e.g., appearance—looks new vs. outdated, ease of use, comfort), or even (f) state what they would like or dislike in their new vehicle.

Interviews can be also conducted with a group of individuals, such as in a focus group, which includes about 8–12 individuals with similar background led by a moderator to brainstorm through a series of questions, and the participants are asked to provide opinions or suggest issues related to the characteristics of one or more products.

Some tools commonly used in communication methods for understanding the importance of customer needs include (1) rating scales: using numeric scales, scales with adjectives (e.g., acceptance ratings and semantic differential scales) and (2) paired comparison–based scales (e.g., Thurstone's Method of Paired Comparisons and Analytical Hierarchy method). These tools are described in Chapter 21.

EXPERIMENTATION METHODS

The purpose of experimental research is to allow the investigator to control a research situation (e.g., selecting a product design, performing a task or a test condition) so that causal relationships between the response variable and independent variables may be evaluated. An experiment includes a series of controlled observations (or measurements of response variables) undertaken in artificial (test) situations with deliberate manipulations of combinations of independent variables to answer one or more hypotheses related to the effect of (or differences due to) the independent variables. Thus, in an experiment, one or more variables (called *independent variables*) are manipulated, and their effect on another variable (called the *dependent* or *response variable*) is measured, while all other variables that may confound the relationship(s) are eliminated or controlled.

The importance of experimental methods is that (a) they help identify the best combination of independent variables and their levels to be used in designing the product, and thus provide the most desired effect on the users, and (b) when the competitors' products are included in the experiment along with the manufacturer's product, the superior product can be determined. To ensure that this method provides valid information, the researcher designing the experiment needs to ensure that the experimental situation is not missing any critical factor related to the performance of the product or the task being studied. Additional information on the experimental methods can be obtained from Kolarik (1995) or other textbooks on design of experiments.

Experiments can be also conducted using computer models with various combinations of input variables (or configurations). The computer modeling methods can be classified as (a) mathematical models, (b) simulation models, (c) visualization or animation models, and (d) prototyping using a combination of hardware and software.

ADDITIONAL METHODS

In addition, many other tools used in fields such as industrial engineering, quality engineering and design for six sigma (DFSS), and safety engineering can be used to analyze and understand customer needs. Some examples of such tools are Pareto charts, process charts, task analysis, arrow diagrams, interface diagrams, matrix diagrams, Pugh analysis, failure modes and effects analysis (FMEA), and fault tree analysis (FTA). The abovementioned tools rely heavily on the information obtained through the methods of communication from the users/customers and members of the multifunctional design teams. Additional information on many of these tools is presented in Chapters 13 through 16 and also in other books, such as Kolarik (1995), Besterfield et al. (2003), Creveling et al. (2003), Yang and El-Haik (2003), and Bhise (2014).

DETERMINING BUSINESS NEEDS: PRODUCT PORTFOLIO, MODEL CHANGES, AND PROFITABILITY

Most auto manufacturers have a range of car and truck products of different brands. Thus, they are continuously looking for opportunities to freshen or update their products to increase their revenues (net sales, topline in the financial sheet) and profits (or net income, bottom line in the financial sheet). Developing a new product is a massive undertaking from the viewpoint of resources (i.e., people, funds, and plant capacity). Therefore, each vehicle line is typically freshened (i.e., minor changes are made) about every 2–4 years, and major changes (e.g., new body-style, new powertrain, new interiors and exteriors) are made about every 4–7 years.

The senior management of the company decides on the overall corporate cycle plan (or product plan) of all vehicles that will be produced over the long term (e.g., in the next 5–10 years). The overall corporate cycle plan attempts to balance the updating needs of every model. The management also decides on which models to terminate and when one or more totally new vehicles should be introduced. The overall corporate cycle plan is usually developed by an advance product planning department with the help of market research, financial planning, engineering, and design leaders within the company.

The upcoming technological and design changes, along with changes in government regulations in all market segments, are continuously studied by various subject matter experts, and their recommendations are reviewed by the senior decision makers during the cycle planning meetings.

GOVERNMENT REQUIREMENTS FOR SAFETY, EMISSIONS, AND FUEL ECONOMY

GOVERNMENT SAFETY REQUIREMENTS

The National Highway Traffic Safety Administration (NHTSA) has a legislative mandate under Title 49 of the United States Code, Chapter 301, Motor Vehicle Safety, to issue Federal Motor Vehicle Safety Standards (FMVSS) and Regulations to which manufacturers of motor vehicle and equipment items must conform and certify compliance (NHTSA, 2015). These Federal safety standards are regulations written in terms of minimum safety performance requirements for motor vehicles or items of motor vehicle equipment. These requirements are specified in such a manner "that the public is protected against unreasonable risk of crashes occurring as a result of the design, construction, or performance of motor vehicles and is also protected against unreasonable risk of death or injury in the event crashes do occur."

The FMVSS can be accessed through the NHTSA website (NHTSA, 2015). The FMVSS are numbered by the following categories in Title 49 Code of Federal Regulations (CFR): (a) crash-avoidance standards (FMVSS 101 to 133), (b) crashworthiness standards (FMVSS 201–224), (c) post-crash standards (FMVSS 301–500), and (d) other regulations included in parts 531–591.

EPA's GREENHOUSE GAS (GHG) EMISSIONS AND NHTSA's CORPORATE AVERAGE FUEL ECONOMY (CAFE) STANDARDS

On October 15, 2012, the U.S. Environmental Protection Agency (EPA) announced the final GHG emissions standards for model years 2017–2025. And the NHTSA has announced the final corporate average fuel economy (CAFE) for MYs 2017–2021 and augural standards for MYs 2022–2025 (EPA and NHTSA, 2012). These standards apply to passenger cars, light-duty trucks, and medium-duty passenger vehicles (i.e., SUVs, cross-over utility vehicles, and light trucks). These standards apply to each manufacturer's fleet (total numbers of vehicles produced by different size and type) and not to an individual vehicle.

It is estimated that the national program will save approximately four billion barrels of oil and reduce GHG emissions by the equivalent of approximately two billion metric tons over the lifetimes of those light-duty vehicles produced in MYs 2017–2025. Although the agencies estimate that the technologies used to meet the standards will add, on average, about $1800 to the cost of a new light-duty vehicle in MY 2025, consumers who drive their MY 2025 vehicle for its entire vehicle lifetime will save, on average, $5700–$7400 (based on 7% and 3% discount rates, respectively) in fuel, for a net lifetime savings of $3400–$5000.

Rationale behind Footprint-based Standard

The requirements in these standards are illustrated in Figures 3.1 through 3.4 for vehicles with different model years (MY). The requirements are based on the footprint of the vehicle. The footprint is the area covered under the four tire touch points on the ground, and it is defined as the product of the wheelbase and the tread width. With this footprint-based standard approach, EPA and NHTSA continue to believe that the rules will not create significant incentives to produce vehicles of particular sizes, and thus, there should be no significant effect on the relative availability of different vehicle sizes in the fleet. These standards will also help to maintain consumer choice during the MY 2017 to MY 2025 rulemaking time frame.

Figures 3.1 and 3.2 present fuel economy requirements for passenger cars and light trucks, respectively. These figures show that a smaller-footprint vehicle will need to have higher fuel economy relative to a larger-footprint vehicle (when both vehicles have a comparable level of fuel efficiency improvement technology).

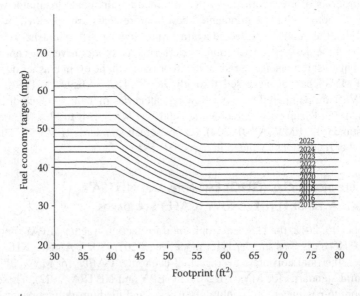

FIGURE 3.1   Passenger car fuel economy requirements. (Redrawn from EPA and NHTSA, *Federal Register*, 77, 199, 2012.)

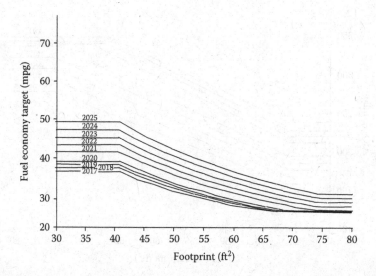

FIGURE 3.2 Light truck fuel economy requirements. (Redrawn from EPA and NHTSA, *Federal Register*, 77, 199, 2012.)

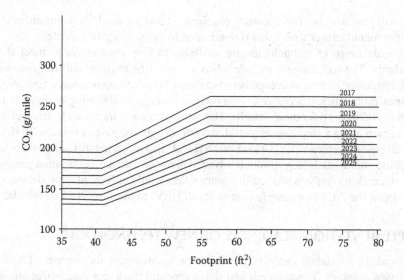

FIGURE 3.3 Passenger car emission requirements. (Redrawn from EPA and NHTSA, *Federal Register*, 77, 199, 2012.)

Conversely, Figures 3.3 and 3.4 (for passenger car and light truck emission requirements) show that a smaller-footprint vehicle will have lower CO_2 emissions relative to a larger-footprint vehicle (when both vehicles have a comparable level of fuel efficiency improvement technology). The standards apply to a manufacturer's overall passenger car fleet and overall light truck fleet, not to an individual vehicle. Thus, if one of a manufacturer's fleets is dominated by small-footprint vehicles, then that

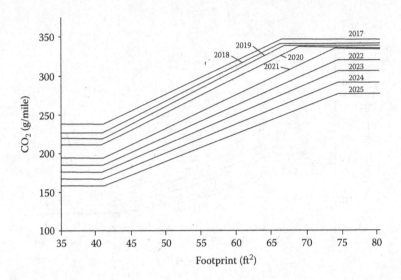

FIGURE 3.4   Light truck emission requirements. (Redrawn from EPA and NHTSA, *Federal Register*, 77, 199, 2012.)

fleet will have a higher fuel economy requirement and a lower CO_2 requirement than another manufacturer's fleet that is dominated by large-footprint vehicles.

A wide range of technologies are available to help automakers to meet these standards. The technologies include advanced gasoline engines and transmissions, vehicle mass reduction, improved aerodynamics, lower rolling resistance tires, diesel engines, more efficient accessories, improvements in air-conditioning systems, and so forth (see Chapter 6 for more details). The automakers will increase electric technologies, such as start-stop systems, mild and strong hybrids, plug-in hybrids (PHEVs), and all electric vehicles (EVs). However, NHTSA also projected that automakers will meet the standards largely through advancements in internal combustion engines. The rulemaking analysis showed that automakers would only need to produce about 1%–3% of the 2025 new vehicle fleet as EVs/PHEVs to meet the 2025 standards.

IMPLEMENTATION READINESS OF NEW TECHNOLOGIES

Automakers constantly evaluate possible new technologies for implementation in their new vehicles. The research and developmental work (e.g., analytical studies, hardware and software development, and testing) requires resources and time, and the probabilities of success in implementation of new technologies are constantly evaluated by all design, engineering, and manufacturing organizations. Based on their evaluations, technologies are selected for implementation in future vehicle programs. The areas of new technologies being considered for future implementation, discussed in more detail in Chapter 6, are (a) engine development, (b) safety technologies, (c) driver information interface technologies, (d) communication technologies (e.g., connected vehicle technologies), (e) lightweight materials, and (f) aerodynamic drag reduction advances.

VEHICLE FEATURES: "WOW," "MUST HAVE," AND "NICE TO HAVE" FEATURES

Every planned vehicle program team is consulted to develop lists of "wow," "must have," and "nice to have" features. According to the Kano model of quality (Yang and El-Haik, 2003; Bhise, 2014; Zacarias, 2016), each product must have (a) features that the customer expects (without which the customer would be dissatisfied—these are classified here as "must have" features [or the product "must be" that way]), (b) features that the customer wants more of (without which the customer will not be satisfied—these can be classified as "nice to have" features [or performance quality]), and (c) features that the customer never expected and is delighted to discover in the product, thus creating a "wow" response [i.e., attractive features]).

Every vehicle must have many features that customers want. Some of these features are the "must have" features that the customers want for sure, without which they would not consider purchasing the vehicle. Such features are necessary for the customers, and the customers have found them to be very useful in maintaining their lifestyle. Some examples of such features are a remote key-fob, disc brakes, seats with adjustable tracks, distance to empty fuel tank display, radio, climate control, cruise controls, and power windows. Other features that the customers would like "more of" can be classified as "nice to have" features. Some examples of such features are rain sensor wipers, heated seats, and a heated steering wheel. Customers have probably found such features to have limited usefulness (or to be very much needed in only some situations), but are willing to forgo them and trade them for some other important feature of the vehicle. Finally, the "wow" features are ones that the customers never thought of getting, and they are totally surprised and delighted to find that the vehicle possesses these features.

The list of features in each of the three categories will be different depending on the type of vehicle, the geographic region where the vehicle is used, and the users' familiarity with the features in their previous vehicles.

There is no universal set of features that can be classified into these three categories. A "must have" feature in a vehicle in a particular market segment may not appeal to, or be found to be necessary by, customers in another market segment, and vice versa. In general, the presence of many features in a luxury vehicle, as compared with "economy" vehicles, will be expected by its customers. Further, Garvin's work suggests that that the perception of the quality of a product will be dependent on the number of features it possesses (Garvin, 1987).

GLOBAL CUSTOMERS AND SUPPLIERS

Variations of a baseline vehicle model are often sold in different countries. Changes are made to conform to local regulations and driver needs to suit local conditions such as (a) the customer population's characteristics, habits, and cultural expectations, (b) country-specific customer driving conditions (road, traffic, and climatic conditions), (c) fuel prices, and (d) the state of the economy (i.e., vehicle owning costs vs. earning power).

Modifications to the vehicles are typically developed by the manufacturer's local product development offices to meet local content and local government regulations. The vehicles are also assembled in local assembly plants using a local workforce, and local suppliers are encouraged to supply a number of components or systems for the vehicles. Special versions of vehicles are also created for some specific markets, for example, Chinese and Indian markets.

COMPARISON OF VEHICLES BASED ON CUSTOMER NEEDS

Auto manufacturers use various approaches to understand the customer needs of future vehicles. The most typical approach is to form a multifunctional team involving designers, engineers, product planners, and marketing and finance personnel, and then to conduct a number of brainstorming sessions. The team members also conduct extensive benchmarking comparisons of a number of existing vehicles. They also use a number of methods, such as observations of customers using their vehicles, personal interviews and focus groups with potential customers, and market research clinics involving a number of vehicles and their features under static and dynamic conditions, to understand customer needs and market trends.

Chapters 23 through 25 provide examples of outputs of design teams organized by the author as part of his graduate-level course in automotive systems engineering. These three chapters describe outputs such as customer characteristics, customer needs, market segment, vehicle benchmarking data, description of the proposed vehicle, and Pugh diagrams (which provide comparative ratings of the proposed vehicle and its competitors in relation to a datum reference vehicle based on customer needs, vehicle attributes, and vehicle systems) for the development of a new car, a pickup truck, and an SUV.

CONCLUDING REMARKS

While many factors that are both internal and external to the automotive manufacturer affect the success of the vehicle program, meeting customer needs, conforming to all government requirements, and meeting business needs should be given the highest priority. Satisfying the external customers who purchase and use the vehicles is probably the most important factor. Meeting government requirements allows the manufacturer to sell the vehicle. Thus, it is a mandatory factor. The business needs of the corporation should be aligned toward customer satisfaction, which can be only achieved through the development and production of quality products. Satisfying the customers requires the right levels of attributes and trade-offs between attributes. Many aspects of these issues are covered throughout this book.

REFERENCES

Besterfield, D. H., C. Besterfield-Michna, G. H. Besterfield, and M. Besterfield-Scare. 2003. *Total Quality Management.* Third Edition. Upper Saddle River, NJ: Prentice Hall.
Bhise, V. D. 2014. *Designing Complex Products with Systems Engineering Processes and Techniques.* Boca Raton, FL: CRC Press.

Creveling, C. M., J. L. Slutsky, and D. Antis, Jr. 2003. *Design for Six Sigma: In Technology and Product Development*. Upper Saddle River, NJ: Prentice Hall.

EPA and NHTSA. 2012. 2017 and Later Model Year Light-Duty Vehicle Greenhouse Gas Emissions and Corporate Average Fuel Economy Standards. *Federal Register*, Vol. 77, No. 199, October 15, 2012, Pages 62623–63200. Environmental Protection Agency, 40 CFR Parts 85, 86, and 600. Department of Transportation National Highway Traffic Safety Administration, 49 CFR Parts 523, 531, 533, 536, and 537. [EPA–HQ–OAR–2010–0799; FRL–9706–5; NHTSA–2010–0131]. RIN 2060–AQ54; RIN 2127–AK79.

Garvin, D. A. 1987. Competing on the Eight Dimensions of Quality. *Harvard Business Review* (November–December 1987). 101–109.

Kolarik, W. J. 1995. *Creating Quality: Concepts, Systems, Strategies, and Tools*. New York, NY: McGraw-Hill.

NHTSA. 2015. *Federal Motor Vehicle Safety Standards and Regulations*. U.S. Department of Transportation. Website: http://www.nhtsa.gov/cars/rules/import/FMVSS/ (Accessed: June 14, 2015)

Yang, K. and B. El-Haik. 2003. *Design for Six Sigma*. New York: McGraw-Hill.

Zacarias, D. 2016. *The Complete Guide to Kano Model*. Website: http://foldingburritos.com/kano-model/ (Accessed: January 19, 2016)

4 Role of Benchmarking and Target Setting

INTRODUCTION

Benchmarking and breakthrough are two methods that are generally used during the very early stages of a product development program. From the information gathered during the benchmarking exercises, the product designers can realize the "gaps" between the characteristics and capabilities of the products of their competitors and their new product concepts, whereas the breakthrough approach forces the design teams to look beyond the existing products and technologies and thus develop a totally new product or new features in the proposed product to achieve major improvements over the existing product designs. This chapter will cover both these methods, with more emphasis on benchmarking, and illustrate how targets are set for the development of future products.

BENCHMARKING

Benchmarking is a process of measuring products, services, or practices against the manufacturer's toughest competitors or those companies recognized as the industry leaders. Thus, it is a search for the industry's best products or practices that can lead to superior performance.

A multifunctional team (or the design team of the product program) within the product development community is usually selected to perform the product benchmarking activities. The benchmarking exercise typically starts with identifying the toughest competitors (e.g., very successful and recognized brands as the industry leaders) and their products (models) that serve similar customer needs to the manufacturer's proposed product. The selected competitor products are used for comparison with the target product. The target product is the product considered by the manufacturer to be its future product (or an existing model of the future product).

The team gathers all important competitive products and information on the products and compares the competitors' products with their target product through a set of evaluations (e.g., measurements of product characteristics, tearing down [disassembling] the products into their lower-level entities for close observation, evaluations by experts [e.g., performance, capabilities, unique features], materials and manufacturing processes used by the competitors, tests and measurements, and estimates of costs to produce the benchmarked products). The information gathered from the comparative evaluations is usually very detailed. However, the depth of evaluations included in benchmarking can vary between problem applications and companies. For example, the benchmarking of an automotive disc brake may involve comparisons based on part dimensions, weights, materials used, surface characteristics, strength characteristics, heat dissipation characteristics, processes needed for its production, estimated production costs, features that would be "liked very much" by customers, features that might be "hated" by customers, features that create a "Wow" reaction among the team members and potential customers, special performance tests such as part temperatures during severe braking torque applications, brake "squealing" sound, and so on. In addition, digital pictures and videos can be taken to help visualize differences (by side-by-side comparison) from the benchmarked products.

The gathered information is generally summarized in a tabular format with product characteristics listed as rows and different benchmarked products represented in columns. Table 4.1 presents an example of a table created to compare the dimensions (or parameters) of four different sports utility vehicles (SUVs). The dimensions are

TABLE 4.1

Benchmarking of 2015 MY Mid-Size SUVs and Target Setting for a 2020 MY SUV

Vehicle Characteristic	2015 Jeep Cherokee (Trailhawk)	2015 Ford Escape (4WD Titanium)	2015 Honda CR-V	2020 Jeep Cherokee (Target)
Wheelbase (in)	106.3	105.9	103.1	104
Length (in)	182	178.1	179.4	180
Height (in)	67.8	66.3	65.1	67
Track width (in)	63.5	61.6	62.2	63
Weight (lb)	4,016	3,645	3,624	3,400
Min ground Clearance (in)	8.8	7.9	6.8	8.8+
Engine	DOHC I4	Turbocharged I4	DOHC I4	Turbocharged I4
Displacement (L)	2.4	1.6	2.4	2.0
Horsepower	184@6,250	173@5,700	185HP@6,400	220@6,250
Transmission	Automatic	Automatic	Automatic	Automatic
Tire size	245/65/17	235/50/18	225/60/18	245/65/17. Must allow for larger aftermarket tires, up to 35" total diameter.
Drivetrain	4×4	4×4	AWD	4×4
Fuel economy (EPA)	19/25/22	22/29/25	26/33/28	25/40/38
Approach angle	29.9	22	28	30+
Construction	Unibody	Unibody	Unibody	Unibody
Towing capacity (lb)	2,000	2,000	1,500	2,000+
MSRP	$30,095	$31,485	$32,895	$30,000–$35,000

Note: MSRP, manufacturer's suggested retail price.

described in the left-hand column of the table, and values of the dimensions of the four SUVs are provided in the subsequent columns. Such data are useful in establishing ranges of values of different parameters and to set target values for each of the parameters of the product being designed.

The data from the benchmarking studies can also be entered into computers as a relational database and later sorted for use by different functional areas as subsets. The information is used to determine the "gaps" between characteristics of the competitors' products and the manufacturer's target product. An action plan can be developed to close the gaps by determining how the target product can be improved by implementing many of the good ideas used in the competitors' designs and avoiding the problems uncovered in the poor designs. Thus, benchmarking can reveal one or more best competitors' products that can be used as reference products during the subsequent product development phases. In addition, some innovative changes can be made to further improve the best design found during the benchmarking exercises.

Peters and Waterman (1982) called the benchmarking exercise "creative swiping from your best competitor." Many companies have benchmarking laboratories, where a number of products produced by different competitors are collected, and the products (or their systems, subsystems, or components) are displayed along with the gathered information. Such laboratories are excellent learning tools for designers, engineers, and product planners. The greatest advantage of benchmarking is that it allows the team members to understand the competitors and learn from their products and the processes needed to create the products within a very short period of time.

Thus, benchmarking can help in reducing the gaps between the manufacturer's target product and its best competitors. However, merely designing as well as the best competitors is not sufficient, because these best competitors will be also continuously improving their future products. Thus, the manufacturer's target product should have capabilities that extend well beyond the best benchmarked products. Simply selecting the best design based on the best set of characteristics among the benchmarked products may not produce an overall best product, because trade-offs exist between different product characteristics related to issues such as costs, performance, customer preferences, manufacturing methods, and so forth.

Benchmarking is an important technique for quality improvements, and it is recognized in the National Institute of Technology and Standards (NIST) Baldrige Quality Award criteria (NIST, 2016) and International Organization for Standardization (ISO) 9000 requirements (ISO, 2016). A quality producer must continuously compare its products with its best competitors and use the information to improve its processes and products continuously. Interestingly, the process of benchmarking is not new—it is like keeping up with the "Joneses" (friends, neighbors, or colleagues). We find out what others have done (e.g., observe or ask others, or conduct literature surveys), compare our situation with the findings reported by others, and then decide on the next course of action.

AN EXAMPLE: MID-SIZE CROSS-OVER SUV

A benchmarking study was conducted to design a mid-size SUV for the author's class project. The students were asked to develop specifications for the 2020 MY

Jeep Cherokee by benchmarking three existing 2015 MY mid-size SUVs: the current Jeep Cherokee, the Ford Escape, and the Honda CRV.

The major customer needs of the SUV generated for the project included the following:

1. Vehicle must look good; styling must convey a sense that the vehicle is a "workhorse."
2. Fuel economy must be at least competitive with other vehicles in the segment, and not vastly different from that of sedans and smaller cars.
3. Vehicle needs to be comfortable during normal road/highway driving.
4. Vehicle needs modern conveniences/infotainment, such as Bluetooth and navigation.
5. Vehicle must be able to accelerate up to highway speeds within an appropriate amount of time.
6. Vehicle must have towing capacity and cargo room consistent with the segment.
7. Vehicle must be able to go off road without incurring expensive damage.

Table 4.1 presents basic vehicle characteristics of the three current 2015 MY SUVs along with the characteristics of the proposed 2020 MY Jeep Cherokee. The tabular format is an effective method of displaying benchmarking data. It allows side-by-side comparison of all the products presented in the columns by considering the vehicle characteristic presented in each row of the table. The team decided to reduce the overall exterior dimensions (wheelbase, overall length, width, height, and track width) of the 2020 MY Cherokee from the existing vehicle to meet the upcoming Environmental Protection Agency/National Highway Traffic Safety Administration (EPA/NHTSA) fuel emissions and fuel economy requirements (see Chapter 3). The target vehicle weight was reduced to 3400 lb from 4016 lb by the use of lightweight materials and a slight reduction in overall dimensions. The present 2.4 L engine was replaced with a 2.0 L turbo-boost engine.

Additional examples of benchmarking studies are provided in Chapters 23 through 25.

PHOTO-BENCHMARKING

Photo-benchmarking is a simple but powerful tool to visualize differences and similarities between products. Here, photographs of current benchmarked products are taken from various selected viewpoints and displayed in a tabular format to allow side-by-side comparisons. Figure 4.1 presents comparisons of three mid-size vehicles from six different viewpoints.

The photographs in Figure 4.1 helped the design team to understand differences between design details of the three vehicles, such as locations of outside door handles, front and rear overhangs, size of C-pillar, grill-opening dimensions, front and rear fascia, center console and center stack design, rear trunk opening, and so forth.

FIGURE 4.1   Illustration of photo-benchmarking for comparison of three mid-size vehicles.

BREAKTHROUGH

The breakthrough approach involves throwing away all the existing product designs and manufacturing processes and conducting brainstorming sessions to develop a totally new design to obtain huge potential gains in terms of styling, performance, costs, and customer satisfaction. Breakthrough designs typically require a radically new thought process and dimensions. This leads to the adoption of new technologies. Thus, the implementation of a breakthrough design creates new problems in systems integration, manufacturing, and project management.

Some examples of breakthrough product designs are (a) SUV styling with off-road driving capability as compared with the traditional station wagon body-style, (b) conventional backlit displays replaced with thin film transistor (TFT) and organic light-emitting diodes (OLED) displays with touch screens, (c) tungsten light sources replaced by compact fluorescent lamps (CFL) and LED lamps to gain a several-fold decrease in electric power consumption, (d) traditional ignition switches with keys replaced with wireless keyless key-fobs and push-button starting (eliminating the need to find, orient, insert and turn the key in the ignition switch), (e) cap-less fuel filler design eliminating fuel filler caps, and (f) fuel systems of gasoline engines with carburetors replaced by direct fuel injection along with turbo-boost technologies to achieve more engine power, meet stricter fuel economy and emissions requirements, and reduce overall powertrain weight.

DIFFERENCES BETWEEN BENCHMARKING AND BREAKTHROUGH

The differences between benchmarking and breakthrough can be highlighted as follows (Kolarik, 1995):

1. Benchmarking is a quick/short-term process to seek ideas for product (or process) improvements from other existing designs, whereas the breakthrough process takes longer for its implementation.
2. Benchmarking generally has a narrower focus (over a smaller set of changes) than breakthrough, which involves expanded focus (or a complete redesign).
3. Benchmarking produces smaller improvements as compared with breakthroughs, which involve dramatically improved performance and radically new dimensions that usually require new technologies.
4. Benchmarking should be conducted before brainstorming to generate the breakthrough designs.

Creative breakthrough thinking, the theory of inventive problem-solving (Altshuller, 1997), the Ideal Design of Effective and Logical Systems (IDEALS) concept (Nadler, 1967; Nadler and Hibino, 1998), and reengineering processes (Champy, 1995) have suggested a number of approaches and principles for developing improved products and processes. Such approaches can be very useful in developing breakthrough product concepts.

BENCHMARKING COMPETITORS' VEHICLES: AN EXAMPLE

A benchmarking exercise can provide a quick start in developing specifications for a new vehicle. An example of how benchmarking can be conducted is provided in this section. A group of graduate students in the author's automotive systems engineering course were asked to select a recent model year (2015 MY) vehicle sold in the U.S. market and two other recent vehicles that currently competed with the reference vehicle in the same market segment. The objective of the benchmarking exercise was to create a set of preliminary specifications for development of a future (2020 MY) model of the reference vehicle. The students were asked to collect data on vehicle dimensions and features from the following: Internet search (e.g., vehicle manufacturer's brochures, articles in automotive magazines and journals), visiting dealerships, attending the Detroit Auto Show, and using their own measurements and photographs of the vehicles and their chunks and systems for side-by-side comparisons.

Table 4.2 presents a benchmarking table comparing the reference vehicle with its two benchmarked comparators based on the exterior and interior dimensions and characteristics of their corresponding systems. Table 4.3 presents a comparison of available standard and optional features in the four vehicles.

In these tables, the last column provides the specifications of the target vehicle. The tables allow comparisons of the existing models with the proposed vehicle by considering a number of vehicle parameters and features in different categories related to the vehicle attributes. The benchmarking tables serve as important reference documents to enable comparison of the existing product with the benchmarked products on a number of parameters and features to decide on the strengths and weaknesses of each product, and thus, the data help in positioning the new product in relation to future models of the benchmarked products. It should be realized that the target vehicle must compete with the 2020 MY vehicles of the benchmarked products.

In the planning process, the benchmarking data are used along with the basic needs of the customers, company needs, and government requirements that the target product must meet.

The data provided in the tables helped to determine where the current vehicle was deficient by comparing the exterior and interior dimensions and features of the existing reference vehicle with those in the benchmarked vehicles. In general, the strategy in developing the new vehicle was to provide more interior space, increase fuel economy, decrease emissions, and provide more new technology features. Comparing the existing Escape with the two benchmarked vehicles, the new vehicle team decided to

1. Decrease the overall vehicle weight to 3500 lb from existing vehicle weight of 3769 lb.
2. Decrease the overall vehicle height to 65.7 in.
3. Increase the ground clearance to make it closer to the competitors.
4. Decrease turning radius to 34.5 ft by decreasing wheelbase slightly to 103 in.
5. Increase interior dimensions by increasing headroom, legroom, and shoulder room for both front and rear rows.
6. Incorporate nine-speed transmission.
7. Adopt all-wheel drive (AWD) to provide the same capability as the two competitors.
8. Increase fuel economy and reduce greenhouse gas emission targets to meet the upcoming fuel economy and emissions requirements.
9. Add new features such as adaptive front lighting system, daytime running lamps, adaptive cruise control.
10. Make previously optional equipment standard, such as forward sensing system and moon roof.

TABLE 4.2
Benchmarking Comparisons of Vehicle Characteristics

Vehicle Parameter/ Characteristic	Reference Vehicle 2015 Ford Escape	Benchmark Vehicle #1 2015 Subaru Forester	Benchmark Vehicle #2 2015 Mazda CX-5	Target Vehicle 2020 Ford Escape
Base MSRP ($)	30,585	33,095	29,220	32,500
Trim	2.0L Titanium with AWD	2.0XT Touring	Grand Touring	2.0L Titanium with AWD
Exterior				
Wheelbase (in)	105.9	103.9	106.3	103.0
Ground clearance (in)	7.9	8.7	8.5	8.5
Length (in)	178.1	180.9	179.3	178.1
Width (no mirror) (in)	72.4	70.7	72.4	72.4
Track (front) (in)	61.5	60.9	62.4	61.5
Track (rear) (in)	61.6	61.1	62.5	61.6
Height (in)	66.3	68.2	65.7	65.7
Curb weight (lb)	3,769.0	3,651.0	3,560.0	3,500.0
Wheel size (in)	18.0	17.0	19.0	18.0
Turning radius (ft)	38.8	34.8	36.7	34.0
Interior				
Headroom (1st row) (in)	39.9	40.0	39.0	40.0
Headroom (2nd row) (in)	39.0	37.5	39.0	39.0
Legroom (1st row) (in)	43.1	43.0	41.0	43.5
Legroom (2nd row) (in)	37.3	38.0	39.3	39.3
Hip room (1st row) (in)	54.8	53.9	55.2	55.2
Hip room (2nd row) (in)	52.4	53.0	53.7	53.7
Shoulder room (1st row) (in)	56.0	57.0	57.5	57.5
Shoulder room (2nd row) (in)	55.3	56.6	55.5	56.6
Capacities				
Seating	5	5	5	5
Passenger volume (cu ft)	98.1	103.3	102.3	103.5
Cargo volume behind 1st row (cu ft)	67.8	68.5	65.4	68.5

(Continued)

TABLE 4.2 (CONTINUED)
Benchmarking Comparisons of Vehicle Characteristics

Vehicle Parameter/ Characteristic	Reference Vehicle 2015 Ford Escape	Benchmark Vehicle #1 2015 Subaru Forester	Benchmark Vehicle #2 2015 Mazda CX-5	Target Vehicle 2020 Ford Escape
Cargo volume behind 2nd row (cu ft)	34.3	31.5	34.1	34.5
Fuel tank (gal)	15.5	15.9	15.3	15.5
Towing (max trailer weight) (lb)	2,000	1,500	2,000	2,100
Powertrain				
Engine type	Gas turbo direct injection, DOHC I4	Gas turbo direct injection, DOHC H4	Direct injection, DOHC I4	Gas turbo direct injection, DOHC I4 with variable displacement, start/stop
Displacement (cc)	1,999.0	1,999.0	2,488.0	1,999.0
Horsepower (hp)	240@5,500	250@5,600	184@5,700	210@5,500
Torque (ft-lb)	270@3,000	258@2,000	185@3,250	260@3,000
Fuel type	Gasoline	Gasoline	Gasoline	Gasoline
Air induction system	Turbo	Turbo	Naturally aspirated	Turbo
Compression ratio	9.3:1	10.6:1	13.0:1	9.3:1
Transmission type	6-speed select shift automatic	CVT w/ 8-speed select shift	6-speed select shift automatic	9-speed select shift automatic
Drive type	4WD	AWD	AWD	AWD
Performance				
0–60 mph (s)	6.8	6.2	8.1	6.2
Quarter mile	15.2 s @ 88.8 mph	14.8 s @ 95.8 mph	16.3 s @ 84.5 mph	15.0 s @96 mph
Braking, 60–0 mph (ft)	123	111	125	115
Lateral acceleration	0.85	0.79	0.79	0.85
Range (miles)	338.8/462.0	381.6/508.8	367.2/459.0	385/520
EPA rating (mpg)	21/28	23/28	24/30	35/46
CO_2 emissions (g/ mile)	371.9	353.8	317.5	175
NHTSA front driver crash test	4	5	5	5
NHTSA front passenger crash test	4	4	5	5

(Continued)

TABLE 4.2 (CONTINUED)

Benchmarking Comparisons of Vehicle Characteristics

Vehicle Parameter/ Characteristic	Reference Vehicle 2015 Ford Escape	Benchmark Vehicle #1 2015 Subaru Forester	Benchmark Vehicle #2 2015 Mazda CX-5	Target Vehicle 2020 Ford Escape
		Chassis		
Front suspension	Independent MacPherson strut with stabilizer bar	Independent MacPherson strut with stabilizer bar	MacPherson strut with stabilizer bar	Independent MacPherson strut with stabilizer bar
Rear suspension	Independent multilink with stabilizer bar	Independent, double wishbone with stabilizer bar	Multilink with stabilizer bar	Independent multilink with stabilizer bar
Brakes	4-Wheel disc with antilock brake system (ABS)	4-Wheel disc with antilock brake system (ABS)	4-Wheel disc with antilock brake system (ABS)	4-Wheel disc with antilock brake system (ABS)
Steering	Electric power-assisted steering (EPAS)	Electric power steering	Rack and pinion, EPAS	Electric power-assisted steering (EPAS)
Tires	P235/50HR18 BSW All-season tires	P225/55HR18 BSW All-season tires	P225/55VR19 BSW All-season tires	P235/50HR18 BSW All-season tires

Thus, the benchmarking data helped in supporting the above changes.

Table 4.4 presents a Pugh diagram showing how each vehicle compares with the reference vehicle (used as the datum) for each vehicle subattribute. (Note that +, S, and − symbols, respectively, indicate that the vehicle in the column is better than, the same as, or worse than the reference vehicle) (datum). The sum of number of subattributes receiving + symbols minus the sum of subattributes receiving − symbols provides a measure of improvement in each vehicle over the datum vehicle. Thus, the 2020 Ford Escape received a total score of 23, which is higher than the corresponding scores of the other benchmarked vehicles.

TABLE 4.3
Benchmarking Comparisons of Vehicle Features

Vehicle Features		Reference Vehicle	Benchmark Vehicle #1	Benchmark Vehicle #2	Target Vehicle
Category	Description	2015 Ford Escape	2015 Subaru Forester	2015 Mazda CX-5	2020 Ford Escape
Lighting and visibility	HID headlights	Optional	Standard	Optional	Optional
	Halogen headlights	Standard	N/A	Standard	Standard
	Fog lights	Standard	Standard	Standard	Standard
	Daytime running lights	N/A	Standard	Standard	Standard
	Adaptive front lighting system	N/A	N/A	Optional	Optional
	Heated side mirrors	Standard	Standard	Standard	Standard
	Auto-dimming center mirror	Standard	Optional	Optional	Standard
	Rear view camera	Standard	Standard	Standard	Standard
	Blindspot mirror	Standard	N/A	N/A	Standard
	Blindspot detection	Optional	N/A	Standard	Optional
Safety and security	Front air bags	Standard	Standard	Standard	Standard
	Driver knee air bag	Standard	Standard	Standard	Standard
	Front seat-mounted side air bags	Standard	Standard	Standard	Standard
	Side curtain air bags	Standard	Standard	Standard	Standard
	Belt minder	Standard	N/A	N/A	Standard
	Tire-pressure monitoring system	Standard	Standard	Standard	Standard
	Passive anti-theft system/ immobilizer	Standard	Standard	Standard	Standard
Driver assistance	Forward sensing system	Optional	N/A	N/A	Standard
	Reverse sensing system	Standard	N/A	Optional	Standard
	Brake assist	Standard	Standard	Standard	Standard
	Active park assist	Optional	N/A	N/A	Optional
	Hill start assist	Standard	Standard	Standard	Standard
	Adaptive cruise control	N/A	Optional	N/A	Standard
	Cruise control	Standard	Standard	Standard	Standard
	Lane-keeping system	N/A	Optional	N/A	Standard
Comfort and convenience	Heated front seats	Standard	Standard	Standard	Standard

(*Continued*)

TABLE 4.3 (CONTINUED)
Benchmarking Comparisons of Vehicle Features

Vehicle Features		Reference Vehicle	Benchmark Vehicle #1	Benchmark Vehicle #2	Target Vehicle
Category	Description	2015 Ford Escape	2015 Subaru Forester	2015 Mazda CX-5	2020 Ford Escape
	Power-adjustable driver seat	Standard	Standard	Standard	Standard
	Leather seats	Standard	Standard	Standard	Standard
	60/40 Split rear seats	Standard	Standard	N/A	N/A
	40/20/40 Split rear seats	N/A	N/A	Standard	Standard
	One-Touch folding rear seats	N/A	Standard	Standard	Standard
	Memory feature for driver settings	Standard	N/A	N/A	Standard
	Heated steering wheel	N/A	N/A	N/A	Optional
	Push-button start	Standard	Standard	Standard	Standard
	Remote start	Standard	Optional	Optional	Standard
	Remote keyless entry	Standard	Standard	Standard	Standard
	Invisible keypad entry	Standard	N/A	N/A	Standard
	Leather steering wheel	Standard	Standard	Standard	Standard
	Rain-sensing wipers	Optional	N/A	Standard	Optional
	Power one-touch window, driver	Standard	Standard	Standard	Standard
	Power one-touch window, all	Standard	N/A	N/A	Standard
	Dual zone climate control	Standard	Standard	Standard	Standard
	Interior ambient lighting	Standard	Optional	N/A	Standard
	Additional storage space under trunk	N/A	Standard	N/A	Standard
Infotainment and connectivity	Large display touch screen	Standard (8")	Standard	Standard (5.8")	Standard
	Audio system	390 W, 10-speaker	440 W, 8-speaker	9-speaker	390 W, 10-speaker
	Voice commands	Standard	Standard	Standard	Standard
	Bluetooth connectivity	Standard	Standard	Standard	Standard
	USB connectivity	Standard	Standard	Standard	Standard

(Continued)

TABLE 4.3 (CONTINUED)
Benchmarking Comparisons of Vehicle Features

Vehicle Features		Reference Vehicle	Benchmark Vehicle #1	Benchmark Vehicle #2	Target Vehicle
Category	Description	2015 Ford Escape	2015 Subaru Forester	2015 Mazda CX-5	2020 Ford Escape
	12 V Auxiliary power ports	Standard (4)	Standard (2)	Standard (3)	Standard (4)
	110 V Power outlet	Standard (1)	Optional	N/A	Standard (1)
	Navigation	Optional	Optional	Optional	Optional
Exterior	Power liftgate	Standard	Standard	N/A	Standard
features	Foot activated liftgate	Standard	N/A	N/A	Standard
	Moon roof	Optional	Standard	Standard	Standard
	Roof rails	Standard	Standard	Optional	Standard
	Trailer hitch	Optional	Optional	Optional	Optional
	Dual chrome exhaust tips	Standard	Standard	Standard	Standard

Note: HID, high-intensity discharge; N/A = Feature not available.

EXAMPLES OF SYSTEM, SUBSYSTEM, AND COMPONENT-LEVEL BENCHMARKING

Electrical systems in different vehicles can be benchmarked and compared to understand how different competitors have configured their electrical systems and subsystems. For example, alternators, electronic control modules, wiring harnesses, electric motors, actuators, sensors, and even connectors can be benchmarked to improve their functional capabilities and reduce product costs.

Brake system engineers benchmark braking systems in different vehicles by comparing all components within the system, such as brake pads, disc rotors, master cylinders, boosters, brake pedals, brake lines, and so forth, to understand the differences in designs, configuration, and integration with other vehicle systems.

Similarly, body engineers benchmark vehicle bodies of different vehicles to understand how the vehicle bodies are designed in terms of use of different materials (e.g., mild steel, high-strength steels, aluminum); the dimensions of various components, such as cross members, floor panels, roof panels, door panels; and the method of welding, sealing, painting, and so forth.

Ergonomics engineers commonly evaluate controls and displays mounted in instrument panels, doors, and center consoles to study ergonomic considerations related to identifying controls and displays, the legibility and use of colors in displays, and reaching, grasping, and operating controls, for every function in each of the benchmarked vehicles. Such exercises often produce lists of examples of good and poor ergonomic features. The generated information is used by designers as a guide in developing future driver interfaces.

TABLE 4.4

Pugh Diagram Comparing Vehicle Subattributes of Benchmarked and Target Vehicles with Existing Vehicle (as Datum)

Vehicle Attribute	Subattribute	Reference Vehicle 2015 Ford Escape	Benchmark Vehicle #1 2015 Subaru Forester	Benchmark Vehicle #2 2015 Mazda CX-5	Target Vehicle 2020 Ford Escape
Cost	Cost	Datum	−	+	−
	Warranty	Datum	S	S	S
Vehicle performance	Horsepower	Datum	+	−	−
	Torque	Datum	−	−	−
	0–60 mph acceleration	Datum	+	−	+
	60–0 mph braking	Datum	+	−	+
	All-wheel drive system	Datum	+	S	+
	Lateral acceleration	Datum	−	−	S
	Vehicle weight	Datum	+	+	+
	Towing capability	Datum	−	S	+
	Turning radius	Datum	+	+	+
Fuel efficiency and emissions	Hybrid option	Datum	S	S	+
	Range	Datum	+	S	+
	Gas mileage	Datum	+	+	+
	Emissions	Datum	+	+	+
Safety	NHTSA front driver crash test	Datum	+	+	+
	NHTSA front passenger crash test	Datum	S	+	+
	Daytime running lights	Datum	+	+	+
	Blind spot side mirror	Datum	−	−	S
	All air bags	Datum	S	S	S
	Belt minder	Datum	−	−	S
	Tire-pressure monitoring system	Datum	S	S	S
	Brake assist	Datum	S	S	S
	Adaptive front lighting system	Datum	S	+	+
Package/space	Number of passengers	Datum	S	S	S
	Passenger volume	Datum	+	+	+
	Cargo volume (behind 1st row)	Datum	+	−	+

(Continued)

TABLE 4.4 (CONTINUED)
Pugh Diagram Comparing Vehicle Subattributes of Benchmarked and Target Vehicles with Existing Vehicle (as Datum)

Vehicle Attribute	Subattribute	Reference Vehicle 2015 Ford Escape	Benchmark Vehicle #1 2015 Subaru Forester	Benchmark Vehicle #2 2015 Mazda CX-5	Target Vehicle 2020 Ford Escape
	Cargo volume (behind 2nd row)	Datum	−	−	+
	Additional storage under trunk	Datum	+	−	+
	Split rear seats	Datum	S	+	+
	One-touch folding rear seats	Datum	+	+	+
Comfort and convenience	Remote start	Datum	−	−	S
	Remote keyless entry	Datum	−	−	S
	Heated front seats	Datum	S	S	S
	Power-adjustable driver seat	Datum	S	S	S
	Leather seats	Datum	S	S	S
	Heated steering wheel	Datum	S	S	+
	Leather steering wheel	Datum	S	S	S
	Rain-sensing wipers	Datum	−	+	S
	Power one-touch window, all	Datum	−	−	S
	Dual zone climate control	Datum	S	S	S
	Interior ambient lighting	Datum	−	−	S
	Foot-activated power liftgate	Datum	−	−	S
	Moon roof	Datum	+	+	+
	Roof rails	Datum	+	−	S
	Trailer hitch	Datum	+	+	S
Driver assist features	Forward sensing system	Datum	−	−	+
	Reverse sensing system	Datum	−	−	S
	Rear view camera	Datum	S	S	S
	Active park assist	Datum	−	−	S
	Hill start assist	Datum	S	S	S

(Continued)

TABLE 4.4 (CONTINUED)

Pugh Diagram Comparing Vehicle Subattributes of Benchmarked and Target Vehicles with Existing Vehicle (as Datum)

Vehicle Attribute	Subattribute	Reference Vehicle 2015 Ford Escape	Benchmark Vehicle #1 2015 Subaru Forester	Benchmark Vehicle #2 2015 Mazda CX-5	Target Vehicle 2020 Ford Escape
	Adaptive cruise control	Datum	+	S	+
	Cruise control	Datum	S	S	S
	Lane-keeping system	Datum	+	S	+
Infotainment and connectivity	Large display touch screen	Datum	–	–	S
	Audio system	Datum	+	–	S
	Voice commands	Datum	–	–	+
	Bluetooth connectivity	Datum	S	S	S
	USB connectivity	Datum	S	S	S
	12 V Auxiliary power ports	Datum	–	–	S
	110 V Power outlet	Datum	–	–	S
	Navigation	Datum	S	S	S
Sum of (–)			20	24	3
Sum of (+)			21	15	26
Sum of (S)			21	23	33
Total score			1	–9	23

CONCLUDING REMARKS

Benchmarking is a very useful tool in the development of new or improved products. Proper selection of products to benchmark allows the entire team to become familiar with the designs created by other manufacturers. It thus educates the design team on all design-related issues within a short time period. Nowadays, the use of benchmarking has become very commonplace in the auto industry.

REFERENCES

Altshuller, G. S. 1997. *40 Principles: TRIZ Keys to Technical Innovation*. Translated by Lev Shulyak and Steven Rodman. Technical Innovation Center, Worcester, MA 01606.
Champy, J. 1995. *Reengineering Management*. New York, NY: Harper Business Books.
ISO. 2016. Quality Management. Website: www.iso.org/iso/home/standards/management-standards/iso_9000.htm (Accessed: May 11, 2016)

Kolarik, W. J. 1995. *Creating Quality: Concepts, Systems, Strategies, and Tools*. New York, NY: McGraw-Hill.

Nadler, G. 1967. *Work Systems Design: The IDEALS Concept*. Homewood, IL: Irwin.

Nadler, J. and S. Hibino. 1998. *Breakthrough Thinking: The Seven Principles of Creative Problem Solving*. Roseville, CA: Prima.

NIST. 2016. Baldrige performance excellence program. Website: www.nist.gov/baldrige/ (Accessed: May 11, 2016)

Peters, T. and R. H. Waterman, Jr. 1982. *In Search of Excellence*. New York, NY: HarperCollins.

5 Business Plan Development and Getting Management Approval

INTRODUCTION

Undertaking a new vehicle development program requires large amounts of resources, such as capital, manpower, tools, equipment, and facilities for development and production of the vehicles. The vehicle program also involves a number of risks: for example, more time and budget may be required to develop the vehicle, the vehicle design may not be perceived by the customers to be better than many of its competitors, and subsequently, the company may not able to generate the projected sales volume. Therefore, the senior management of the auto company needs sufficient information about the proposed vehicle program to decide whether to approve the program or not. A business plan for the vehicle program is prepared to provide necessary information on the proposed program and is presented to the senior management to aid their decision-making process. This chapter describes the process of preparing a business plan, its content, and the risks involved in implementing the business plan.

BUSINESS PLAN

What Is a Business Plan?

A business plan is a proposal for creating or developing a new product. It is an essential roadmap for business success. This document generally projects three or more years ahead and outlines the route a vehicle program will take to meet its objectives. It is usually prepared internally within the auto company to obtain concurrence from the top management to approve the vehicle program. The business plan is thus a document prepared to describe details of a proposed vehicle, vehicle program timing, corporate resources needed to develop the vehicle, and future revenues and income that the vehicle will generate. It is typically prepared jointly by the company's product planning, engineering, marketing, and finance activities.

Contents of the Business Plan

The business plan is a decision-making tool for the senior management of the company to help decide whether the proposed vehicle program should be approved. Its contents should help management to think through all the relevant business issues that the automotive company will face in launching the vehicle program and continuing it till the product is discontinued. Thus, it should include the following:

1. *Description of the Proposed Product*
    a. Product configuration (e.g., for an automotive product, the body-style of the vehicle such as a sedan, a coupe, a crossover, a sports utility vehicle [SUV], a pickup, or a multi-passenger vehicle [MPV] and its variations). It should be noted that some large vehicle programs include a new line of vehicles with various versions (models). These vehicles most likely will share (i.e., use) a common vehicle platform (i.e., many common parts and manufacturing and assembly equipment) to create different models for one or more brands.
    b. Model year of the vehicle.
    c. Size class. U.S. Environmental Protection Agency (EPA) size classes: (a) sedans: mini-compacts, subcompact, compact, mid-size/intermediate, and large; (b) station wagons: small, mid-size and large; and (c) pickup trucks: small and standard (EPA, 2016). Euro market segments: A-segment minicars, B-segment small cars, C-segment medium cars, D-segment large cars, E-segment executive cars, F-segment luxury cars, S-segment sports coupes, M-segment multipurpose cars, J-segment sports utility cars (EAFO, 2016).
    d. Market segment (e.g., ultra-luxury, luxury, entry-luxury, or economy vehicles of different body-styles).
    e. Markets where the product will be sold and used (countries).
    f. Energy use characteristics (e.g., gasoline, diesel, electric power, hybrid, plug-in hybrid), target fuel economy and emissions levels.
    g. Powertrain type and configuration (e.g., types of engines and transmissions, left-hand drive or right-hand drive, front-wheel drive [FWD], rear-wheel drive [RWD], or all-wheel drive [AWD]).
    h. Types of suspensions, brakes, and tire sizes.
    i. Manufacturer's suggested retail price (MSRP) and price range with different optional equipment.
    j. Production capacity and estimated annual sales volumes over the product life cycle.
    k. Makes/brands, models, and prices of leading competitors in the market segment of the proposed vehicle.
2. *Target Attribute Rankings* (i.e., how the product would be positioned in its market segment, such as best-in-class, above the class average, average in the class, or below average, by considering each product attribute).
3. *Pugh Diagram* showing product attributes (and/or vehicle systems) and changes in the proposed concept with respect to the datum (selected reference product) and competitors.
4. *Dimensions, Major Changes, and Options*
    a. Overall exterior dimensions (e.g., product package envelope with length, width, height, wheelbase, and cargo/storage volume)

 b. Interior dimensions (e.g., people package with legroom, headroom, shoulder room, number of seating locations, and so forth)

 c. Curb weight and gross vehicle weight

 d. New government requirements affecting the program

 e. Major changes in the vehicle systems as compared with the systems in the previous (outgoing) model, e.g., styling changes, weight reduction, drive options (FWD, RWD, AWD), powertrains (types, sizes, and capacities of engines/motors and transmissions), descriptions of unique features (e.g., type of suspension), technologies to be introduced, and standard and optional features/equipment.

5. *Short Description*: A one-paragraph description of the proposed product with several adjectives to describe its image, stance, and styling characteristics (e.g., futuristic, traditional, retro, fast, dynamic, aerodynamic, tough, or chunky—rugged like a Tonka truck).

6. *Program Schedule*

 a. Program kick-off date, timings of major milestones

 b. Job#1 date (i.e., date when first production unit will come out of the assembly plant)

 c. Vehicle life-cycle events: future minor and major changes and end of vehicle life cycle (expected date when the vehicle production will be discontinued)

7. *Market Analysis and Projected Sales Volumes*

 a. Descriptions of competitors to the proposed products

 b. Results of market research (e.g., percentage of survey participants who liked the proposed vehicle)

 c. Probable scenarios of competitors' plans and market share

 d. Quarterly (or yearly) sales estimates of each model in each market segment

8. *Organization and Manpower Needs*

 a. Manpower needed for product development, manufacturing, and marketing and sales

 b. Unique expertise needed

9. *Make versus Buy Analysis and Supplier Plan*

 a. Make versus buy analysis of new product entities

 b. Supplier needs and capabilities

10. *Proposed Plant Location, Vehicle Production Capacity, and Plant Investments*

11. *Product Life Cycle*

 a. Estimated life span

 b. Possible product refreshments, future models, and variations and changes

 c. Recycling of the plant, equipment, and products

12. *Financial Analysis*

 a. Curves of estimated cumulative costs and revenues during product life cycle for different scenarios (best case, average, and worst case)

 b. Anticipated quarterly funding needed during product development and during revenue buildup

    c. Anticipated date of breakeven point

    d. Estimated quarterly net profit over the product life cycle and return on investments

13. *Risks in the Proposed Product Program*: list of risks the company will face in developing (e.g., during new technology implementation) and marketing the vehicle. Level of major risks (probability and consequences of each major loss)

14. *Program Justification*: Reasons for undertaking the proposed plan versus other alternatives. Effect of the vehicle program on other existing and future vehicle programs (e.g., sharing of entities and resources)

PROCESS OF PREPARING A BUSINESS PLAN

The process of preparing a business plan requires a lot of brainstorming, forecasting trends in design and technologies, projecting competitors' abilities to produce a similar and better product in the same market segment, and undertaking some pilot or advanced concept design work. A team of experienced product planners, market researchers, engineers, and designers is usually assigned to come up with an early specification of the proposed vehicle and early designs of the vehicle concepts.

Typical steps involved in business plan development include

1. Study trends in automotive markets, technologies, and sales history of existing products of the company and its competitors.
2. Identify opportunities and brainstorm product ideas.
3. Benchmark a number of existing competitors' vehicles.
4. Create early vehicle concept ideas, sketches, and drawings.
5. Develop package drawings of promising vehicle concepts to understand design and engineering challenges.
6. Develop vehicle description, specifications, and list of standard and optional features.
7. Develop a list of major vehicle development tasks, estimate time and manpower required to complete the tasks, and create a program timing plan and a technology implementation plan.
8. Conduct financial analysis and prepare financial plan.
9. Prepare a draft documentation of the business plan.
10. Request key experts in product engineering offices and management personnel to review the draft of the business plan, provide comments, and suggest changes to the engineering feasibility of the proposed vehicle, including a list of their major concerns.
11. Resolve major concerns, make necessary changes to the draft, and issue a final draft of the business plan.
12. Distribute the final draft to members of the senior management.
13. Schedule a meeting of the senior management personnel to present and discuss the business plan.

RISKS IN PRODUCT PROGRAMS

One of the important issues in the business plan that the management closely examines is the risks involved in the proposed program. The risks include potential losses that the company may face if the program does not progress according to the proposed plan. The potential losses can occur due to many different reasons. Some of the reasons may be under the control of the program management (e.g., failure to take a decision at a required time), whereas many other events that can lead to losses may be outside the program management's control (e.g., earthquake, adverse weather, unforeseen changes in the economic situation, or loss of experienced professionals).

The risks can be categorized by considering major areas such as program timing, technical issues, economic situations, and manpower issues. It should be realized that many of the issues are interrelated; for example, technical problems can affect program timings and costs. For example, failure to meet a given level of fuel economy can be related to a variety of technical reasons, such as inability to design durable engine bearings or inability to develop components with lightweight materials (e.g., difficulty in manufacturing certain carbon-fiber body parts). Many of the risks may be interrelated due to a combination of causes.

Important risk categories and some examples in each of the categories are (Blanchard and Fabrycky, 2011)

1. *Technical risks* occur due to one or more technical problems involved with the design of the vehicle. Inability to develop a required technology within the allocated time, a flaw occurring in the design of a system, a nonfunctional subsystem, and a failed component in an electrical unit are examples of technical problems. The correction of a flaw in the product may require additional analyses, experimentation, modification of tools' and equipment, and so forth. Sometimes, the planned implementation of new technologies may not be production ready within the required time constraints. Additional tasks to incorporate changes would require additional time and resources, and thus, introduce uncertainty in meeting the program timing goals.

   Additional examples of technical risks are
   a. Performance of the product did not meet required specification; for example, the braking system may be unable to stop the vehicle from 60 to 0 mph within the stopping distance specified in the braking system design specification.
   b. Vehicle failed to meet required fuel economy (miles per gallon or liters per kilometer) goal.
   c. Vehicle failed to meet a crashworthiness requirement in a federal standard.
2. *Cost risks* include additional expenditures required to correct the problems. The rework of the product design is also one of the major cost risks. In some cases, the problem is with the incorrect cost assumptions used during the product planning process. In such situations, the design complexity

is usually not understood, and thus, the time and/or resources needed to develop the product are underestimated and underallocated. Similarly, the sales forecasts may be too optimistic. Cost and revenue estimates are central to any business plan for deciding the viability of the planned program. But costs are often underestimated and revenues overestimated, which results in larger cost overruns, revenue shortfalls, and possibly nonviability of the program.

3. *Schedule risks* are related to inability to meet the scheduled deadlines (e.g., gateways) for a number of different reasons. Proper scheduling of the project activities is critical to ensure that the engineers and designers get enough time to properly design and test all the systems and subsystems of the vehicle before it is launched on the market. If a certain task takes longer than specified, this can lead to delays in the production of the vehicle, and there will be a delay in reaching certain revenue targets. There are numerous possible causes for schedule delays: machine and assembly line downtime due to breakdown, a strike by the workforce, delay in components arriving from suppliers, and so forth. Changes made to the plans at the last moment can cause delays in receiving raw materials, which also causes schedule risks.

4. *Programmatic risks* are related to the product development aspects of the program. Any major change in the program, such as modification in the program, cancellation of projects related to the program, or delay to the program due to late decisions, can cause programmatic risks. This type of risk is closely associated with scheduling risks but is more related to management issues, wherein frequent and major changes in the program due to rescheduling of projects, budget constraints, and requirement for approval of an increase in budgets could lead to programmatic delays. This type of risk is prevalent in product development programs due to frequent changes in budgets, program requirements, and management directions, and even program cancellations. Program budgets calculated early, before the changes to the program, may not be applicable to the new changed program. Dropping of a project due to insufficient funds is also a programmatic risk.

Additional examples of program-related risks are

a. The vehicle development program failed to meet the required Job#1 date.

b. The vehicle program could not be contained within the allocated budget.

c. The required manpower was not available to perform the technical and/ or management tasks.

d. An exterior lamp (e.g., a headlamp) failed to meet a photometric specification included in the Federal Motor Vehicle Safety Standard 108.

e. The number of components in a rear suspension system was increased to meet weight and performance targets (e.g., changing from solid axle to independent rear suspension).

5. *Supplier-related risks* are related to failure by the suppliers to meet the assembler's vehicle build timings. The suppliers may fail to provide the required quality and/or quantity of entities (systems, subsystems, or components) according to the agreed delivery plan. (Note: suppliers provide a high percentage of automotive systems, subsystems, and components to automotive companies.)

6. *Risks due to external causes* that are outside the control of the program management. For example, projects can be delayed due to political impasse. For example, a VW crossover vehicle plant in the United States was delayed due to a tug-of-war between Mexico and Tennessee over financial incentives in June 2014. Unexpected changes in economic conditions, changes in oil prices, earthquakes, and weather conditions (e.g., unexpected floods) can also have a substantial effect on the plant operations and hence the sales volumes of the planned vehicles.

MAKE VERSUS BUY DECISIONS

The business plan should include a list of entities that will be outsourced (i.e., designed, manufactured, and delivered by suppliers). Some important considerations related to the decision on whether to make or buy are

1. Lack of in-house manufacturing capability and capacity
2. Availability of reliable and low-cost suppliers that can deliver in needed volumes and quality
3. Unavailability of capital required for internal production of the needed entities
4. Need to maintain confidentiality of competitive information on future product designs or specialized knowledge on unique processes needed to produce certain entities within the organization to retain competitive advantage

CONCLUDING REMARKS

A business plan is a useful product planning and communication tool. It summarizes all important information about the vehicle specifications along with its competitors and information on business issues such as timings, resource needs, and financial analysis, including costs, revenues, and return on investments. A good business plan will also help the design team organize their ideas, set priorities, and see how their model may play out financially. It also enables the product planning team to share their business concept with everyone affected by the vehicle program. Unless you want to explain it over and over again, sharing your business plan with others in the organization provides a clear vision with facts, timing plan, and financial picture. Thus, it promotes better decision making all around. Chapters 23 through 25 provide examples of business plans for a sedan, a pickup truck, and an SUV, respectively. Additional information on financial analysis included in the business plan is provided in Chapter 19.

REFERENCES

Blanchard, B. S. and W. J. Fabrycky. 2011. *Systems Engineering and Analysis*. Upper Saddle River, NJ: Prentice Hall PTR.

EAFO. 2016. European vehicle categories. Website: www.eafo.eu/content/european-vehicle-categories (Accessed: June 9, 2016)

EPA. 2016. Classes of comparable automobiles. Website: www.gpo.gov/fdsys/pkg/CFR-2015-title40-vol30/xml/CFR-2015-title40-vol30-sec600-315-08.xml (Accessed: June 9, 2016)

6 New Technologies, Vehicle Features, and Technology Development Plan

INTRODUCTION

New advances in technologies, changes in government requirements, and fierce competition between auto manufacturers are requiring auto manufacturers to incorporate new features with improved functionality and safety in their new vehicles. There is also pressure to reduce costs, weight, emissions, and fuel consumption. This chapter reviews new technology applications that could be introduced in future automotive products and covers technology implementation issues. Each new automotive product must have a technology plan that provides a summary of all the new features that will be planned for the vehicle along with an assessment of technology implementation challenges and an action plan to meet the challenges. The chapter also provides an example of a technology plan for a future automotive product.

IMPLEMENTING NEW TECHNOLOGIES

Implementing new technologies always creates a lot more work. The teams involved in the applications of new technologies to improve the performance or capabilities of one or more vehicle systems first have to clearly understand a number of issues and their effects on all other interfacing vehicle systems and all vehicle attributes. The important issues that need be considered are

1. Level of improvement in performance or capability of the vehicle that could be achieved (note: proving that an improvement is achievable and assessing the level of improvement generally requires a considerable amount of research)
2. Ability and willingness of the customers to adapt to the new changes, accept the changes, and maintain/service/upgrade the new features as needed
3. Effect of the change on other vehicle systems and resulting trade-offs between various affected vehicle attributes
4. Ready availability of technical resources (e.g., availability of specialists, analysis techniques, and test equipment) to analyze effects of changes in the vehicle design

5. Availability of packaging space in the vehicle to incorporate the hardware needed to incorporate the changes
6. Effect on the overall cost of the vehicle
7. Effect on the curb weight of the vehicle
8. Effect on the fuel consumption characteristics of the vehicle
9. Time and costs associated with implementing the new technology (i.e., making sure that the new technology is effective and is ready for implementation)
10. Effect on vehicle quality in the short and long term (i.e., to ensure that the new technologies do not introduce defects into the vehicle)
11. Effect of "make or buy" decisions related to new entities on the company's production resources versus the capabilities of potential suppliers of the new or affected vehicle systems

MAJOR REASONS FOR CHANGES AFFECTING FUTURE VEHICLE DESIGNS

1. Meeting government requirements (e.g., the National Highway Traffic Safety Administration [NHTSA]'s corporate average fuel economy [CAFE] and the Environmental Protection Agency [EPA]'s greenhouse emissions [GHE] requirements; see Chapter 3 for more details)
2. Advances in technologies related to vehicle attributes and features that can provide many advantages, such as improved functionality, efficiency, safety, comfort, convenience, and packaging space, weight reduction, and cost reduction. Some examples of implementation areas are
   a. Incorporation of advanced driver aids and safety technologies
   b. Improvements in driver comfort and convenience
   c. Incorporation of advanced driver information and communication systems
   d. Advances in new lightweight materials
   e. Improvements in manufacturing and assembly methods and equipment (e.g., material joining techniques, robotics systems, material handling systems)
   f. Advances in global communication, sourcing, and project management methods
   g. Improvements in vehicle reliability, durability, and quality

Another major consideration in implementing the new technologies is to ensure that the new vehicle will be perceived by its customers to be "improved" (or advanced) as compared with the older vehicle that will be replaced by the new vehicle. This is especially important if the major competitors of the vehicle have already adopted or are in a process of adopting many of the new technologies.

CREATING A TECHNOLOGY PLAN

The early assumptions and description of the new vehicle will generally include a list of improvements (e.g., features and technologies) to be incorporated. A technology plan should be initiated in parallel with the development of specification and attribute requirements. The technology plan should consider the inclusion of every major vehicle system and its subsystems, and it should include descriptions of changes and technologies, risks in incorporating the changes, and major open issues. A technology plan is discussed later in this chapter (see Table 6.1). Additional technology plans for a sedan, a pickup truck, and a sports utility vehicle (SUV) are included in Chapters 23 through 25, respectively.

RISKS IN TECHNOLOGY IMPLEMENTATION

Implementing any new technology involves risks. It is important for the vehicle development team members to consider all types of risks and make sure that the higher management of the company understands the risks, consequences, and challenges facing the company. The risks can be classified into the following three categories:

1. *Technical risks*: Technical risks result from the inability of the auto company to develop the technology or perfect it to create the required functionality within the available time or resources. The technology readiness may be overestimated during the development of the business and technology plans. The new features may not be debugged thoroughly to remove all possible errors or defects, which may require costly product recalls to fix the problems and/or defending liability situations.
2. *Schedule risk*: The vehicle program may be delayed to allow for the required technology development and its implementation.
3. *Cost overruns*: The costs to develop the required technology and its implementation may increase well above the budgeted amounts.

Cost sharing with other vehicle models or with other vehicle manufacturers in joint development projects (e.g., Ford and GM jointly developed new nine- and ten-speed transmissions [Wernle and Colias, 2013]) is one of the possible approaches considered in undertaking new technology development projects.

NEW TECHNOLOGIES

This section provides descriptions of leading design trends and technologies considered during the early phases of the vehicle development process.

DESIGN TRENDS IN POWERTRAIN DEVELOPMENT

Smaller, Lighter, and More Fuel-Efficient Gasoline Engines

The average vehicle engine size and engine weight are decreasing due to a combination of improved engine technologies and use of lightweight materials.

The current improvements in automotive engines involve

1. *Forced induction/turbo charging/turbo-boost*: Forced induction is the process of delivering compressed air to the intake port of an internal combustion engine. A forced induction engine uses a gas compressor (e.g.,

TABLE 6.1
Illustration of a Technology Plan for an Automotive Product

Vehicle System	Major Changes Planned	Brief Description of Major Technological Challenges	Comments
Body system	Use lightweight and recyclable materials for the exterior body and for interior when permissible	Lightweight materials are expensive, and their properties, such as strength and stiffness, need improvements. Lighter-weight vehicle can be less safe. Reduce aerodynamic drag.	Using lightweight materials such as aluminum, magnesium, or glass fiber-reinforced polymer composites will increase cost of the vehicle. Recyclable materials are environmentally friendly. Most parts of the body are cheaper to make internally than buy from outside suppliers. Incorporate improved exterior shape and underbody panels and reduce vehicle height.
Electric powertrain system	160 kW permanent magnet electric traction motor 60 kWh lithium-ion battery or more advanced replaceable battery	Increasing motor power may increase its power consumption, thus increase the vehicle cost.	Some body modifications will be required to accommodate the new motor and the replaceable battery. It is recommended to buy the motor and the battery from suppliers. If major breakthrough happened in battery industry in the future, the electric powertrain would be more appealing to the customers.
Chassis system	Low-friction bearings Low-rolling resistance tires Quick change battery exchange (release and installation)	Safety must be considered because low-rolling resistance tire may increase braking distance and time response for braking. Also, adding new battery releasing and installing system needs special consideration about how the new system will interface with all other systems.	Low-rolling resistance tire and low-friction bearings reduce the energy used to roll tires, thus increasing the range. Some modifications are required to accommodate the battery releasing and installing system.

Climate control system	Automatic climate control system	More sensors and wiring needed to measure sun load at different body locations.	This climate control system will adjust the interior temperature according to the sun illumination and sun angle. The automatic climate control system will help to reduce the power consumption by adjusting the internal environment according to the outside environment.
Entertainment system	Voice recognition, voice controls, and voice displays Reconfigurable displays and multifunction controls Navigation systems with real-time traffic alerts Text-to-speech conversion systems Projected displays	Since there are many voice-activated systems, highly sophisticated software and hardware (e.g., voice filters) are needed to differentiate human voice versus road noise. Incorporation of many advanced systems in vehicle will increase costs.	Some modifications are required to accommodate all these systems. Buying these systems from supplier is recommended.
Steering system	Tilt/telescoping steering column	May increase the overall price of the vehicle.	The cross-car beam around the steering column needs to be reconfigured.
Braking system	Electronic braking system (EBS)	All-weather reliability needs to be evaluated.	This braking system activates all components electronically, and brake force distribution adapts to load distribution. It is recommended to get this system from suppliers.
Lighting system	Smart lighting and visibility systems LED lamps, fiber optics, smart headlamp Night vision system	Adding new systems to the vehicle will increase costs.	Smart lighting will help to reduce power used for lighting by turning the lighting on and off according to the environment around the vehicle.
Solar power system	Solar panels or more advanced technology for reduced battery load	Solar panels and additional electronics will add more weight and cost to the vehicle.	Modifications needed on the roof to accommodate the solar panels. Solar panels must be curved to be compatible with the body-style of the vehicle. Explore breakthrough in solar panel construction.

(Continued)

TABLE 6.1 (CONTINUED)
Illustration of a Technology Plan for an Automotive Product

Vehicle System	Major Changes Planned	Brief Description of Major Technological Challenges	Comments
Safety system	Smart air bags and belts	Additional sensors may increase costs. Need more extensive verification tests with large range of dummies to ensure protection to increased percentage of population.	Additional verification testing recommended.
	Drowsy driver or alertness monitoring and lane-departure warning systems	Adding all these safety system will increase the overall price of the vehicle. Accuracy and reliability need additional verification tests.	
	Collision avoidance systems	Higher costs, false alarms, and reliability concerns.	
	Driver assistance systems	Higher costs, false alarms, and reliability concerns.	
	Blind spot warning	Higher cost to incorporate sensors and warning displays.	

a turbo charger, which is an exhaust-powered or electric motor–driven turbine) to increase the pressure, temperature, and density of the air. An engine without forced induction is considered a naturally aspirated engine. Turbo charging has helped in downsizing engines and maintaining or even increasing their output. For example, many of the currently available turbo-boost gasoline engines are providing about 120 hp/L output as compared with about 80–100 hp/L outputs provided by naturally aspirated gasoline engines. Turbo chargers also help recycle exhaust energy and reduce the energy loss when hot exhaust gases are released into the atmosphere. The energy loss is typically about 25%–30% of the energy in the fuel consumed.

It should be noted that a supercharger does not work off the exhaust gas, as it is attached to and powered by the engine, and thus, it spins with the crankshaft. When the crankshaft spins the supercharger, it forces air into the engine. The turbo is more efficient, as it does not require engine power to spin it, so it generates more power per boost. A supercharger also does not create full boost till the redline (near the top end of engine speed), which is when the engine is spinning the supercharger as fast as possible.

Thus, with the implementation of forced induction techniques, the number of cylinders used in automotive engines has been decreasing, which has resulted in an increase in the percentage of four-cylinder engines with turbo-boost, and eight-cylinder engines are being replaced by six-cylinder engines and turbo-boost. Electric assist from motor-generators attached to the turbine shafts can further assist in recovering the electrical energy.

2. *Direct fuel injection versus carburetor-based engines*: Fuel injection engines are more efficient and reduce emissions as compared with engines with carburetors. The carburetor contains jets that inject the fuel (e.g., gasoline) into the combustion chambers. The amount of fuel that can flow through these jets depends completely on the amount of air that can be pulled into the carburetor intake. The main disadvantage with obtaining the best performance using a carburetor is that it cannot adjust the air-to-fuel ratio for each individual cylinder. Fuel injection systems, which can inject a precise amount of fuel into the engine, are now more popular for obtaining the best performance from engines.

There are two different versions of fuel injection: port fuel injection and direct injection. Port fuel injection is the most commonly used, and direct fuel injection is the latest-developed fuel injection system. Both systems use computer-controlled electric injectors to spray fuel into the engine, but the difference is where they spray the fuel. Port injection sprays the fuel into the intake ports, where it mixes with the incoming air. The injectors are often mounted in the intake manifold runners, and the fuel sits in the runners till the intake valve opens and the mixture is pulled into the engine cylinder. Port injection systems are much cheaper to manufacture than injectors mounted in the cylinders. The port injectors are not exposed to the high heat and pressure of the combustion chamber, and they do not have to handle the high fuel pressures. Port injection systems typically operate in the 30–60 PSI range, which is dramatically lower than direct injection systems. Support systems such as fuel pumps are also cheaper, because fuel pressures are lower.

In direct injection, the injectors are mounted in the cylinder head, and the injectors spray fuel directly into the engine cylinder, where it then mixes with the air. Only air passes through the intake manifold runners and past the intake valves with direct injection. Direct injection can meter the amount of fuel exactly into each cylinder for optimum performance, and it is sprayed in under very high pressure—up to about 15,000 PSI in some vehicles—so the fuel atomizes well and ignites almost instantly. With current computer controls, the injectors can be pulsed several times for each combustion stroke, so the fuel can be injected over a longer time frame to maximize the power out of the cylinder.

Thus, the main advantage of using direct injection is that the amount of fuel and air can be precisely injected into the cylinder according to the engine load conditions. The electronics used in the system will calculate this information and constantly adjust the timings of the fuel injection. The controlled fuel injection results in a higher-power output, greater fuel efficiency, and much lower emissions. Improvements of about 15% are not uncommon just by changing from port to direct injection.

The disadvantages of direct injection are its cost and complexity. Because the injector tips are mounted right into the combustion chamber, the materials in the injector have to withstand both high temperatures and high pressures, and thus, they are more expensive. Also, the high pressure needed to inject fuel directly into the cylinders means that more expensive high-pressure fuel pumps are required. These are typically mechanically driven from the engine, and thus, they increase the engine complexity.

3. *Cylinder deactivation*: This method involves deactivation of some cylinders (typically two to four cylinders in six- to eight-cylinder vehicles) when the vehicle is cruising at constant speed and the demand for power is lower than when the vehicle is accelerating. Under light driving load conditions, cylinder deactivation will reduce pumping losses from deactivated cylinders and thus improve fuel economy.

4. *Stop/start*: The stop/start method involves stopping the internal combustion engine when the vehicle comes to a full stop and restarting it immediately when the driver presses the gas pedal to accelerate the vehicle. The system requires a larger starter motor and battery capacity to handle frequent stop/start cycles. The stop/start method can reduce energy consumption in city traffic conditions where the vehicle makes frequent stops in traffic and at intersections.

5. *Alternate fuel sources*: To reduce demand for gasoline, engines using a number of alternative fuel sources have been developed. These include (a) natural gas (compressed natural gas [CNG] and liquefied natural gas [LNG]), (b) diesel (e.g., turbo-diesel), (c) biomass fuels, and (d) hydrogen (i.e., hydrogen-powered fuel cell vehicles).

Each of the alternate sources has some disadvantages and advantages relative to gasoline-powered internal combustion engines. For example, since the energy density of CNG is much lower than that of gasoline, large onboard CNG tanks are required. To carry LNG, a refrigeration unit is needed to store the fuel at a low temperature in its liquefied state before use. Diesel engines are more expensive than gasoline engines. The biomass fuels (developed from organic materials, e.g., lumber, crops, manure) are not very common and are not standardized. A hydrogen-powered vehicle would need a large hydrogen tank or to carry a hydrogen fuel cell to generate hydrogen.

6. *Hybrid powertrains*: The hybrid powertrains involve an internal combustion engine along with one or more electric motors. In series configuration, the drive wheels are powered by an electric motor, and the internal combustion engine drives an alternator, which charges the battery. The electric motor is driven by the battery through an electronic module. In parallel powertrain configuration, both the internal combustion engine and the electric motor provide power to the drive wheels. Some hybrid powertrains have two or more electric motors (e.g., each wheel motor directly drives a wheel). The hybrid powertrain consumes less fuel, because energy is supplied by the electric motor more efficiently than by the internal combustion engine. Further, during vehicle deceleration, the electric motor acts like a generator, recovering the dynamic energy of the vehicle and using it to recharge the battery.
7. *Electric vehicles*: Vehicles driven purely on electric power through energy stored in batteries or electric power generated by onboard sources (e.g., hydrogen fuel cells) are available in steadily increasing numbers. Future advances in the ability to increase energy storage capabilities, and reduction in battery weight and volume, will increase driving distance range, and thus, accelerate their market share.

Higher-Efficiency Transmissions

The share of 8–10-speed transmissions, continuous variable transmissions (CVT), and duel clutch transmissions is slowly increasing. These transmissions can improve fuel efficiency by about 2%–10% over five- or six-speed transmissions. The added weight and complexity of these newer higher-speed transmissions, however, can increase the cost and may not provide substantial improvements in fuel economy. However, several manufacturers have produced vehicles with such complexity and claimed improvements in fuel consumption.

DRIVER AIDS AND SAFETY TECHNOLOGIES

These features are incorporated in vehicles to perform certain functions to aid drivers in performing driving tasks safely. Such features typically include sensors that monitor the vehicle motion and other variables related to road, traffic, and weather, and warn the driver and activate vehicle controls (e.g., braking or steering the vehicle) to avoid drivers getting into unsafe situations. The features are

1. *Lane-Departure Warning Systems*: A lane-departure warning system provides a warning to a driver when his vehicle begins to move out of its lane (unless a turn signal is activated in that direction of the lane deviation) on freeways and arterial roads (typically while driving over about 40 mph). These systems are designed to minimize run-off-the-road accidents by addressing the main causes of collisions: driver error, distractions, and drowsiness. There are two main types of system: (a) systems that warn the driver (lane-departure warning [LDW]) if the vehicle is leaving its lane by providing visual, audible, and/or vibratory warning (e.g., vibrating the steering wheel), and (b) systems that warn the driver and, if no action is taken, automatically take steps to ensure that the vehicle stays in its lane.

2. *Driver Monitoring or Alertness Warning Systems*: If the driver is not paying attention to the road ahead and a dangerous situation is detected, the system will warn the driver by flashing lights, warning sounds, and/or a vibratory warning. If no action is taken by the driver, the vehicle will apply the brakes (e.g., a warning alarm will sound, followed by a brief automatic application of the braking system).

3. *Adaptive Cruise Control System*: Adaptive cruise control (also called autonomous or radar cruise control) is an optional cruise control system that automatically adjusts the vehicle speed to maintain a safe distance from vehicles ahead. The control is based on sensor information from onboard sensors (radar or laser based). Most systems provide steering wheel–mounted controls for setting maximum cruising speed and safe headway distance from the leading vehicle.

4. *Automated Braking System*: This system applies vehicle brakes when the sensor and processor in the vehicle determine that the vehicle is headed on a collision course with a stationary or moving object. The unit applies brakes automatically if the vehicle is on a collision course and the driver has not executed a collision avoidance maneuver.

5. *Backup Camera System*: A backup camera is a special type of video camera that is produced specifically for attaching to the rear of a vehicle to aid in backing up, and to alleviate the rear blind area. The backup camera is alternatively known as the *reversing camera* or *rear view camera*. It is specifically designed to avoid a backup collision by providing the driver with a view of the projected path of the vehicle with color-coded distance markers in the rear camera view. The rear-facing video camera is typically mounted at the vehicle centerline, above the rear license plate or near the top or bottom edge of the backlite (rear window). During backing maneuvers (as soon as the gear shifter is placed in reverse gear), the camera output, along with the projected color-coded markings, is displayed in a screen located in the center stack. The red, yellow, and green color-coded zones, respectively, indicate that an object to the rear of the vehicle is very close, somewhat close, or far from the vehicle. In some vehicles, the rear camera display with the color-coded zones is integrated inside the rearview mirrors.

6. *Blind Spot Monitoring System*: A blind spot monitor is a vehicle-based sensor device that detects other vehicles located on both sides of the driver (i.e.,

in adjacent lanes) and to the rear. Warnings can be displayed via visual, audible, vibrating, or tactile signals. While driving in the forward direction, the most common warning signal is activation of an amber-colored light-emitting diode (LED) warning lamp mounted near the outboard edge of each outside mirror. In backup-sensing mode, the system provides auditory warning beeps when the detected targets are approaching close to the collision zone. The system can also be integrated with the backup camera system.

7. *Night Vision System*: This system allows the driver to see further than he or she would be able to see with the vehicle headlamp system during night driving. The night vision system typically uses an infrared camera that can detect objects on the roadway far beyond what a driver can see with the low beam of the vehicle headlamps. The output of the infrared camera is provided to the driver through a separate display in the front of the driver or in an augmented screen of a head-up display. The detected objects are typically shown as augmented superimposed images on the view of the forward road scene captured by the cameras.

8. *Adaptive Forward Lighting Systems*: Adaptive forward lighting systems offer the greatest potential for improving night driving safety performance. The system monitors the forward road scene for oncoming drivers and road features such as curves, grades, and intersections (e.g., through an integrated global positioning system [GPS] and map database system) and alters the beam pattern to provide more illumination in target areas and reduce glare illumination into the oncoming driver's eyes. Some of these functions are already permitted under Federal Motor Vehicle Safety Standards (FMVSS) 108 by allowing a portion of the emitted light to move within a compliant headlamp beam and/or through an automatic re-aim of a headlamp beam pattern (NHTSA, 2016).

9. *Active Rollover Protection/Stability System*: Active rollover protection (ARP) systems involve sensors and microprocessors to recognize impending rollover and selectively apply brakes to resist the rollover. ARP builds on an electronic stability control and its three chassis control systems: the vehicle's anti-lock braking system, traction control, and yaw control. ARP adds another function: detection of an impending rollover. Excessive lateral force, generated by excessive speed in a turn, may result in a rollover. ARP automatically responds whenever it detects a potential rollover. ARP rapidly applies the brakes with a high burst of pressure to the appropriate wheels, and in some situations, decreases the engine torque to interrupt the rollover before it occurs.

10. *Advanced Automatic Collision Notification System*: The system is also known as *advanced automatic crash notification* (AACN) and is the successor to the automatic collision notification (ACN) system. It alerts emergency medical responders, and the real-time crash data from the AACN vehicle telematics system and similar systems can be used to determine whether injured patients need care at a trauma center. By using a collection of sensors, a vehicle telemetry system such as AACN sends crash data

to an advisor if a vehicle is involved in a moderate or severe front, rear, or side-impact crash. Depending on the type of system, the data include information about crash severity, the direction of impact, air bag deployment, multiple impacts, and rollovers (if equipped with appropriate sensors). Advisors can relay this information to emergency dispatchers, helping them to quickly determine the appropriate response involving a combination of emergency personnel, equipment, and medical facilities.

11. *Inflatable Seat Belts*: This technology enables a tubular air bag to inflate during a crash from the seams of the seat belt across the occupant's chest. Inflatable seat belts have two advantages. First, they spread the crash force over a wider area of the body, potentially reducing the risk of injury to the chest. Second, deployment of the bag tightens the belt and reduces the forward movement of the occupant. Thus, it reduces the potential for head injury. Inflatable seat belts were introduced by Ford for outboard rear occupants.

12. *Tire-Pressure Monitoring System*: A tire-pressure monitoring system (TPMS) is an electronic system designed to monitor the air pressure inside the pneumatic tires on various types of vehicle. TPMS reports real-time tire-pressure information to the driver of the vehicle via a gauge, a pictogram display, or a simple low-pressure warning light.

13. *Automated Lighting Systems*: These systems monitor the ambient light environment, activate vehicle exterior and/or interior lamps, and adjust their intensity levels for the driver's safety and convenience (e.g., a perimeter lighting system). The system is activated via proximity sensors, for example, a driver with a wireless key-fob approaching the vehicle.

DRIVER INFORMATION INTERFACE TECHNOLOGIES

These involve the implementation of combinations of the following control- and display-related technologies and features:

1. Steering wheel–mounted controls: Controls such as push buttons, rocker switches, or rotary (e.g., thumb wheels) mounted in the steering wheel spokes.

2. Touch screens: Touch controls on touch screens (controls activated by finger touch on the display surface).

3. Touch pads: Touch pads located in center consoles or center stack regions. The inputs to the touch pad are generally shown on a display screen.

4. Touchless controls (or proximity and/or motion sensors): The sensors can detect proximity of a body part (e.g., finger or hand) and/or its movement and operate a control based on the direction of the movement of the body part.

5. Multifunction controls (a single control can control different functions depending on its selected mode). These controls can be programmed to activate different functions, or they can be moved in different directions to

control many functions (e.g., a rotary control that can also be moved like a joystick and can be pushed in to select its action, like a mouse "click").

6. Haptic controls (controls that provide tactile [or force] feedback during the movement of a control): Controls that can provide different tactile feel/ feedback during the operation of a selected control mode (e.g., turning a rotary switch can provide feeling of different characteristics of detents by changing different level of crispness [change in force or torque with switch movements] and change in force/torque levels during movements for different functions activated by the same switch).

7. Controls with haptic feedback displays and augmented reality enhancements (e.g., touch controls on a display with additional superimposed details/targets such as locations of other vehicles, pedestrians, and/or animals on the road view). For example, the location on the touch display where a target is detected can provide visual and vibratory feedbacks when a finger is moved over it.

8. Gesture-based controls: Body movement–based controls (e.g., proximity sensors or camera-based sensor to detect and identify certain hand/finger or body movements as control actions).

9. Eye gaze–operated controls: Controls activated by eye fixation on a selected location in the driver's visual scene.

10. Voice controls (controls activated by voice commands): Voice command can be recognized in different languages, and a control action corresponding to the recognized command will be made. The accuracy with which a voice command can be correctly recognized inside a noisy moving vehicle, the response time to activate the control after the voice command is given, and the willingness of the driver to use the voice commands are three important considerations in developing and evaluating voice recognition systems.

11. Digital displays: High-resolution changeable message/graphics with enhancing features such as use of colors, touch control areas, haptic feedback in touch controls, legibility under bright sunlight, and display size are important considerations in developing digital displays. The displays can vary in size from very large (e.g., Tesla has a 17 in portrait-type display) to small displays (e.g., 3–5 mm high changeable message displays that can be incorporated in the knobs or buttons as illuminated labels). The number of such displays is expected to increase in future vehicles due to cost reductions and new display technologies (e.g., LED, organic light-emitting diode [OLED]).

12. Auditory displays: Sound feedback for activation of controls; tones, beeps, or spoken or synthesized voice commands are examples of auditory displays. The advantage of auditory displays is that the driver can be free to use his/her eyes to look at any other visual scene or visual display and can acquire the auditory message without turning his or her head in a given direction. The auditory or voice displays can be used to provide a message or the status of one or more vehicle functions. Additional features such as language and the speaker's accent can be selected by the user.

13. Tactile displays: In these displays, the information is presented through the motion of actuators whose movement or vibrations can be sensed by the user's skin touching the actuator surface (e.g., pins moving above control surfaces [control knobs], vibrations provided through touch or grasp surfaces of screens, steering wheel, pedals or seats).

CONNECTED VEHICLES OR VEHICLE-TO-X (V2X) TECHNOLOGIES

These wireless technologies (called *vehicle-to-X* [V2X]) allow two-way communication between the vehicle and other entities (X) outside the vehicle, such as

1. V2V = Vehicle-to-vehicle: The subject's vehicle communicates its position, motion, and state of control activations (e.g., turning, accelerating, decelerating) to other vehicles.
2. V2H = Vehicle-to-home: The subject's vehicle communicates information with his/her home-related programming or controlling of functions related to the vehicle (e.g., charging of an electric vehicle) or home systems (e.g., security system, appliances).
3. V2I = Vehicle-to-infrastructure: The subject's vehicle can communicate with roadside infrastructure, such as traffic signals at intersections, the state of the road, or traffic conditions.
4. V2P = Vehicle-to-person or pedestrian communication: The subject's vehicle can communicate with nearby persons or pedestrians (e.g., by sending them warning messages through wireless devices about the vehicle location, direction of approach, or arrival time).
5. V2C = Vehicle-to-cloud-based data sources: The driver can access information from other cloud-based databases for personal needs (e.g., looking for the nearest bank, gas station, or restaurant).

Thus, V2X technologies allow connected vehicles to wirelessly communicate with each other and other locations. The communicated information can be used to assist the driver by providing warning messages related to different unsafe situations or even initiating certain maneuvers to avoid accidents. Some examples of information that can be communicated to drivers include

1. Warnings of approaching intersection and traffic signal mode
2. Warnings of approaching vehicles during left turning
3. Warnings of approaching work zone and sudden slowing or stopping of vehicles ahead
4. Warnings of approaching road hazards (e.g., a pavement defect [uneven surface, pothole], a downed power line, an accident)
5. Warnings of speed when approaching curves
6. Warnings while in visually obstructed (blind zones) or low sight distance areas
7. Warnings of bicycles and pedestrians
8. Warnings of sudden slowing or stopping of vehicles ahead

In a report on V2V applications, Harding et al. (2014) state the following:

1. V2V communications represent an additional step in helping to warn drivers about impending danger as compared with the benefits of "vehicle-resident" crash-avoidance technologies.
2. V2V communications use onboard dedicated short-range radio communication devices to transmit messages about a vehicle's speed, heading, brake status, and other information to other vehicles and receive the same information from the messages, with range and "line-of-sight" capabilities that exceed current and near-term "vehicle-resident" systems—in some cases, nearly twice the range. This longer detection distance and ability to "see" around corners or "through" other vehicles helps V2V-equipped vehicles to perceive some threats sooner than sensors, cameras, or radar can and warn their drivers accordingly.
3. V2V technology can also be fused with vehicle-resident technologies to provide even greater benefits than either approach alone. V2V can augment vehicle-resident systems by acting as a complete system, extending the ability of the overall safety system to address other crash scenarios not covered by V2V communications, such as lane and road departure. A fused system could also augment system accuracy, potentially leading to improved warning timing and reducing the number of false warnings.

For a discussion of NHTSA's views on how the various levels of vehicle automation will play an important role in reducing crashes and how onboard systems may someday work cooperatively with V2V technology, see NHTSA's Preliminary Statement of Policy on Vehicle Automation (NHTSA, 2013).

With such warning information, the driver can take action to reduce the severity of the collision or avoid it completely. The NHTSA estimates that this technology could be a "game changer," potentially addressing 80% of vehicle crashes involving nonimpaired drivers (NHTSA, 2013).

Harding et al.'s (2014) report also includes preliminary estimates of safety benefits showing that two safety applications—left turn assist (LTA) and intersection movement assist (IMA)—could prevent up to 592,000 crashes and save 1083 lives per year. Thus, the V2V technology could help drivers avoid more than half of these types of crashes that would otherwise occur by providing advance warning. LTA warns drivers not to turn left in front of another vehicle traveling in the opposite direction, and IMA warns them if it is not safe to enter an intersection due to a high probability of colliding with one or more vehicles. Additional applications could also help drivers avoid imminent danger through forward collision, blind spot, do not pass, and stop light/stop sign warnings.

Several major automakers and numerous technology providers have been working with NHTSA and researching the potential safety benefits of V2V. Since 1999, the Federal Communication Commission (FCC) has set aside a separate band (5.9 GHz frequency band) of airways for V2V wireless communications. Currently, the FCC is exploring whether this spectrum can be shared with unlicensed Wi-Fi devices, a decision that automakers believe should not be made till it can be proven that there will be no interference.

SELF-DRIVING VEHICLES

A number of vehicle manufacturers have demonstrated vehicles that have capabilities to drive without any inputs or interventions from the drivers. These vehicles have sensing capabilities to continuously monitor the roadway and traffic situations and take the necessary lateral (steering) and longitudinal (accelerator and brake pedal actions) control actions. With integrated GPS support, the vehicles can also select routes and reach preprogrammed destinations. With the implementation of such technologies, the vehicle becomes "autonomous" (i.e., acting separately from other things or people; having the power or right to govern itself).

Many of the currently available driver assistance systems, such as automatic braking, adaptive cruise control, and lane-keeping systems will be integrated over time to create the self-driving cars.

The future of such technologies is currently debated, because drivers may not be ready to trust such systems. Further, the problem of hacking into such cars needs to be solved to the highest degree of confidence, because if hackers can get into the electronic systems of such vehicles, they can alter the output actions of a vehicle. It is expected that in the near future, automakers will integrate many of the driver assistance capabilities and offer vehicles with limited capabilities (semi-automated and not fully automated self-driving vehicles), such as (a) adaptive cruise control with lane-changing capabilities, (b) self-parking vehicles, and (c) autopilot features that allow drivers to take their hands off the wheel under certain preapproved conditions (Naughton, 2015).

Self-driving trucks are another important application area for this technology. Many commercial applications, including those of the army, can use self-driving trucks. Sedgwick (2016) describes how the army can benefit from having a convoy of self-driving trucks that can follow a lead truck with a human driver. The potential for reducing driver workload and number of human drivers is also very appealing for many commercial delivery applications. The self-driving trucks can operate over long distances with much fewer breaks (no coffee breaks and only stopping to refuel) and thus, can transport cargo in shorter delivery times.

LIGHTWEIGHTING TECHNOLOGIES

Lightweight materials and new structural optimization technologies are used to reduce the weight of the vehicle. Reducing vehicle weight results in less power being required to accelerate and maintain a given speed of the vehicle, and thus, weight reduction reduces the fuel consumption of the vehicle. During the early stages of product development, all vehicle systems are studied to evaluate weight reduction possibilities. Automakers have been experimenting for decades with lightweighting technologies, but the effort is gaining urgency with the adoption of tougher gas mileage standards (EPA and NHTSA, 2012). To meet the government's goal of nearly doubling average fuel economy to 45 mpg by 2025, light vehicles need to lose some weight.

The weight reduction possibilities generally involve combinations of the following approaches: (a) use of different lightweight materials (e.g., high-strength steels, aluminum, magnesium, composites/plastics/carbon fiber), (b) new structural designs and mechanisms (e.g., space-frame designs with composite body panels and hollow coil springs), (c) different production techniques (e.g., hydro-formed body and chassis components, titanium suspension links, spray-painted metal circuits), (d) joining methods (e.g., riveting of steel and aluminum body parts, adhesives, laser welding of dissimilar materials), and (e) smaller, lower-weight, and more efficient fuel-saving powertrains. Many technological advances, such as turbo-boost engines, eight-speed transmissions, stop-start, and cylinder deactivation have been attempted to improve the fuel-saving capabilities of powertrains. Hybrid and electric power plants provide improved fuel economy; however, the need to carry heavy batteries generally increases the weight of the vehicle. All these approaches generally increase costs and development time and add challenges to maintaining high levels of reliability and durability while achieving the desired performance levels.

There follows a short summary of various materials currently used in the auto industry (Helms, 2014).

1. *High-strength steel (HSS)*: HSSs are lighter and stronger steels, composed of mixtures with other elements, such as nickel and titanium. Currently, HSS makes up at least 15% of the car's weight. Some newer vehicles (e.g., the 2014 Cadillac ATS) are using nearly 40% HHSs. HSS costs about 15% more than regular steel, but less than aluminum. HSS weighs more than aluminum. However, with continuing advances in structural designs, the vehicle weight can be further reduced with HSS.

2. *Aluminum*: The typical vehicle already contains around 340 lb of aluminum, which makes up 10% of the weight of a mid-size car. The 2013 Range Rover dropped around 700 lb with its all-aluminum body, while the 2014 Acura MDX shed 275 lb with increased use of HSS, aluminum, and magnesium. The 2015 F-150 pickup has reduced its weight by up to 700 lb as compared with its earlier version. Aluminum is most commonly used in engines, wheels, hoods, and trunk lids. Aluminum is lighter than steel and easy to form into a variety of parts. It is also more corrosion resistant than steel. The supply of steel is many times greater than that of aluminum, and will be for many years. Aluminum costs about 30% more than conventional steel, and a rapid increase in demand could make aluminum prices volatile. Some projections have estimated that the use of aluminum in the auto industry will triple by 2030.

3. *Carbon fiber*: This is a high-strength material made from woven fibers. The specific weight of carbon fiber is about half that of steel. Carbon fiber is resistant to dents and corrosion, and it offers high design flexibility, as it can be formed into a variety of shapes as compared to the stamped steel. However, the high cost of carbon fiber and the longer part-forming (manufacturing) times are substantial drawbacks for the auto industry. Carbon-fiber parts are made from petroleum-based strands, which must go through several stages before they are woven into carbon fiber. After that, it takes about 5 min to form the material into a part, compared with about 1 min for steel or aluminum. Carbon fiber is about five to six times more expensive than steel.

Many organizations are experimenting with cheaper materials for the fibers and faster-curing resins that could shorten the time and costs of forming parts. Carbon fiber is expected to be used in limited amounts on low-volume or luxury cars till major advances occur. The 2014 Chevrolet Corvette Stingray has a carbon-fiber hood and roof, and the 2014 BMW i3 electric car is built around a carbon-fiber frame.

AERODYNAMIC DRAG REDUCTION

Many improvements in aerodynamics are constantly developed and incorporated into new vehicle designs. The improvements range from coming up with a more aerodynamic basic vehicle shape to introducing active aerodynamic elements to alter air flow around the vehicle (Gehm, 2015). Some recently introduced aerodynamic improvements include

1. Lowering vehicle height at higher driving speeds by use of adjustable suspensions.
2. Lightweight underbody panels to reduce underbody turbulence.
3. Active shutters in grills and front bumpers to reduce and deflect air into the engine compartment.
4. Active deflector elements that move outward and rearward to reduce drag around wheels.
5. Flush (noncupped) wheels or active wheel rims (e.g., the intelligent aerodynamic automobile (IAA) concept introduced by Mercedes-Benz, which changes their cupping from 50 mm to zero—from five-spoke to flat-disc wheels).
6. Extendable rear ends and spoilers.
7. Low-profile or flush-mounted door handles.
8. Smaller rear view mirrors or replacing outside mirrors with rear-facing video cameras displaying their view on a screen mounted in the instrument panel. Note: FMVSS 111 requires flat (unit magnification) inside and left outside mirrors (NHTSA, 2016).

TECHNOLOGY PLAN

Table 6.1 illustrates how changes in various vehicle systems and subsystems are planned to meet the weight reduction and fuel economy objectives of a future electric vehicle.

CONCLUDING REMARKS

Advances in technology are influencing vehicle designs substantially through developments in powertrains, aerodynamics, materials, electronics, driver interfaces, and so forth. Fuel economy and cleaner vehicles are being developed along with alternate energy sources. Ongoing research studies on autonomous or driverless vehicles have now demonstrated that even higher levels of passenger comfort, convenience, and safety are possible. The technological developments have also changed how vehicles are being developed by increasing use of the computer-assisted technologies in design, engineering, and management activities covered in other chapters of this book.

REFERENCES

EPA and NHTSA. 2012. 2017 and Later Model Year Light-Duty Vehicle Greenhouse Gas Emissions and Corporate Average Fuel Economy Standards. *Federal Register*, Vol. 77, No. 199, October 15, 2012, Pages 62623–63200. Environmental Protection Agency, 40 CFR Parts 85, 86, and 600. Department of Transportation National Highway Traffic Safety Administration, 49 CFR Parts 523, 531, 533, 536, and 537. [EPA–HQ–OAR–2010–0799; FRL–9706–5; NHTSA–2010–0131]. RIN 2060–AQ54; RIN 2127–AK79.

Gehm, R. 2015. Active in Aero. *Automotive Engineering*, November 2015. (Published by the SAE International, Warrendale, PA.)

Harding, J., G. R. Powell, R. Yoon, J. Fikentscher, C. Doyle, D. Sade, M. Lukuc, J. Simons, and J. Wang. 2014, August. *Vehicle-To-Vehicle Communications: Readiness of V2V Technology for Application.* (Report No. DOT HS 812 014.) Washington, DC: National Highway Traffic Safety Administration.

Helms, J. H. 2014. Advanced engineered material technologies for a challenging environment. *SAE Off-Highway Engineering.* Website: http://articles.sae.org/13054/ (Accessed: June 16, 2015).

Naughton, K. 2015. Self-driving cars are a lot closer than you think. *Automotive News.* Website: www.autonews.com/article/20150507/OEM06/150509895/self-driving-cars-are-a-lot-closer-than-you-think (Accessed: May 7, 2015).

NHTSA. 2013. NHTSA's preliminary statement of policy on vehicle automation (May 2013). Website: www.nhtsa.gov/staticfiles/rulemaking/pdf/Automated_Vehicles_Policy.pdf (Accessed: January 22, 2014).

NHTSA. 2016. *Federal Motor Vehicle Safety Standards and Regulations.* Website: http://www.nhtsa.gov/cars/rules/import/FMVSS/ (Accessed: November 8, 2016).

Sedgwick, D. 2016. Army marches with self-driving trucks. *Automotive News*, February 6, 2016. Website: www.autonews.com/article/20160206/OEM06/302089997/army-marches-forward-with-self-driving-trucks (Accessed: February 6, 2016).

Wernle, B. and M. Colias. 2013. Ford, GM work together on new nine-, 10-speed transmissions. *Autoweek*, April 14, 2013. Website: http://autoweek.com/article/car-news/ford-gm-work-together-new-nine-10-speed-transmissions (Accessed: October 23, 2014).

7 Relation of Vehicle Attributes to Vehicle Systems

INTRODUCTION

After collection of all the necessary background information (covered in the preceding chapters) needed to design a vehicle, the challenge is to come up with a list of detailed design specifications for the vehicle. The detailed specification must include all the functions that must be performed by the vehicle and its systems. Detailed benchmarking of existing vehicles provides useful information in understanding the configuration, construction, and functions of all vehicle systems within the selected vehicles. The information facilitates brainstorming of how the new vehicle should be designed. The vehicle design generally begins with the overall vehicle specifications (vehicle type and exterior dimensions) and allocation of the vehicle functions to different vehicle systems, creating the configuration (i.e., the arrangement of all the entities within each of the systems) and allocating space for each of the systems within the overall vehicle envelope. Therefore, this chapter describes the tasks that need to be performed and the relationships that must be met between customer needs, vehicle attributes, vehicle functions, and system design details.

OVERVIEW OF TASKS AND RELATIONSHIPS BETWEEN CUSTOMER NEEDS AND SYSTEMS DESIGN

A new vehicle program begins with the desire and direction of the higher management of the company to develop a new vehicle (or a set of new vehicles). Figure 7.1 presents a flow diagram of activities, beginning with the management direction and ending with the evaluation of the new vehicle concept. The need for a new vehicle program is generally presented to and discussed with the upper management of the company by the advanced product planning department in a series of advanced product planning meetings. If the management is convinced of the need, it directs the vehicle program management to develop a concept of the proposed vehicle and evaluate the concept by showing it to customers in market research clinics. The flow diagram of activities undertaken to follow the management directions is shown in Figure 7.1. The flow of activities is iterative and involves a number of professionals from different disciplines. The process and activities shown in Figure 7.1 are

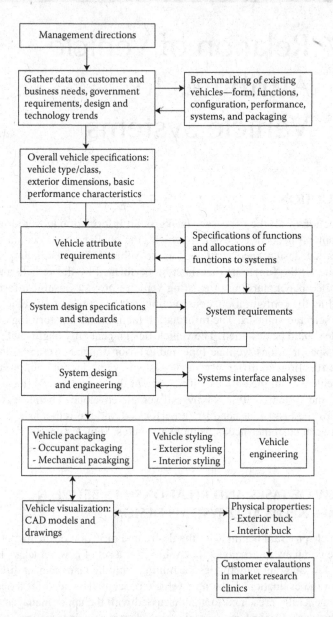

FIGURE 7.1   Flow diagram of activities relating management directions to customer evaluation of new vehicle concept.

dependent on many factors, such as the state of sales of the existing products of the company and its competitors, technological advances, and the availability of resources to develop a new vehicle. Thus, the process of activities varies greatly within different vehicle programs of an automotive company and also between different automotive companies.

Important points to understand from Figure 7.1 are

1. *Overall vehicle specifications*: These specifications must be decided very early in the vehicle program. Everyone in the program must understand the type of vehicle and overall vehicle characteristics such as number of passengers, payload, exterior dimensions, and curb weight. Thus, these specifications are the starting point of the program. These specifications are determined from careful studies of (a) customer needs, (b) business needs, (c) government requirements that the vehicle must meet, (d) design and technological trends, and (e) detailed benchmarking of leading competitors' vehicles.

2. *Vehicle attribute requirements*: All the required attributes of the proposed vehicle and the requirements for each attribute must be defined (see Chapter 2 for details). The target values for each attribute requirement at the vehicle level must be carefully determined to ensure that the vehicle designed to meet these requirements will meet all the customer, business, and regulatory needs.

3. *System requirements*: The system requirements specify how each system should be designed and what it should do. The system requirements used for designing each vehicle system must be developed (or cascaded) from the vehicle attribute requirements and careful allocation of functions (i.e., all the functions the vehicle must perform) to each of the vehicle systems. It should be noted that the functions that the vehicle must perform are developed to meet the attribute requirements. In most automotive companies, internal system design specifications (or standards) are developed and made available to expedite the process of creating (or selecting) design requirements for each system of the proposed vehicle. These standards are generally very comprehensive, including technical information on details such as recommended configurations, construction and packaging considerations, design requirements, interfaces of the system with other systems (i.e., an interface diagram showing subsystems of the system and other interfacing vehicle systems; see Chapter 8), design and installation guidelines, test procedures, test equipment, and so forth.

4. *Overall vehicle packaging and design*: This task involves the integration and consideration of design, engineering, and manufacturing issues by the coordinated efforts of package engineers, designers, and engineers and experts from all the different attribute departments (see the widest box in Figure 7.1). Their basic task is to create the vehicle design, which can be easily visualized and evaluated using computer-assisted design (CAD) models, drawings, and physical bucks. The vehicle design illustrates the form of the vehicle (e.g., its shape, size, and the proportions of different vehicle dimensions), the spaces and locations of all major vehicle systems, and some features of constructional details.

5. *Vehicle concept evaluation*: Once the overall vehicle concept is designed in sufficient detail to allow visualization and design reviews of the proposed vehicle, it should be evaluated to ensure that it will meet the needs

of customers. Market research clinics are generally conducted to evaluate the vehicle concept. In the market research clinic, representative samples of customers are invited and led by an interviewer to observe the vehicle concept and provide answers to a series of questions. The vehicle concept is shown to the customers using methods such as computer visualization, CAD simulations, vehicle views, and/or physical bucks illustrating interior and exterior details of the vehicle. The market research clinics are described in Chapter 21.

Additional descriptions of the concepts and details involved in these tasks are provided in the following sections of this chapter.

ALLOCATION OF ATTRIBUTE REQUIREMENTS TO VEHICLE SYSTEMS

This section describes the following tasks involved in the process of allocation of vehicle attribute requirements to each vehicle system: (1) development of overall vehicle specification, (2) development of attribute requirements for the proposed vehicle, (3) refinement of vehicle attribute requirements, and (4) cascading vehicle attribute requirements to vehicle systems.

DEVELOPMENT OF OVERALL VEHICLE SPECIFICATIONS

The vehicle specifications involve (a) identifying the major vehicle parameters and (b) assigning target values to the identified parameters. The major vehicle parameters include key exterior and interior dimensions, capacities, and capabilities to perform major vehicle functions. The specifications of the vehicle should include

1. Vehicle type, body-style, size (vehicle class), passenger and cargo/luggage carrying capacity (total number of occupants, seating configuration [number of seat rows and occupants in each row]), trunk/cargo volume, and vehicle weight
2. Major exterior dimensions: Overall vehicle length, width and height, cowl and deck point locations, windshield and backlite (rear window) slope angles, tumblehome angle, wheelbase, front and rear overhangs, wheel size, front and rear tread (or track) widths, ground clearance, approach and departure and ramp break-over angles (these exterior dimensions are defined in Chapter 20)
3. Major interior dimensions: Height of the seating reference point (SgRP) from the ground and vehicle floor, accelerator heel point to SgRP, seat track length, leg room, shoulder room, head room, and hip room for each occupant row, and couple distance (longitudinal distance between front and second row SgRPs) (see Chapter 20 for definitions and more details on the above interior dimensions)
4. Longitudinal motion performance: (a) Time to accelerate to a given speed (e.g., 0–60 mph time in seconds), (b) time and maximum attainable speed in a given distance (e.g., time taken to travel ¼ mile and maximum speed attained in ¼ mile), and (c) 60–0 mph stopping distance.

5. Steering and lateral motion performance: (a) Minimum turning radius and (b) maximum lateral acceleration (e.g., maximum velocity while going around a 300 ft diameter circle on a dry pavement)
6. Fuel consumption and range: (a) fuel type (gasoline, diesel, natural gas [compressed natural gas, CNG, or liquefied natural gas, LNG], hydrogen, or electric power), (b) travel distance per fuel volume (miles per gallon or kilometers per liter) in city, highway, and combined driving situations, or electric power consumption (kilowatt hours per kilometer), and (c) maximum travel distance per full fuel tank and fully charged battery (if hybrid or electric vehicle)
7. Vehicle systems details: Powertrain type, engine size and output characteristics, front and rear suspension characteristics
8. Towing capacity: Towing load (pounds or kilograms) for truck and sports utility vehicle (SUV) products
9. Power and comfort features and their characteristics: For example, power windows and lock, power seats, power steering, tilt/telescopic steering column, cruise control, dual zone climate controls, rearview camera, and parking aids
10. Entertainment and communication systems and their characteristics: Audio, navigation and information/communication systems
11. Safety and driver assistance systems and their characteristics: Crashworthiness features (seat belts and air bags), crash-avoidance features (e.g., braking, steering and stability systems, exterior lighting systems, blind area detection and rear vision systems), and driver assistance features (e.g., lane-departure warning and forward braking systems).

The items included in the vehicle specifications may vary somewhat depending on vehicle type, vehicle program, manufacturer, and reporting organization.

It should be noted that the design specifications are generic; that is, they only specify what to achieve—they do not specify how systems should be designed to achieve the specifications (that part comes later).

Defining Attribute Requirements for the Proposed Vehicle

After the initial vehicle design specifications are agreed on by the program management, the product planning team begins development of the attribute goals and requirements for the vehicle. The attribute targets (goals) define how well the vehicle should be positioned or compare with other vehicles in its class (or market segment). Table 7.1 illustrates the attribute targets of a mid-size passenger car (e.g., Ford Fusion or Toyota Camry).

The attribute requirements must be drafted from the targets to ensure that they will meet customer needs and create the overall image of the vehicle as fitting into the specified market segment and brand and possessing other key characteristics, for example, body-style, performance, and feature content.

Benchmarking of leading products in the market segment is conducted. The data gathered are used to prepare a Pugh diagram. A Pugh diagram is very useful in

TABLE 7.1

Attribute Targets of a Mid-Size Passenger Car

Serial No.	Vehicle Attribute	Subattributes	Targets (Goals)
1	Package	Occupant seating package, entry and exit, luggage/cargo package, fields of view, powertrain package, suspensions and tire package, other mechanical and electrical package	The vehicle should have at least 40 mm more rear leg room than its existing model. The overall vehicle package rating must be in the top quartile of the vehicles in its market segment.
2	Ergonomics	Locations and layouts, hand and foot reach, visibility and legibility, posture comfort, operability	The vehicle should be among the leaders within its class.
3	Safety	Front impact, side impact, rear impact, rollover and roof crush, air bags and seat belts, sensors and ECMs, other safety features (visibility, active safety)	The vehicle must meet all federal safety standards and receive 5 star ratings in all impact categories.
4	Styling and appearance	Exterior—Shape, proportions, stance, and so on, Interior—Configuration, materials, color, texture, and so on.	The vehicle should be rated as the best among its class.
5	Thermal and aerodynamics	Aerodynamics, thermal management, water management	The vehicle should get ratings better than the class average.
6	Performance and drivability	Performance feel, fuel economy, long range capabilities, drivability, manual shifting, trailer towing	The vehicle should get ratings among the top quartile of the vehicles in its class. The vehicle must also meet federal fuel economy requirements for its footprint size for MY 2022.
7	Vehicle dynamics	Ride, steering and handling, braking	The vehicle should get ratings better than the class average. The vehicle should meet federal motor vehicle safety standards on braking systems.
8	Noise, vibrations, and harshness (NVH)	Road NVH, powertrain NVH, wind noise, electrical and mechanical systems NVH, brake NVH, squeaks and rattles, passerby noise	The vehicle should get ratings better than the class average.

(continued)

TABLE 7.1 (CONTINUED)
Attribute Targets of a Mid-Size Passenger Car

Serial No.	Vehicle Attribute	Subattributes	Targets (Goals)
9	Interior climate comfort	Heater performance, air-conditioning performance, water ingestion	The vehicle should get ratings better than the class average.
10	Weight	Body system weight, chassis system weight, powertrain weight, electrical system weight, fuel system weight	The vehicle should meet 3400 lb max curb weight.
11	Security	Vehicle theft, contents/ component theft, personal security	The vehicle should get ratings better than the class average.
12	Emissions	Tailpipe emissions, vapor emissions, onboard diagnostics	The vehicle must meet federal emissions requirements for its footprint size for MY 2022.
13	Communication and entertainment	Internet connectivity, within-vehicle coactivity, vehicle-to-infrastructure communication, vehicle-to-vehicle communication, audio reception	The vehicle should be rated as the best among its class.
14	Costs	Cost to the customer, cost to the company	The vehicle cost should be near the average of the vehicles in its class in MY 2020.
15	Customer life cycle	Purchase and service experience, operating experience, life stage changes, system upgradability, disposal and recyclability	The vehicle should attract 20% more customers than the existing vehicle.
16	Product and process complexity	Commonality, reusability, carryover, product variations, plant complexity, tooling and plant life-cycle changes	The vehicle should use at least 30% carryover components and must not exceed current complexity levels in the final assembly plants.

Note: ECM, electronic control module.

understanding the level of each attribute needed in defining the vehicle. The Pugh diagram is used to further improve different vehicle concepts (see Chapter 18). Promising concepts are further evaluated in market research clinics to select a leading concept for development of the proposed vehicle.

REFINEMENT OF VEHICLE ATTRIBUTE REQUIREMENTS

During benchmarking, concept development, and concept selection, a number of decisions are iteratively made about how the proposed vehicle should be configured in terms of locations, functionality, and performance characteristics of major vehicle systems. Constant communication (in daily or weekly team meetings) between various design teams at vehicle and systems levels and design reviews facilitate developing a balanced set of attribute requirements (by achieving trade-offs between different attributes, functions to be allocated to various systems, and their configuration as represented in the vehicle package and CAD models).

SPECIFICATION OF VEHICLE FUNCTIONS FROM VEHICLE ATTRIBUTE REQUIREMENTS AND ALLOCATION OF FUNCTIONS TO VEHICLE SYSTEMS

For a vehicle to possess the stated attributes, it must meet the requirements for each of the vehicle attributes. To meet the attribute requirements, the vehicle must perform many functions. Vehicle functions are actions and outputs that the vehicle as a whole must perform to achieve its specifications and attribute requirements. The primary vehicle goal is to safely transport passengers and luggage/cargo on roads within the specified market under all daytime, nighttime, and weather conditions. The braking, accelerating, and steering functions to be accomplished by the vehicle for the transportation must be established in terms of customer expectations such as stopping distances and speeds to be achieved in a specified time. The level of occupant comfort to be met during ride and handling must be specified. Similarly, the seating and interior configuration must be specified (e.g., in terms of maximum levels of vertical, longitudinal, and lateral space and acceleration) to provide the required level of safety, comfort, and convenience.

The vehicle-level functions are met by assigning (or allocating) one or more of the vehicle functions to each of its vehicle systems. Each vehicle system must therefore perform certain operations to perform its assigned functions. The functions to be performed by each vehicle system thus become the targets (goals) in designing and configuring each vehicle system. For example, to meet the vehicle function of accelerating, the powertrain system must function to generate a specified relationship between engine speed and output torque.

The process of assigning functions to vehicle systems requires many design iterations, as there are many different ways the vehicle systems can be configured and designed. Finding a unique combination of designs of all the vehicle systems that meets all the vehicle-level requirements is challenging, as it requires a number of trade-offs to be made between vehicle attributes. For example, if a very powerful engine is selected to meet the vehicle acceleration capability, it may not meet the requirements for the fuel economy and vehicle weight attributes. Similarly, to incorporate a high-performance braking system, a larger brake system may not be able to meet the requirements for its cost and weight attributes.

The functions of each vehicle system must be defined by considering (a) attribute requirements, (b) the relationship of the attribute to the system, and (c) interfaces of the systems with other systems.

For example, the basic functions of the vehicle body system are to

1. Provide the basic vehicle structure (framework) to position and hold all other major vehicle systems, such as powertrain, chassis, fuel, and electrical system

2. Position and protect occupants from wind, precipitation, and debris
3. Enhance vehicle appearance (styling) and reduce aerodynamic drag
4. Absorb crash energy during accidents
5. Provide safe and comfortable operating environment (includes occupant space, lighting and visibility, and driver interface) during all driving conditions
6. Provide for comfortable entry/egress (e.g., seat height and door openings)
7. Reduce corrosion and protect all vehicle systems
8. Provide space for luggage/cargo, spare tire, and tire-changing tools

Functional analysis can be performed using many different techniques. Functional analysis systems technique (FAST) and integration definition of function modeling (IDEF) are two commonly used techniques to define and organize functions to ensure that all the needed functions are included during requirement development (Bytheway, 2007; Colquhoun et al., 1993).

Figure 7.2 presents a flow diagram of vehicle functions created using FAST. FAST involves brainstorming all possible functions of the product or a system being designed by the use of pairs of verbs and nouns, for example, transport people, accommodate people. The functions are arranged from left to right such that the most basic functions are on the left side and other secondary functions that need to be performed to achieve the basic functions are listed on the right side. Additional functions of the same level can be listed above or below other functions. A multi-disciplinary team is involved in creating the lists of functions and organizing them within the scope of the problem between the two dotted lines to the left and right, which are labeled as "How?" and "Why" (i.e., How should the product function? And why is a secondary function created?). After all the possible functions have been listed and joined by arrows to indicate the flow, entities such as vehicle systems and their lower-level entities (e.g., subsystems, sub-subsystems down to component level) can be proposed (i.e., functions are allocated to entities). The proposed entities can be arranged in many possible ways within the vehicle space till a balanced and acceptable vehicle configuration is created.

CASCADING VEHICLE ATTRIBUTE REQUIREMENTS TO VEHICLE SYSTEMS

All vehicle attribute requirements must be cascaded to all vehicle systems. This means that all the requirements for each vehicle system must be developed to meet one or more of the vehicle-level attribute requirements. Thus, each vehicle system requirement can be traced up to one or more vehicle-level requirements. Thus, no vehicle system should be designed without making sure that it helps meet at least one vehicle-level attribute requirement. Otherwise, the system may perform some functions that are not needed to fulfill customer needs.

The relationship matrix between vehicle attributes and vehicle systems presented in Table 2.3 also provides more useful insights. For example, Table 2.3 shows that many attributes are related to many vehicle systems; that is, the related attributes affect the design of many vehicle systems. Further, a single attribute (e.g., weight) can be related to the design of many vehicle systems.

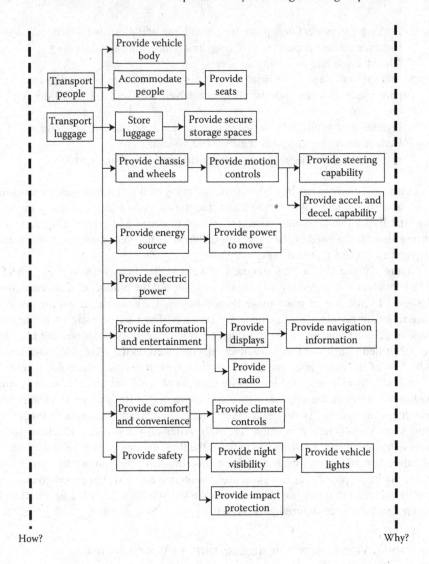

FIGURE 7.2   Illustration of function diagram for a vehicle.

The following two observations from Table 2.3 illustrate these points:

1. Package attribute: The package engineering will provide space for all vehicle systems. Thus, the locations of all vehicle systems must be studied to determine the spaces required to develop interfaces (e.g., physical attachment points and attachment mechanisms, electrical cables, fuel lines, coolant pipes, and brake fluid lines) between various systems.

2. Ergonomics attribute: All user (driver, passenger, and installer) interface equipment, such as controls and displays involved in operation of each of the systems, must be designed to meet the requirements for the ergonomics attribute (see Bhise, 2012 for more detail on ergonomic issues).

This topic of cascading attribute requirements to vehicle systems is covered in greater detail in Chapter 9.

Further, since many vehicle systems are connected to other vehicle systems to perform their functions, the interfaces between different vehicle systems need to be analyzed and designed to meet the interface requirements. The topics related to interface design and requirements are covered in Chapter 8. In designing any vehicle system, understanding its interfaces with other vehicle systems is very important. The interfaces must be designed to ensure that together they meet all the vehicle-level attribute requirements. Table 8.1 presents an interface matrix of major vehicle systems. The matrix shows the following:

1. All vehicle systems are interfaced with five to seven other vehicle systems.
2. The vehicle body system, the powertrain system, the electrical system, and the driver interface system have the greatest number of interfaces to other systems. Thus, these systems have a substantial effect on the overall vehicle performance, drivability, and driver comfort and convenience.

Additional information on the interface matrix is provided in Chapter 8. Bhise (2014) also provides additional information on interface analysis.

SYSTEM DESIGN SPECIFICATIONS

System design specifications must be created for each vehicle system. The system design specifications help in reducing the effort and time needed to develop requirements for each system. The system design specification should include descriptions and details on

1. The objective of the system
2. Functions to be performed by the system
3. Interfaces of the system with other vehicle systems, including interface diagrams and interface matrices
4. Descriptions of possible configurations of the system, including its subsystems, and interfaces between the subsystems and other vehicle systems
5. Design specifications, requirements, and guidelines for the system
6. Test procedures and test equipment for verification of the functions of the system
7. Special requirements: Mandatory government requirements, requirements specific to certain market segments related to vehicle body-style, level of luxury, comfort, and so forth
8. Additional information on customer feedback and lessons learned from implementation of similar systems in other existing products
9. Additional reference information (e.g., benchmarking data, research studies and reports)

CONCLUDING REMARKS

The challenge is to design the vehicle using a top-to-bottom (vehicle level down to lower levels) approach, so that all requirements and design work are traceable to the beginning step involving customer needs, government requirements, and company needs. The engineering competence and skills come from the design team's capabilities to cascade the requirements down to the lowest level of systems to create a balanced vehicle design that meets all the attribute requirements by selecting acceptable trade-offs between various requirements. Chapters 8 and 9 provide additional information on interfaces between systems, interface requirements, and cascading vehicle attribute requirements to vehicle systems requirements.

REFERENCES

Bhise, V. D. 2012. *Ergonomics in the Automotive Design Process*. Boca Raton, FL: The CRC Press.

Bhise, V. D. 2014. *Designing Complex Products with Systems Engineering Processes and Techniques*. Boca Raton, FL: The CRC Press.

Bytheway, C. W. 2007. *FAST Creativity and Innovation: Rapidly Improving Processes, Product Development and Solving Complex Problems*. Fort Lauderdale, FL: J. Ross.

Colquhoun, G. J., R. W. Baines, and R. Crossley. 1993. A State of the Art Review of IDEF0. *International Journal of Computer Integrated Manufacturing* 6.4: 252–264.

8 Understanding Interfaces between Vehicle Systems

INTRODUCTION

An automotive product contains many vehicle systems. The vehicle systems must be interfaced with other vehicle systems such that the systems work together to perform all the functions of the vehicle. Automotive designers work with studio engineers and package engineers to create exterior and interior surfaces to form envelopes. All vehicle systems are packaged within their respective envelopes. For the systems to work with other systems, the interfaces (i.e., connections) between the systems must be designed to ensure that all systems fit within the vehicle space and perform their allocated functions. In this chapter, we will review types of interfaces, interface diagrams, and interface matrices used to understand interface design tasks and requirements for the interfaces.

INTERFACES

WHAT IS AN INTERFACE?

An interface can be defined as a "joint" whereby two (or more) entities (e.g., systems, subsystems, or components) are linked together to serve their allocated functions. Thus, the interface affects the design of both the entities and the parameters defining the joint (i.e., the configuration of connecting elements at the interface). The joint or the interface between the two entities must be compatible; that is, the values of the parameters (e.g., dimensions of the interfacing portions) of the two interfacing entities defining their capabilities must match. An interface can involve (a) physical connection (or attachment), (b) sharing of space (i.e., packaged close to each other), (c) exchange of energy (e.g., transfer of mechanical, hydraulic, electric, thermal, or luminous energy), (d) exchange of material (e.g., oil, coolant, gases), and/or (e) exchange of data (e.g., digital and/or analog signals).

Knowing the type of interface and its characteristics is important to ensure that the two interfacing entities work with each other to perform their allocated functions. During the early design phases of the product, as the functions and their requirements are allocated and the systems are identified, the interfaces between different entities and their parameters must be identified. As the design progresses further, the parameters that define each interface in terms of its characteristics (e.g., dimensions, strength of physical attachment forces, amount of current or data flow passing through the interface) and their level of strength or capacity must be determined and controlled during subsequent detailed design activities. The engineers involved in designing both the interfacing entities must know how the two entities work with each other and how and what the interface must exchange, communicate, or share to enable the two entities to work together and perform their intended functions.

It should be realized that since each system in a product performs one or more functions, all the systems in a product must work together for the product to function. Thus, each interface must be carefully designed to ensure that both interfacing systems are compatible.

TYPES OF INTERFACE

Interfaces between the systems, subsystems, or components of a product and other external systems that affect the operation of the product and their components (e.g., parts, subassemblies, human operators, software) need to be studied and designed to ensure that the product can be used by its customers. Interfaces can be categorized by considering many engineering characteristics and user needs of the product (Lalli et al., 1997). Some commonly considered types of interfaces are described in the following list.

1. *Mechanical or Physical Interface*: This type of interface ensures that any two interfacing components perform as follows: (a) they can be physically joined together (e.g., by the use of bolts, rivets, threads, couplings, welds, or adhesives), (b) their linkages (or joints) can be fixed or allow a range of movements (e.g., through pins or hinges), (c) they can transmit forces between entities using elements, such as a link, spring, damper, or frictional element (e.g., the interface between a brake drum and a brake shoe pad), and (d) they have the required strength or transfer capabilities (e.g., for transfer of materials, heat, or forces) and durability (i.e., ability to work under many work cycles involving loads, vibrations, temperatures, and so forth).

2. *Fluidic or Material Transfer Interface*: A fluidic or material transfer interface (for the transfer of fluids, gases, or powdered/granular materials) can be considered as a different type of interface, or it can be considered as a mechanical interface involving pipes, tubes, hoses, ducts, seals, and so on. The fluidic interface will enable the flow of fluids, gases, or powdered/granular materials, with characteristics including flow rates, purity, pressures, temperatures, insulation, sealing, corrosion resistance, and so forth.

3. *Packaging Interfacing Entities*: Physical space is required to package or accommodate the two interfacing entities. The required space can be determined from (a) the sizes/volumes and shape of spaces (i.e., three-dimensional envelopes) occupied by the two interfacing entities and their interfaces, (b) clearance spaces required around the entities to account for vibrations, movements of parts/linkages, air passages for cooling, and hand/finger or tool access space for assembly/service/repair, and (c) consideration of minimum and maximum separation distances required for their operations. Some examples of packaging interfaces are (a) the engine is packaged within the engine compartment, (b) the engine and the radiator are packaged together, and (c) the occupants are packaged within the passenger compartment.

4. *Functional Interface*: In some cases, if it is necessary to provide one or more functions, one or more of the above types of interfaces may be combined and defined as a functional interface. For example, an automotive suspension system forms a unique functional interface (involving physical links and their relationships with relative movements with energy transfer) between the sprung and unsprung masses of the vehicle.

5. *Electrical Interface*: An electrical interface ensures that two interfacing entities can form an electrical connection/coupling (e.g., with connectors, pins, screws, soldering, or spring-loaded contacts/brushes) that can carry the required electrical current or signals, provide the necessary insulation protection, and/or enable data transfer, and may have other characteristics, such as resistance, capacitance, electromagnetic fields, or interferences.

6. *Software Interface*: A software interface ensures that when data are transferred from one entity (with a software system) to another, the format and transmission characteristics of the coded data through the two interfacing entities are compatible so as to facilitate the required amount and rate of data transfer.

7. *Magnetic Interface*: A magnetic interface generates the required magnetic fields for operation of devices such as solenoids/relays, electric machines (motors, generators), levitation devices, and so on.

8. *Optical Interface*: An optical interface (e.g., fiber optics, light paths, light guides, light piping, mirrors or reflecting surfaces, lenses, prisms, and filters) allows the transfer of light energy between adjoining entities through luminous or nonluminous (e.g., infrared) energy transmission and reflection. The interface may also function to prevent radiant energy transfer by shielding, baffling/blocking, or filtering of unwanted radiated energy.

9. *Wireless Interface*: This type of interface can communicate signals or data without wires via radio frequency communication, microwave communication (e.g., long-range line-of-sight via highly directional antennas, or short-range communication), infrared (IR) short-range communication, Bluetooth, and so forth. The interface applications may involve point-to-point communication, point-to-multipoint communication, broadcasting, cellular networks, and other wireless networks.

10. *Sensor or Actuator Interface*: A sensor has a unique interface that converts certain sensed energy or object characteristic (e.g., light, motion, touch, distance or proximity to certain objects, pressure, and temperature) into an electrical output or signal. For example, a float or floating sensor device can sense fluid levels and convert them into electrical signals, whereas an actuator produces an output (e.g., movement of a control or mechanical links) by converting an input from one type of modality to a different output modality. For example, a stepper motor produces a precise angular movement for each electrical pulse input.

11. *Human Interface*: When a human operator is involved in operating, monitoring, controlling, or maintaining a product, the human–machine or human–computer interface (commonly referred to as the HMI or HCI, respectively) will include devices such as human accommodating or

positioning devices (e.g., chairs, seats, armrests, cockpits, standing platforms, steps, foot rests, handles, access doors), controls (e.g., steering wheel,
gear shifter, switches, buttons, touch controls, stalks, levers, joysticks, pedals, and voice controls), tools (e.g., hand tools, powered tools), and displays
(e.g., visual, auditory, tactile, and olfactory displays).

INTERFACE REQUIREMENTS

To design an interface, an engineer must first understand the overall requirements
of the product and the allocated functions and characteristics of both the entities
attached or linked at the interface. The requirements for the interface should specify
the following: (a) the functional performance of both the entities, (b) configuration of
the entities, (c) available space to create the interface, (d) environmental conditions
for the operation of the product and comfort of the human operators, (e) durability
(minimum number of operational cycles for which the product must function), (f)
reliability and safety considerations in performing the required functions, (g) human
needs (e.g., viewing and reading needs, hearing needs (sound frequencies and levels),
lighting and climate control needs, and product-operating needs), and (h) electromagnetic interference. In addition, the requirements should include any other special
constraints (e.g., weight requirements, aerodynamic considerations, and operating
temperature ranges) that must be met.

Steps involved in the interface requirements development process generally use
an iterative approach (with a series of steps and loops as shown in Figures 2.2 and
2.3) unless a previously developed requirements document (or a standard) is available. The series of steps typically involves the following:

1. Gather information to understand how the interfacing entities work, fit into
   the product, and support the overall functionality, performance, and requirements of the product (e.g., review existing system design documents and
   standards). Draw an interface diagram (described in the next section). Meet
   with the design team members of the interfacing entities (e.g., core engineering functions such as body engineering, powertrain engineering, electrical
   engineering, and climate control engineering for an automotive product) and
   product design teams to understand issues and trade-off considerations with
   the product attributes (e.g., packaging space, safety, maintenance, and costs).
2. Document all design considerations, such as inputs, outputs, constraints,
   and trade-offs, associated with the interface and its effects on other entities (e.g., develop a cause-and-effect diagram; conduct a failure effects and
   mode analysis (FEMA) (see Chapter 18 and Bhise [2014])).
3. Study existing designs of similar interfaces and compare them by benchmarking competitors' products (see Chapter 4 for benchmarking technique).
4. Study existing and new technologies that could be implemented to improve
   the interfaces.
5. Create an interface matrix (described in the next section) of selected systems
   to understand all interfaces (between the selected systems, their subsystems and other vehicle systems), their types, and their characteristics.

6. Create a preliminary set of requirements on how each interface should function.
7. Translate requirements into design specifications (use of the quality function deployment [QFD] technique can help in this step; see Chapter 18 and Bhise [2014]).
8. Brainstorm possible verification tests (or obtain available test methods from existing standards) that need to be performed to demonstrate compliance with the requirements.
9. Develop alternate interface concepts/ideas.
10. Review alternate concepts and ideas with subject matter technical experts (i.e., conduct design iterations; see Chapter 2).
11. Select a leading design by analyzing all other entities that are functionally linked to the entities associated with the interface (develop a Pugh diagram to aid in decision making [see Chapter 17]).
12. Modify and refine interface diagram and interface matrix.
13. Iterate Steps 1–12 till an acceptable interface design is found (see Chapter 2).

The iterative workload described in this process can be reduced if an internal (company) design guide or standard for designing the entities being interfaced can be used as a starting document along with the product-level requirements. Experts and other knowledgeable people in the organization can provide information on valuable lessons learned during the development of similar interfaces from past product programs.

VISUALIZING INTERFACES

REPRESENTING AN INTERFACE

An interface between any two entities (which could be systems, subsystems, or components) can be represented by a simple arrow diagram, as shown in Figure 8.1.

The arrow (between the two entities) indicates a link (or relationship) between the two entities, namely entity A and entity B. The arrow representing the link can denote any of the following (see Figure 8.1):

1. Output of entity A is an input to entity B.
2. Entity A is mechanically attached to entity B.
3. Entity A is functionally attached to entity B (i.e., function of A is required by B to perform its function).
4. Entity A provides information to entity B.
5. Entity A provides energy to entity B.
6. Entity A transmits or sends signals, data, or material (e.g., fluids, gases) to entity B.

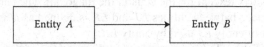

FIGURE 8.1 Interface between two entities.

FIGURE 8.2   Interfaces between the interior door trim panel and the vehicle body.

For example, in an automobile, the interior door trim panel (on which the door armrest and door switches are mounted) is physically attached (using plastic press-fit studs) to the sheet metal door frame, and the door frame in turn is physically mounted (bolted using screws) to the vehicle body via door hinges (see Figure 8.2). (Note: the letter P placed above the arrows in Figure 8.2 indicates a physical connection.)

INTERFACE DIAGRAM

An interface diagram is a flow diagram (or an arrow diagram) showing how different systems, subsystems, and components of a product, shown in blocks (or rectangles in the flow diagram), are interfaced (i.e., joined or linked) by arrows. It provides a visual representation of the product or a portion of the product showing where the interfaces occur. It should also show the type of each interface by the use of letter codes, such as P for a physical connection, E for an energy transfer, M for material/ fluid transfer, or D for data transfer placed next to the arrow.

An interface diagram is a useful tool in understanding how various systems, subsystems, and components are interfaced with each other. The diagram can be created at any level: at the product (vehicle) level, showing all the systems of the product; at a system level, showing all the subsystems of the system; at a subsystem level, showing all the components of the subsystem; or at a mixed level, showing a system, its subsystems, and also other major systems of the product. Two examples of the interface diagram are shown in later sections of this chapter. Figure 8.4 presents an interface diagram of vehicle systems in a car, and Figure 8.5 presents an interface diagram for an automotive braking system.

INTERFACE MATRIX

An interface matrix is a commonly used method to illustrate the existence and types of interfaces between different entities (i.e., systems, subsystems, or components). All the entities involved in the analysis are represented in the matrix. The entities are shown as headings for both the rows and the columns of the matrix. The headings for the rows are placed on the left side of the matrix, and the headings for the columns are placed above the matrix. Each cell of the matrix is defined by the intersection of its row and column, represented by the two interfacing entities. The description of the interface is shown in the cell by one or more applicable letter codes to indicate the type(s) of the interface.

Figure 8.3 presents the output-to-input relationships between six entities in a 6×6 interface matrix (except for the cells in the diagonal). The entities are labeled *E1–E6*. The interface between entities *EJ* and *EK* is defined as *IJK*. Thus, *IJK* represents the output of entity *EJ* used by entity *EK*.

The contents of the cells (i.e., the representation of *IJK*) of the interface matrix typically include letter codes that define descriptors of the types of interface between

Input to entities

	E1	E2	E3	E4	E5	E6
E1		I12	I13	I14	I15	I16
E2	I21		I23	I24	I25	I26
E3	I31	I32		I34	I35	I36
E4	I41	I42	I43		I45	I46
E5	I51	I52	I53	I54		I56
E6	I61	I62	I63	I64	I65	

Output of entities

Note: E1 to E6 = Entities
IJK= Interface between jth and kth entities

FIGURE 8.3 Output-to-input relationships of entities indicated by the cells of the interface matrix.

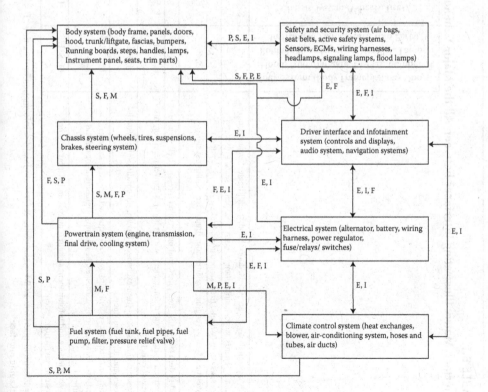

FIGURE 8.4 Interface diagram of vehicle systems.

TABLE 8.1
Interface Matrix Illustrating Interfaces between All the Major Vehicle Systems

System	Body System (Body Frame, Panels, Doors, Hood, Trunk/Liftgate, Fascias, Bumpers, Running Boards, Steps, Handles, Lamps, Instrument Panel, Seats, Trim Parts)	Chassis System (Wheels, Tires, Suspensions, Brakes, Steering System)	Powertrain System (Engine, Transmission, Final Drive, Cooling System)	Fuel System (Fuel Tank, Fuel Pipes, Fuel Pump, Filter, Pressure Relief Valve)	Electrical System (Alternator, Battery, Wiring Harness, Power Regulator, Fuse/Relays/Switches)	Climate Control System (Heat Exchanges, Blower, Air-Conditioning System, Hoses and Tubes, Air Ducts)	Safety and Security System (Air Bags, Seat Belts, Active Safety Systems, Sensors, ECMs, Wiring Harnesses, Headlamps, Signaling Lamps, Flood Lamps)	Driver Interface and Infotainment System (Controls and Displays, Audio System, Navigation Systems)
Body System (body frame, panels, doors, hood, trunk/liftgate, fascias, bumpers, running boards, steps, handles, lamps, instrument panel, seats, trim parts)		S, F, M	P, S	P,S	P, S, E	P, S, E	P, S, E, I	P, S, E, I
Chassis system (wheels, tires, suspensions, brakes, steering system)	M, F, S		F, P, S	P, S,			E, I	E, I
Powertrain system (engine, transmission, final drive, cooling system)	S, F	S, F, M		F, M	E, I	M, P, E, I	E, I	E, I
Fuel system (fuel tank, fuel pipes, fuel pump, filter, pressure relief valve)	S	P	M, F		F, E			I
Electrical system (alternator, battery, wiring harness, power regulator, fuse/relays/switches)	F, E, M	F	F, E	E		E, I	E, F	F, E, I
Climate control system (heat exchanges, blower, air-conditioning system, hoses and tubes, air ducts)	M, F, S		F, P, E		E, I			E, I
Safety and security system (air bags, seat belts, active safety systems, sensors, ECMs, wiring harnesses, signaling lamps, flood lamps)	M, S, I		F, E	I	E, I			E, F, I
Driver interface and infotainment system (controls and displays, audio system, navigation systems)	M, S, I	I	F, E, I	I	E, I	E, I	I	

E = Electrical interface; ECM = electronic control module; I = information or data flow; F = functional interface; M = material flow; P = physical interface; S = spatial-packaging interface.

each outputting entity and the receiving entity. The codes typically include P = physical interface, S = spatial-packaging interface, E = energy transfer, M = material flow, I = information or data flow, and 0 (or a blank cell) = no relationship.

Thus, an interface matrix (a) captures the existence of all interfaces, (b) shows the output-to-input relationships between any two entities (see Figure 8.3), and (c) presents the type(s) of interface between any two entities. Examples of interface matrices are provided in the next section. The interface matrix is also called an *interaction matrix* in some organizations.

The interface diagram and interface matrix are both very useful tools in visualizing relationships and documenting the presence of the interfaces (NASA, 2007; Sacka, 2008). These tools make the design team aware of the presence of many interfaces and the types of these interfaces in the product. The next step is to understand the connection configuration details and the functional requirements of the interfacing entities, and to develop requirements for these interfaces to ensure that the interfacing entities can be designed to work together to perform their allocated functions.

EXAMPLES OF INTERFACE DIAGRAM AND INTERFACE MATRIX

VEHICLE SYSTEMS INTERFACE DIAGRAM AND INTERFACE MATRIX

Figure 8.4 illustrates an interface diagram for all the major systems in a vehicle. All the eight major vehicle systems presented in Table 1.1 are shown in the blocks in the interface diagram. The arrows between the blocks show interfaces between the systems, and the letter codes above or on the right side of each arrow indicate the type of interface. The interface diagram shows that every system in the vehicle is attached to several other systems of the vehicle. For example, all vehicle systems are attached to the body system (which holds and positions all the systems to create the vehicle). The electrical system, which provides the electrical power, is also interfaced to all the other vehicle systems.

Table 8.1 presents an interface matrix illustrating the interfaces between all the major vehicle systems shown in Figure 8.4. The advantage of the interface matrix over the interface diagram is that it presents the interface information in an easy-to-follow format. One can look across each row to determine how the outputs of the system represented by the row are linked to other vehicle systems. For example, scanning across all the columns and down all the rows, the matrix shows that the body system, the powertrain system, the electrical system, and the driver interface system have the highest number of interfaces to the other vehicle systems. Thus, the engineers working on these systems must be in constant communication with other vehicle systems engineers to ensure that all the identified interfaces are designed to meet their respective requirements. Similarly, scanning horizontally across all columns indicates the interfaces that receive the inputs from the systems in the respective rows.

VEHICLE BRAKE SYSTEM INTERFACES

The automotive brake (or braking) system illustrated in this section is for a vehicle with front disc brakes, rear drum brakes, antilock braking system (ABS) capability, and a hand parking brake that applies the rear drum brakes. The braking system was decomposed into four subsystems, and the major components in each of the subsystems were assumed to be

1. Hydraulic subsystem
   a. Brake pedal
   b. Vacuum booster
   c. Vacuum pump
   d. Master cylinder
   e. Brake fluid reservoir
   f. Brake lines
   g. ABS solenoid valves
2. Mechanical subsystem
   a. Calipers with pistons
   b. Brake pads
   c. Brake rotors (disc)
   d. Wheel hubs
   e. Spindle/axle
3. Parking brake subsystem
   a. Parking handbrake
   b. Parking brake cables
   c. Cams and brake pads
4. ABS subsystem
   a. ABS computer/controller
   b. ABS warning light
   c. Wheel speed sensors

Other vehicle systems that interface with the braking system are (1) the body system, (2) the electrical system, (3) the suspension system, and (4) the powertrain system. The braking system can also be considered as a subsystem of the vehicle safety system. It can also be interfaced with the vehicle exterior lighting system through the operation of the stop lamps.

The interface diagram of the braking system is presented in Figure 8.5. The interfaces between components of the subsystems and other vehicle systems are shown by arrows, and letters placed above or to the right side of each arrow indicate the type of interface. The letter codes used are P = physical attachment, S = Spatial—sharing of space, F = functional interface, E = electrical interface, and M = material transfer (e.g., brake fluid).

Table 8.2 presents an interface matrix of the braking system, its subsystems and components, and other interfacing systems. The systems, subsystems, and components are also identified by the codes S = system, SS = subsystem, C = component, and OS = other system. The letter codes are followed by numbers to identify the system in the first digit and the serial number in the second digit. The number 0 in the interface matrix shows the diagonal, with 0 code designating no interface (the same as a blank cell).

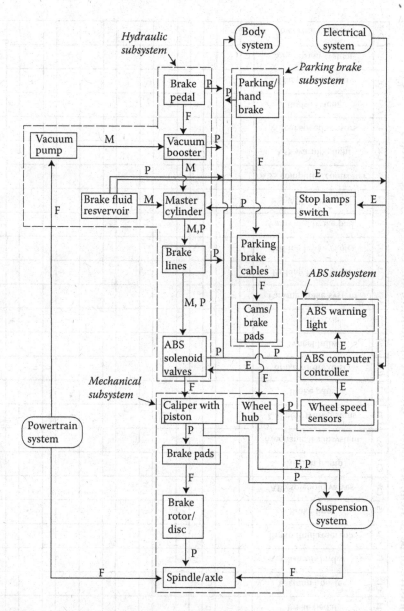

FIGURE 8.5 Interface diagram of an automotive brake system.

A quick visual check of the interface matrix shows that most components are sequentially interfaced to the next component (in each row and column) in the first three subsystems, and most of the components are attached to the vehicle body (see column OS1, labeled "Body system").

During the interface analysis, important issues, trade-off considerations, and other observations were made, which are listed in the following subsections.

TABLE 8.2
Interface Matrix of Automotive Brake System, Its Subsystems and Components, and Other Vehicle Systems

	S	SS1	C11	C12	C13	C14	C15	C16	C17	SS2	C21	C22	C23	C24	C25	SS3	C31	C32	C33	SS4	C41	C42	C43	OS1	OS2	OS3	OS4
S Brake system																								P	E	F	
SS1 Hydraulic subsystem										F						F				E,F				P	E		F
C11 Brake pedal				F																				P			
C12 Vacuum booster			O		M																			P			
C13 Master cylinder				M		O	P,M																	P			
C14 Brake fluid reservoir					O																						
C15 Brake lines						O		P,M																			
C16 ABS solenoid valves							O				F													P	E		
C17 Vacuum pump				M																							P
SS2 Mechanical subsystem																								P	E		
C21 Calipers with pistons												P															
C22 Brake pads											O		F														
C23 Brake rotors (disc)											P	O		O													
C24 Wheel hubs													O		P												
C25 Spindle/axle														O												F	F

Code	Component												
SS3	Parking brake system						0			P	E	P	F
C31	Parking hand brake						F	0		P	E	P	F
C32	Parking brake cables					F	0			F			
C33	Cams and brake pads			F		0							
SS4	ABS system									P	E	P	
C41	ABS computer/controller							0	F	P	E	P	
C42	ABS warning light							0	0	P	E		
C43	Wheel speed sensors			P				E	E				F
OS1	Body system									0			
OS2	Electrical system							E	E	P	0	P	PF
OS3	Suspension system									PS			
OS4	Powertrain system									P	E	P	S

CIJ = ijth component of ith subsystem; OCIJ = ijth component of other system; OSI = ith other system; S = system; SSI = ith subsystem.

Important Interfaces

1. Hydraulic Subsystem

    a. The hydraulic subsystem must interface with the powertrain system via the connection of the brake booster to the intake manifold. The powertrain system also includes an electric vacuum pump that will pump up the brake booster if there is insufficient engine vacuum in the manifold to do so. Poor design of this interface may result in the loss of power-assisted braking.

    b. The hydraulic system also interfaces with the body system. The pedal box needs to be rigidly mounted to the body. The brake booster also needs to attach in a spot where there is enough room, as it is a fairly large component. If these components are not interfaced with the body correctly, the brake system may not work properly.

    c. The hydraulic subsystem also interfaces with the ABS subsystem. If the interface is not done correctly, ABS braking performance may be poor, or complete brake failure may occur.

2. ABS Subsystem

    a. The ABS subsystem interfaces with the electrical system. In most modern cars, many other subsystems may react to ABS braking events (transmission shifting, engine power reductions, etc.), and this information needs to be communicated to other electrical modules to ensure that the entire vehicle reacts appropriately.

    b. The interface with the drivetrain system is necessary to ensure that wheel speed can be appropriately measured at all points. Proper wheel speed is necessary to ensure that the ABS activates when needed.

    c. The interface with the mechanical subsystem is critical to ensure that proper hydraulic pressure is delivered or reduced as needed by the ABS subsystem and that proper braking performance is maintained by the ABS.

3. Mechanical Subsystem

    a. The mechanical subsystem interfaces directly with the drivetrain system to decelerate the vehicle. It is important that all components fit together well to ensure that proper braking torque is delivered to the wheels.

    b. The interface of the mechanical system to the ABS subsystem is important. The ABS is responsible for delivering proper hydraulic pressure to this subsystem so that the vehicle will decelerate without wheel lock-up on low-friction pavements.

    c. The components within this subsystem must interface properly with each other. Improper fit and coordination of components can result in many braking problems, such as rotor warping, premature wear, and noise, vibrations, and harshness (NVH) issues.

Design Trade-Offs

1. An important trade-off is balancing the size of the mechanical subsystem components that interface with the driveline components. Large calipers, pads, and rotors that are specially designed to increase braking friction and

improve heat reduction are critical to meeting brake performance objectives. The wheel hub, wheel, and suspension all need to be designed to incorporate these components to ensure that proper brake performance measurements are met. Larger brake system components (i.e., calipers, brake pads, and rotors) can increase unsprung weight, which can affect vehicle ride and handling performance.

2. The brake pedal and booster need to be rigidly mounted to the vehicle body. The mechanical interface needs to be very robust, that is, not affected by vibrations, corrosion, temperature changes, or high brake pedal actuation forces. This leads to the desire to use a large, heavy brake pedal and linkage to the booster to ensure that the subsystem will not be damaged due to heavy usage by aggressive drivers. This leads to a need for large forces in the attachment hardware. Thus, the space required to provide a robust booster must be considered when trading and allocating space for other components in the engine compartment.

3. There is a trade-off between the electrical system cost and ABS pump performance. When active, the ABS pump represents a significant load on the electrical system. As the pump becomes more powerful, the amperage load is greater, requiring larger cables and an alternator to support the load.

4. There is a trade-off between the capacity of vacuum pump required and the cost and the space required to incorporate it in the engine compartment. The brake booster relies on the engine to provide the vacuum needed for power-assisted braking. Since engines have been becoming more fuel efficient, sometimes the vacuum created is not enough. Thus, an additional vacuum pump may be needed to provide vacuum for the booster in lieu of the engines, especially when operating at higher altitudes. Thus, to provide better braking performance, additional space and electrical load required for the vacuum pump must be considered.

Other Observations

An observation that can be made from this example is that many systems and subsystems are often involved in providing basic vehicle functions. Managing the complexity of these systems and interfaces is always a challenge for systems designers and engineers. Large amounts of data are gathered and used in designing all the interfaces for each component in the selected system. Exercises such as development of the interface diagram and matrix can help the component engineers in organizing and understanding the information needed to develop their components and ensure the required functionality in the vehicle. The information gathered will also be useful in developing the interface requirements used during the interface design process. An interface requirement must specify the characteristics of the two interfacing entities and how they should perform under a given set of operating conditions.

DESIGN ITERATIONS TO ELIMINATE OR IMPROVE INTERFACES

Reducing the number of interfaces will involve (a) reducing vehicle features, (b) increasing the complexity of interfaces by combining two or more interface types (e.g., mechanical attachment also functions as an electrical connector), and (c) changing the modality of an interface (e.g., change from electrical connection to wireless data transfer).

Improving interfaces requires a lot of brainstorming in areas such as new configurations of systems within the product envelope; breakthrough concepts; applications of new technologies; discarding old designs and carryover entities; reducing weight, size, and costs; and investments in new interface designs. For example, current trends in driver interface designs involve reconfigurable driver interfaces with new technology displays and controls, more intuitive touch displays, and multifunction controls—which provide the driver with options to select his or her preferred combinations of displays and controls.

SHARING OF COMMON ENTITIES ACROSS VEHICLE LINES

Sharing an entity (system, subsystem, or component) across a number of vehicle lines involves standardization (or "commonization") of the shared interfaces so that they can work together (e.g., attach and transmit signals or materials such as fluids or gases) with their corresponding mating entities in a different vehicle. This restricts the design (i.e., configuration) of the mating entities, which in turn, may also affect the performance of the involved systems. For example, the use of a common alternator in different vehicle lines would restrict the mechanical and electrical interfacing systems used in the different vehicle lines. The commonization would reduce or even eliminate the design work associated with designing different alternators and their corresponding connectors, which in turn, would reduce design and manufacturing costs; however, it could restrict the overall availability of electric power within the electrical systems of the vehicles that share the same alternator design.

CONCLUDING REMARKS

Interfaces are very important, as they allow different entities to link and function together to perform the required functions of the product. Interface design and the production of interfaces involve expenditure of time, money, and specialized resources. It is important to remember the following point, which wiring harness producers often make: "The major costs increases are not in the increase in length of the wiring harness but they are in the complexity of the 'connectors' at both the ends (i.e., the interfaces) of the harness."

REFERENCES

Bhise, V. D. (2014). *Designing Complex Products with Systems Engineering Processes and Techniques*. Boca Raton, FL: CRC Press. ISBN: 978-1-4665-0703-6.

Lalli, V. R., R. E. Kastner and H. N. Hartt. 1997. *Training Manual for Elements of Interface Definition and Control*. NASA Reference Publication no.1730. Washington, DC: National Aeronautics and Space Administration, Office of Management, Scientific and Technical Information Program.

NASA (National Aeronautics and Space Administration). 2007. *NASA Systems Engineering Handbook*. Report no. NASA/SP-2007-6105 Rev1. NASA Headquarters, Washington, DC 20546. Website: http://ntrs.nasa.gov/archive/nasa/casi.ntrs.nasa.gov/20080008301_2008008500.pdf (Accessed: October 15, 2012) p. 139.

Sacka, M. L. 2008. A Systems Engineering Approach to Improving Vehicle NVH Attribute Management. Master's Thesis for M. S. degree in Engineering Management at the Massachusetts Institute of Technology, Cambridge, MA.

9 Cascading Vehicle Attribute Requirements to Vehicle Systems

INTRODUCTION

Vehicle attribute requirements should be derived from customer needs, business needs, and government requirements enforced in the market where the vehicle will be sold and used. The vehicle attribute requirements are thus the highest level of requirements for the vehicle. These requirements are cascaded down to all lower-level entities to ensure that the lower-level systems within the vehicle are designed to perform their functions to meet the vehicle-level attribute requirements.

This chapter describes the cascading process and provides a few examples and considerations related to specifications of requirements for the lower-level entities.

WHAT IS A REQUIREMENTS CASCADE?

A requirements cascade involves creating requirements for all lower-level entities of a product from each requirement for a higher-level entity of the same product to ensure that the requirement for the higher-level entity is met. Here, an entity is defined by the level of a system—with the product level (or vehicle level) being at the highest, and a component at the lowest, level. For example, a requirement at the vehicle level should be cascaded down to requirements for the vehicle systems. Figure 9.1 presents a schematic representation of consideration of vehicle attributes (shown as A1 to An) for designing vehicle systems (shown as S1 to Sm). Figure 9.2 similarly presents a second-level view of consideration of attributes and subattributes (shown as SA11, SA12, etc.) to vehicle systems and subsystems (shown as SS11, SS12, etc.) It should be noted that the intersection of horizontal and vertical lines in Figures 9.1 and 9.2 illustrate locations of cascaded requirements. The vehicle system requirements for any given attribute can be cascaded down to requirements for its subsystems (see Figure 9.2); the subsystem-level requirements can be cascaded down to its sub-subsystems, and so on. The cascading process thus ensures that a requirement at any level exists only to meet the requirement at its higher-level entity within the product.

Further, it is important to realize that *all* vehicle attributes must be included when developing requirements for any entity within a vehicle. For example, in developing requirements for the braking system, if only the safety attribute is considered (which includes the requirements in the National Highway Traffic Safety Administration [NHTSA] Federal Motor Vehicle Safety Standards [FMVSS] 135 on vehicle braking systems [NHTSA, 2015]), the requirements for the braking system will be

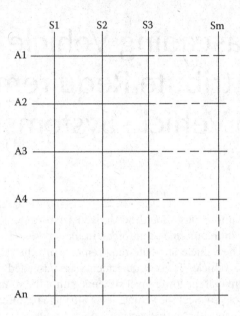

FIGURE 9.1   Considerations of vehicle attributes for vehicle systems design.

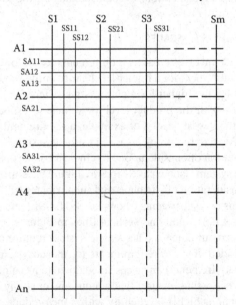

FIGURE 9.2   Considerations of vehicle attributes and subattributes in designing vehicle systems and their subsystems.

incomplete, as the braking system also affects other vehicle attributes such as noise, vibrations, and harshness (NVH), weight, costs, and product and process complexity. Thus, the attribute requirements that can affect an entity must be included in the cascading.

CASCADING ATTRIBUTE REQUIREMENTS TO LOWER LEVELS

To ensure that all issues (or design considerations) related to all product attributes are taken into account during the design stages of the vehicle, each attribute is subdivided into lower levels: subattributes, sub-subattributes, sub-sub-subattributes, and so on. Figure 9.3 illustrates an attribute tree showing that attributes (A1, A2, A3, and A4) are subdivided into their corresponding subattributes such as SA11, SA12, SA21, SA22, ..., SA42. The requirements defining each of the subattributes are shown in Figure 9.3 as R111, R112, ..., R425. The development of an attribute tree helps in progressively dividing the attributes into a manageable number of lower-level attributes so that the requirements for each subattribute can be clearly defined. For each requirement, one or more test procedures (verification tests) along with performance measures and minimum acceptance levels should also be specified to ensure that the requirement can be verified. The attribute tree also helps in maintaining the relationship (or traceability) between lower-level attribute requirements and upper-level requirements, and meeting the lower-level requirements ensures that the upper-level requirements will be met.

EXAMPLE: SUBATTRIBUTES OF VEHICLE ATTRIBUTES

For example, "package and ergonomics" is a vehicle attribute, and one of its subattributes is "easy-to-view displays." Thus, each display in the vehicle (such as a navigation display screen mounted in the center stack) must meet "easy-to-view" requirements for the subattribute. The "easy-to-view" requirements for each display would need to include requirements for (a) display size, (b) display location (e.g., located in an obstruction-free zone), (c) display resolution, (d) display luminance (physical brightness), (e) display color, (f) minimum size in terms of subtended (viewing) angles of displayed letters (or visual details), (g) luminance or contrast of the letters against the display background, (h) display orientation (adjustment) angles, (i) display surface reflectivity, (j) scratch-resistant display surface, and so forth.

Some of the requirements for the subattributes of the display can be specified as follows:

1. The display shall have minimum size of 33 cm (13 in) measured diagonally with a length-to-width ratio of 7:4.
2. The display shall be located in an obstruction-free area (e.g., use Society of Automotive Engineers [SAE] J1050 procedure to determine areas obscured by the steering wheel rim, spokes, and hub [SAE, 2009]).
3. The display shall have a minimum resolution of 4 pixels/mm.
4. The display shall produce minimum luminance of 600 cd/m^2 (or bright white visual details), and the minimum contrast ratio between white and black areas shall be at least 1000:1.
5. The display shall have rotary controls to adjust luminance.
6. The display shall be legible to 65-year-old viewers from 60° to the left to 60° to the right of the display axis (normal to the display surface).

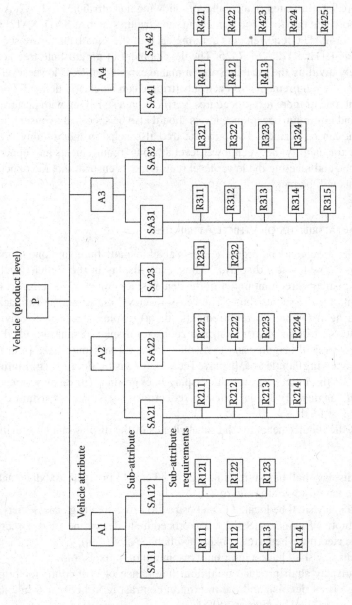

CASCADING ATTRIBUTE REQUIREMENTS TO DEVELOP SYSTEMS DESIGN REQUIREMENTS

To ensure that the vehicle meets all its requirements, the vehicle must possess the attributes that its customers want. The requirements are therefore written for each vehicle attribute. Since most vehicle attributes are complex, they should be further divided into specialized areas called *subattributes* as described in the previous section (see Figures 9.2 and 9.3). The vehicle is divided into systems and subsystems (see Table 1.1). Every system within the vehicle performs certain assigned functions that enable the system to meet one or more sets of attribute requirements. The process of assigning or cascading attribute requirements to systems and subsystems requires a lot of thinking and analysis, and it is usually an iterative process, because the configuration of each system affects the configuration of other systems with which the system interfaces.

CONSIDERATIONS RELATED TO CASCADING ATTRIBUTE REQUIREMENTS FOR VEHICLE SYSTEMS

The problem of cascading requirements from vehicle attribute-level requirements to lower levels of vehicle systems is complex because

1. A vehicle has many attributes.
2. Each vehicle attribute includes many considerations that need to be described (or divided) in terms of lower levels of attributes, such as subattributes, sub-subattributes, sub-sub-subattributes, and so on.
   Table 2.2 provides a list of vehicle attributes and their subattributes. For example, the safety attribute can be decomposed into the following subattributes: (a) front impact, (b) side impact, (c) rear impact, (d) rollover and roof crush, (e) air bags and seat belts, (f) sensors and electronic controls, (g) visibility from the vehicle, (h) visibility of the vehicle, (i) acceleration and deceleration capabilities, (j) steering capabilities, (k) vehicle handling and stability, and (l) driver state monitoring. The sub-subattributes of "visibility from the vehicle" will include (a) forward field of view, (b) rear field of view, (c) side field of view, and (d) indirect field of view available through the inside and outside mirrors. The sub-sub-subattribute of "forward field of view" will include (a) visibility over the hood (or downward visibility), (b) obstructions in the forward field of view caused by the A-pillars, (c) up angle visibility (e.g., visibility of high-mounted signs and signals), (d) forward visibility or road scene and targets (e.g., lane markings) at night (due to the headlamp system), and so forth.
3. Requirements for each level of attributes must be met by determining the functions the vehicle must perform.
4. A vehicle performs many functions. The functions need to be allocated to different vehicle systems.
5. Vehicle systems can vary between different vehicle models due to the use of different technologies (e.g., differences in powertrain technologies, electronic content, and the type of materials used in construction of vehicle components).

6. Many vehicle systems are complex, and they can be decomposed into lower levels (for convenience and management), such as subsystem, sub-subsystem, and so on, till the component level is reached.
7. The vehicle system requirements can be allocated to its lower-level entities.
8. The tasks of decomposition of attributes and vehicle systems are generally accomplished by experts from various engineering offices within each automotive company, depending on factors such as (a) engineering disciplines and specializations, (b) division of engineering design and manufacturing responsibilities and organizational structure, (c) ability of selected suppliers to deliver vehicle systems or their lower-level entities.

For example, in some companies, the body engineering department will be responsible for designing body-in-white and body electrical systems (wiring harnesses, switches, and lighting equipment that fit within the vehicle body), but the electrical engineering department will be responsible for developing the electrical system architecture, which involves electrical power generation, storage (battery), and the control and distribution system, and the electronic subdivision within the electrical system department will be responsible for engineering all the high-tech features such as electronic control modules and driver information, entertainment, warning and assistance systems.

Thus, the process of cascading attribute requirements from the vehicle level down to systems will vary between different vehicle manufacturers and even between different vehicle programs of the same manufacturer.

EXAMPLES OF ATTRIBUTE CASCADING

THE BRAKE SYSTEM AND ITS SUBSYSTEM REQUIREMENTS

This section provides a simplified example illustrating how requirements for an automotive brake system and its subsystems can be developed from the requirements for vehicle attributes. The brake system considered here is described in terms of its interface diagram and interface matrix in Chapter 8 (see Figure 8.5 and Table 8.2).

Table 9.1 illustrates how the vehicle attribute requirements are cascaded into requirements for the brake system and its subsystems. The first column of the table presents the vehicle attributes (see Table 2.2 for more details on subattributes). The second and third columns present subattributes and requirements for the subattributes, respectively. Each row of the table presents a subattribute. (Note that Figure 9.2 schematically illustrates the combinations represented by the format of Table 9.1.)

All the subattributes of each of the attributes must be entered as rows in the cascading table (e.g., Table 9.1) to ensure that no attribute and its subattributes are missed. This is very important, because if any of the subattributes is missed (i.e., not considered in the cascading process), the resulting requirements for the systems and subsystems will be incomplete.

Columns 4 through 8 present requirements for the brake system and its four subsystems. The requirements for the brake system and its subsystems are developed (or cascaded) to meet the subattribute requirements listed in each row. It should be

TABLE 9.1

Cascading of Vehicle Attribute Requirements into Brake System and Its Subsystem Requirements

Vehicle Attributes and Requirements			Brake System Requirements	Subsystems of the Brake System			
Vehicle Attribute: Package	Subattribute	Requirement		Hydraulic Subsystem	Mechanical Subsystem	Parking Brake System	ABS Subsystem
	Occupant accommodation	95% of the user population shall be accommodated with 7 or above rating on 10-point scale	Brake system shall be operable by small 1st percentile female to 99th percentile male	Brake booster and pedal shall not be larger than necessary to support occupant pedal force capability			
	Mechanical packaging	All mechanical equipment shall be accommodated within vehicle hood/body spaces	All brake system components shall be accommodated within vehicle hood/body spaces	Hydraulic system shall be accommodated within the hood compartment with required hand clearance for maintenance	Mechanical subsystem shall be accommodated with required hand/foot clearance for operation and maintenance	Parking brake system shall be accommodated with required hand/foot clearance for operation and maintenance	ABS system shall be protected from water and heat exposure
	Cargo space	The vehicle shall provide at least 25 cubic feet of cargo space	Brake system shall not intrude into cargo space				

(continued)

TABLE 9.1 (CONTINUED)

Cascading of Vehicle Attribute Requirements into Brake System and Its Subsystem Requirements

Vehicle Attributes and Requirements			Brake System Requirements	Subsystems of the Brake System			
Vehicle Attribute:	Subattribute	Requirement		Hydraulic Subsystem	Mechanical Subsystem	Parking Brake System	ABS Subsystem
Ergonomics	Controls and displays	Controls and displays shall be located in expected locations and shall be operable in minimum time. Easy to use rating of 7 or above on 10-point scale	Controls and displays of the brake system shall be operable by small 1st percentile female to 99th percentile male	Brake fluid leak shall activate brake malfunction warning lamp. Brake warning lamp must be located in the instrument cluster in visible areas	Brake pedal shall be located within foot reach distances and easily operated by 99.5% of the driver population	Parking brake control shall be located within reach distances and easily operated by 99.5% of the driver population	Provide ABS operation, and malfunction warning lamp shall be provided
Safety	Crash avoidance	The vehicle shall meet FMVSS 135 requirements. Must be able to stop vehicle within 95% of the FMVSS specified distances without losing vehicle control	The brake system shall meet FMVSS 135 requirements. Must be able to stop vehicle within 95% of the FMVSS specified distances without losing vehicle control	The hydraulic system shall be capable of delivering brake fluid pressure under max load over the life of the vehicle	Brake pads shall be designed to meet fade resistance and stopping distance requirements	The parking brake shall be able to hold the vehicle when parked on a steep (30%) grade	The operation of the ABS shall be able to meet directional stability requirements during braking on low-friction surfaces.

Crash protection	The vehicle shall meet all FMVSS 200 and 300 series applicable requirements. Must obtain 5 star NHTSA crash protection ratings	The braking system must be designed to allow minimal intrusion into cabin during front crash testing	Brake pedal and booster shall be designed to allow minimal intrusion into cabin during front crash testing	Brake pedal and booster shall be designed to allow minimal intrusion into cabin during front crash testing	Parking brake system shall be designed to allow minimal intrusion into cabin during front crash testing	
Visibility	Vehicle shall have lighting systems that meet FMVSS 108 requirements	Brake system shall be capable of providing visual warning messages to the driver	Provide warning lamp signal in case of brake system malfunction. Brake warning lamp shall be located in the instrument cluster in visible areas	Provide warning when thickness of brake pads due to wear reaches limit	Provide parking brake telltale	Provide ABS operation and malfunction warning lamp
Styling and appearance / Appearance	The wheels shall provide appearance to meet the vehicle image	Visual appearance of the brake rotors, pedals, and engine compartment must be consistent with the vehicle design theme	Visual appearance of the brake pedal must be consistent with the interior design theme	Visual appearance of the brake rotors must be consistent with the exterior design theme	Visual appearance of the hand brake lever must be consistent with the interior design theme	Must provide ABS operation and malfunction warning in the instrument cluster

(continued)

TABLE 9.1 (CONTINUED)
Cascading of Vehicle Attribute Requirements into Brake System and Its Subsystem Requirements

Vehicle Attributes and Requirements		Brake System Requirements	Subsystems of the Brake System			
Vehicle Attribute:	Subattribute	Requirement	Hydraulic Subsystem	Mechanical Subsystem	Parking Brake System	ABS Subsystem
Thermal and aerodynamics	Aerodynamics and thermal management			Provide ability to cool brake rotors to avoid excessive temperatures		
Performance and drivability	Fuel economy	The vehicle shall meet 39 mpg combined city and highway driving requirement	Minimum weight to improve fuel economy	Minimum weight to improve fuel economy	Minimum weight to improve fuel economy	Minimum weight to improve fuel economy
	0–60 acceleration	At least 8.5 s				Interaction with launch control system to reduce wheel slip during acceleration

Braking	The vehicle shall meet FMVSS 135 requirements. Must be able to stop vehicle within 95% of the FMVSS specified distances without losing vehicle control	The brake system shall be designed to meet applicable FMVSS requirements	Brake hydraulics need to be powerful enough to ensure brake distance specification is met	Physical brake components must be large and powerful enough to meet brake distance specification	The parking brake system shall meet FMVSS requirements	ABS system shall be designed to ensure that wheels do not lock during brake test, but that stopping distance is minimized
Range	The vehicle shall have gas tank capacity for 500 mile highway range					
Vehicle dynamics	Lateral Gs	The vehicle shall be capable of meeting at least 0.8 g acceleration on curves					Traction control system to reduce wheel slip during traction tests
Driver comfort	Must pass ride and drive testing with benchmarked vehicles		No complaints of hydraulic brake noise during ride and drive	No complaints of brake grind/squeal during ride and drive		

(continued)

TABLE 9.1 (CONTINUED)
Cascading of Vehicle Attribute Requirements into Brake System and Its Subsystem Requirements

Vehicle Attributes and Requirements			Brake System Requirements	Subsystems of the Brake System			
Vehicle Attribute:	Subattribute	Requirement		Hydraulic Subsystem	Mechanical Subsystem	Parking Brake System	ABS Subsystem
Noise, vibrations, and harshness (NVH)	Road noise	No worse than benchmarked vehicles on proving grounds NVH course	No worse than benchmarked vehicles on proving grounds NVH course				
	Powertrain noise	No worse than benchmarked vehicles					
Interior climate comfort	Heater	The interior of the vehicle should be capable of providing 70°F within 5 min from any exterior ambient temperature from −20° to 120°					
Weight	Weight	Completed vehicle weight less than 3700 lb		Minimum size necessary to reduce weight	Minimum size necessary to reduce weight		Minimum size necessary to reduce weight

Security	Vehicle theft/personal security	Lock out vehicle usage under unauthorized usage conditions				ABS controller can lock brakes during unauthorized vehicle use
Emissions	• Tailpipe emissions	Must meet federal and California base emissions standards				
	Onboard diagnostics	Fully compliant with CARB regulation				ABS controller must meet OBD regulations for emissions-relevant signals
Communications and entertainment	Vehicle state information	Provide warning during malfunction	Provide warning during malfunction	Provide warning during malfunction	Provide warning during malfunction	Provide warning during malfunction
Cost	Vehicle price	Priced within 5% of benchmarked vehicles	Priced within 5% of benchmarked vehicles	Must meet cost targets for hydraulic components	Must meet cost targets for mechanical components	Must meet cost targets for ABS components
	Profit	5% profit margin per vehicle		Must meet cost targets for hydraulic components	Must meet cost targets for mechanical components	Must meet cost targets for ABS components
Customer life cycle	System reliability	10 claims/100 overall warranty claims.	Brake system complaints shall be below 2/100 vehicles should be below	Must meet durability requirements for hydraulic components	Must meet durability requirements for mechanical components	Must meet durability components for ABS components

(continued)

TABLE 9.1 (CONTINUED)

Cascading of Vehicle Attribute Requirements into Brake System and Its Subsystem Requirements

Vehicle Attributes and Requirements			Brake System Requirements	Subsystems of the Brake System			
Vehicle Attribute:	Subattribute	Requirement		Hydraulic Subsystem	Mechanical Subsystem	Parking Brake System	ABS Subsystem
	Full warranty	3 year 36,000 miles		Must meet durability requirements for hydraulic components	Pads and shoes must last initial 36,000 miles		Must meet durability components for ABS components
	Powertrain warranty	5 year 100,000 miles					
Product and process compatibility	Carryover components	Check for feasibility and performance improvements through use of carryover components		Share components with other vehicle models	Share components with other vehicle models	Share components with other vehicle models	Share components with other vehicle models
	Platform and plant sharing	Share platform and plant with other product lines	Share platform and plant with other product lines	Share platform and plant with other product lines	Share platform and plant with other product lines	Share platform and plant with other product lines	Share platform and plant with other product lines

Note: ABS, antilock braking system; CARB, California Air Resources Board; OBD, On-Board Diagnostics.

noted that to limit the size of Table 9.1, not all the subattributes listed in Table 2.2 are included. But in a real cascading exercise, all attributes and their subattributes must be listed, and all the system and subattribute requirements needed to completely cascade the subattribute requirements must be meticulously tracked and cascaded. The exercise needs to be conducted with the active participation of engineers managing all vehicle attributes and systems engineers responsible for designing the system, its subsystems, and other interfacing systems (see Figure 8.5).

In performing this analysis, a number of issues need to be simultaneously considered. The important issues to be considered are

1. Specification of the vehicle being developed
2. Overall configuration of the vehicle in terms of its package and initially proposed locations of various systems in the vehicle
3. Interfaces between systems and subsystems
4. Functions allocated to the systems
5. Referring and following design guidelines provided in the systems engineering specifications available within the engineering organizations
6. Benchmarking of package and systems in leading competitors' products
7. Trade-offs between attributes and their subattributes to be considered while deciding Steps 1–4

The above process is iterative, as many changes and trade-offs are considered and made till a balanced overall configuration of the vehicle package with the functional allocation and locations of all major vehicle systems and their subsystems is achieved. The whole process can take many meetings and several weeks, depending on availability of experts (including presentation of many analyses and data) from all related functional and systems areas and the schedule of design reviews. A well-developed and well-documented set of systems design specifications can help expedite the entire cascading process.

CONCLUDING REMARKS

The process of cascading is time consuming and difficult and requires careful consideration of many issues described in this chapter. However, since the vehicle is a complex product and involves the fulfillment of many attributes, the intricate work of complete cascading is necessary. Any shortcuts, approximations, or sloppiness in cascading the attribute requirements to vehicle systems can result in development of vehicle systems that will be incomplete, lack some characteristics or features, and ultimately cause customer dissatisfaction and lower vehicle acceptance.

REFERENCES

NHTSA (2015). Federal Motor Vehicle Safety Standards and Regulations, U.S. Department of Transportation. Website: www.nhtsa.gov/cars/rules/import/FMVSS/ (Accessed: June 14, 2015)
SAE. 2009. *SAE Handbook.* Warrendale, PA: SAE.

10 Development of Vehicle Concepts

INTRODUCTION

This chapter provides information on the activities that take place during the development of a vehicle concept. The process formally begins with the formation of the core team to begin the vehicle program. The team members begin the work by gathering information developed during the pre-program phase (or advanced product planning phase) and conduct additional detailed surveys to obtain information in areas such as trends in technologies, automotive designs, and government regulations. They also benchmark competitive vehicles, visit auto shows, meet and interview customers, and create lists of issues to understand customers, market segments, competitors, new materials, manufacturing processes, and the capabilities of their existing production and assembly plants.

The vehicle concept development work usually begins with creating vehicle sketches, drawings, computer-aided design (CAD) models, and physical properties (mostly nonworking models) to illustrate how the proposed vehicle would look and function. Both the exterior and the interior of the vehicle are illustrated in these drawings and models. Various specialized engineering teams (e.g., body engineering, powertrain engineering, production engineering) review the drawings and properties and conduct their analyses to determine whether the vehicle concept is feasible, that is, whether a functional vehicle with all its attribute requirements can be designed and manufactured within the available resources and timings. Many changes and trade-offs suggested by various engineering teams responsible for delivering different attributes are incorporated with minimum alterations in the overall design (styling) concept.

WHY CREATE A VEHICLE CONCEPT?

The vehicle concept development phase provides a "heads-up" view of the new vehicle. Everyone in the organization gets a better idea about how the future vehicle could look from the properties created to illustrate the concept. The concept vehicle also shows how various vehicle characteristics and new features are to be incorporated in the design. The exercise of creating one or more vehicle concepts also reveals a number of technical issues and problems that need to be solved in creating such a vehicle. The properties created to illustrate the vehicle concepts can be used to conduct design reviews with all key specialists to assess the feasibility and risks associated with creating the vehicle.

Table 10.1 presents a summary of different methods used to communicate vehicle concepts. The vehicle concepts can be compared with other existing vehicles to

TABLE 10.1
Tools Used for Concept Visualization and Evaluations

Type of Tool	Use of the Tool
Sketches of vehicles	To illustrate views from different angles to get a better idea of overall shape, proportions and stance, and important features (e.g., body-style, grill, headlamps, tail lamps, windshield rake angle)
Two-dimensional views of the vehicles (side view, end views, and plan views)	To get a better idea of relative dimensions and spaces available for packaging (e.g., occupant space, wheelbase, overhangs, luggage/cargo space)
Two-dimensional full-size tape drawings	To get a better idea of overall size and vehicle package space. Other vehicle concepts or existing vehicle profiles can be superimposed in different colors to illustrate differences between different designs
Three-dimensional CAD wireframe/mesh models	To visualize available packaging spaces and relative sizes of different spaces (e.g., space for engine, suspensions, tires, occupants, gas tank). Basic occupant packaging/ergonomics tools will be used to check occupant positioning, reach, clearances, field of view, and so forth. Digital human models can be used to illustrate occupant positioning and accommodation
CAE tools	CAE tools will be used to evaluate weight, center of gravity, vehicle stability, powertrain selection, aerodynamic drag, and so forth
Three-dimensional CAD models with color, texture, reflections, shading, and shadow effects	To allow more realistic visualization of the product from various viewing locations and in different backgrounds, such as show room, road, and urban environment
Virtual reality simulations	The CAD model can be viewed through head-mounted displays and/or in computer-assisted virtual reality environments (CAVE), where a subject seated in the vehicle buck can see projected images of the CAD model and other road environments. Useful to evaluate interior spaces and field of view
Foam-core bucks/mock-ups	Quick evaluation of size, space, and clearances between items in the vehicle (e.g., locations of controls and displays). Especially useful for interior evaluations (e.g., instrument panel)
Clay buck (vehicle exterior and/or interior)	Good for quick management appraisal of exterior and interior surfaces. Scale models (e.g., 1/4th or 1/3rd scale) can be made quickly and at less cost as compared with full-size clay models. However, the full-size models make it easier to visualize actual dimensions and volumes as compared with the smaller-scale models. Clay models are not durable and are difficult to transport (heavy and fragile)
Wooden or aluminum extruded frame buck with fiber glass exterior and interior surfaces	Good for transporting long distances for market research clinics or auto shows. Generally used for evaluation of vehicle exterior styling and interior surfaces (e.g., instrument panel, header, roof liner, door trim panels)
Working prototypes	For customer evaluations and showcasing future styling and technology features

Note: CAE, computer-aided engineering.

better understand similarities and differences between various vehicles in its class and with its competitors. The concept vehicles and their features can also be shown to representative customers to assess their desirability and marketability in relation to some of the existing vehicles.

During the early part of the concept development phase, many alternate vehicle concepts are created by the design team. They are usually illustrated in the form of vehicle sketches. Some examples of such sketches are presented in Figure 10.1. The sketches are reviewed by a number of team members, experts, and management personnel and a few concepts (typically two to four) are selected for further development. The selected concepts are refined, and additional analyses are conducted to explore feasibility issues.

FIGURE 10.1   Sketches of the proposed vehicle concept.

Figure 10.2 presents line drawings showing side and plan views of a proposed vehicle concept. Such drawn-to-scale views provide a better understanding of the size and proportions of different exterior characteristics (e.g., ratios of hood length to overall vehicle length, overall height to width of the vehicle, and overall height vs. beltline height).

Figures 10.3 through 10.7 show additional views of the vehicle concept illustrated using CAD models. These CAD models and pictures were created by the University of Michigan-Dearborn students while working on a reconfigurable vehicle development program with the author serving as the faculty advisor. A report on the project is available on the university website (see Gupta, 2009).

In addition, websites can be used to illustrate new vehicle concepts (see Land Rover USA, 2014 website).

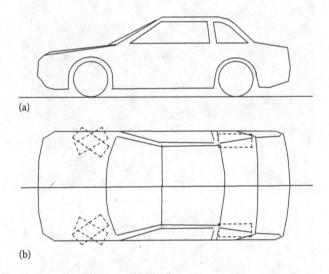

(a)

(b)

FIGURE 10.2   Illustration of early (a) side and (b) plan views of a proposed vehicle.

FIGURE 10.3   Illustration of a clay model of a vehicle concept.

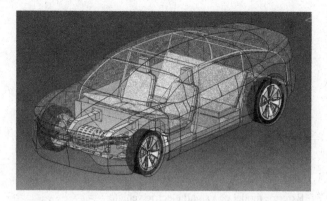

FIGURE 10.4 Illustration of a transparent exterior (wireframe model) showing electric powertrain and seats.

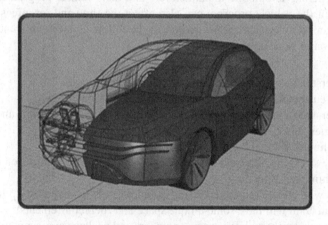

FIGURE 10.5 Partially rendered vehicle concept CAD model.

FIGURE 10.6 Chassis and body frame concept of a small electric vehicle.

FIGURE 10.7   Exterior model of a small electric vehicle.

PROCESS OF DEVELOPING VEHICLE CONCEPTS

The steps involved in developing a vehicle concept will generally vary widely between different automobile manufacturers and their vehicle programs. The size and scope of the vehicle program can also affect the time and resources allocated to the concept development phase. However, the following steps are generally performed in major vehicle programs:

1. Understand objectives and scope of the vehicle program.
2. Visit customers, understand their lifestyles and wants.
3. Understand customer needs, market segment, trends in design and technologies, and government requirements.
4. Benchmark leading competitors' products.
5. Study strengths and weaknesses of characteristics and features of various benchmarked vehicles.
6. Meet with all key functional teams to understand their needs (e.g., the powertrain team will have some unique space and fuel requirements).
7. Work constantly with the studio engineers and package engineers to ensure that engineering and package requirements for all major vehicle systems can be met by the vehicle concepts.
8. Develop sketches of many alternate vehicle concepts.
9. Review the concept vehicle sketches with the design teams to discuss issues related to customer desirability, engineering feasibility, platform sharing (covered in the next section of this chapter), costs, performance, risks, and so forth
10. Modify vehicle concepts based on feedback received.
11. Select a few leading vehicle concepts, incorporate refinements, and develop more detailed sketches.
12. Model exterior and interior surfaces of selected vehicle concepts in 3-D CAD.
13. Incorporate key features and packaging constraints in the CAD models (see Chapter 20).
14. Review CAD models of the vehicle concepts with design teams, experts from various functional areas, and management personnel.

15. Select two or three leading concepts for customer evaluations in market research clinics. (Pugh diagram and other decision-making tools are often used to select the leading concepts. See Chapters 17 and 18.).
16. Prepare rendered exterior and interior views, or fully surfaced and rendered 3-D CAD models, for internal reviews and market research (see Chapter 13).
17. Prepare exterior physical bucks for internal reviews and market research
18. Plan and conduct market research clinics at key cities in major vehicle market regions (see Chapter 21).
19. Review results of the market research clinics with the team members and key management personnel.
20. Select a leading vehicle concept for further development or make decisions to modify the vehicle program, including possible termination of the program if the vehicle concept is not well received.

OTHER ISSUES RELATED TO VEHICLE CONCEPT CREATION

PRODUCT VARIATIONS AND DIFFERENTIATION

The whole purpose of creating a new vehicle is to introduce one or more new vehicle models that will be liked and purchased by the customers. The amount of change between the outgoing and new incoming models needs to be planned based on changes anticipated in customer needs, government regulations, technologies, design trends, cost of resources, and so forth.

One important factor is platform sharing. If the new vehicle can share many of its components and systems with an existing vehicle platform, then the new vehicle development and production costs can be reduced substantially. Building multiple models from a single platform yields economies of scale.

DEFINITION OF A VEHICLE PLATFORM

A vehicle platform is a shared set of common design, engineering, and production efforts, as well as major components, over a number of perceptually different and distinct models that can differ in body-style, vehicle size, and brands. The term *platform sharing* has been used rather loosely in the auto industry, and it can be interpreted in several different ways. There are a number of different methods for model and platform sharing, but the practice is used primarily to lower production costs by reducing the specialization (or increasing commonality) in design and manufacturing efforts.

Very different vehicles can be on the same platform, and very similar ones can be on different platforms. Parts sharing alone does not mean platform sharing, and platform sharing does not mean just parts sharing. If a new vehicle can be built in an existing plant of an existing vehicle, then it can share a number of dimensions, the sequence of manufacturing processes, machines, tools and fixtures, and so forth to reduce costs. The platform can be considered as a set of dimensions. It dictates the physical maximum/minimum size of some parts (such as floor pans, roof panels, chassis parts) of the sharing vehicles. These common dimensions allow fitting/sharing of different vehicles in the manufacturing and assembly equipment, such as conveyors, automated guided vehicles (AGVs), lifts, jigs, fixtures, and tools. It dictates the common locations for various hard-points (or locators) that are used so that the robots can grab, manipulate, position, machine, and weld the various body parts and hoist components into place.

For example, common platforms to share different brands have been used very commonly to create vehicles of the same body-style and size but of different brands (e.g., Ford Fusion and Lincoln MKZ; Ford Edge and Lincoln MKX; Chevrolet Malibu and Buick Regal).

Thus, as a new vehicle concept is being created, the question of platform sharing is addressed to determine how time and costs in vehicle design and production activities can be reduced.

NUMBER OF VEHICLE CONCEPTS AND VARIATIONS

Depending on the scope of the vehicle program (e.g., the number of models and body-styles to be introduced), many variations in concept development for market research are possible, for example, (a) a single concept vehicle, (b) a single concept vehicle with variations in certain exterior or interior characteristics (e.g., different appearance of grill, rear end), or (c) multiple concept vehicles, each with a different exterior and interior styling.

A large vehicle program may include the development of vehicle concepts for a number of body-styles (e.g., sedan, coupe, sports utility vehicle [SUV]) for different brands sold by the auto manufacturer. Early planning before concept development should include specifications for (a) the number of exterior and interior changes and features for brand differentiation and (b) similarities in vehicle size and hardware configuration.

Changes in exterior features considered for variations in concept development typically include changes in overall body shape, changes in proportions, and changes in exterior components, such as grill, headlamps, tail lamps, door handles, wheel caps, exterior mirrors, and trim components. Variations in interiors can be accomplished by considering changes in the shape and materials, with differentiation based on (a) visual appearance (e.g., color, texture, feeling of luxury, genuine vs. fake [simulated, e.g., woodgrains, painted metallic plastics] materials or plastic parts), (b) tactile feel (e.g., texture, compressibility, softness/harshness of surface) characteristics, (c) movement feel (e.g., feeling crispness or mushiness during switch activation), and (d) redesign of major components, such as the instrument panel, the instrument console, the instrument cluster, door trim panels, the steering wheel, and seats.

DESIGNING VEHICLE EXTERIOR AND INTERIOR AS A SYSTEM

In creating new vehicle designs, it is important to coordinate exterior and interior designs of vehicles. Some important considerations for coordination include

1. Common theme and/or brand cues (e.g., feeling of luxury, feeling of sportiness, feeling of ruggedness for a truck-like and/or an off-road vehicle)
2. Look and feel that the interior and exterior were designed by the same designer (using a common theme or a set of design cues)

TABLE 10.2

Pugh Diagram Comparing Three Vehicle Concepts with Four Competitors and the Existing Vehicle as the Datum

Sr. No.	Attribute	Concept #1	Concept #2	Concept #3	Competitor #H	Competitor #T	Competitor #B	Competitor #W	Existing Vehicle (Datum)
1	Package	+	S	−	+	S	+	−	D
2	Ergonomics	S	+	S	−	−	−	−	D
3	Safety	+	+	S	S	S	+	S	D
4	Styling and appearance	+	+	S	−	−	+	−	D
5	Thermal and aerodynamics	S	+	S	+	S	+	S	D
6	Performance and drivability	+	+	S	−	−	S	S	D
7	Vehicle dynamics	+	+	S	−	−	S	+	D
8	Noise, vibrations, and harshness (NVH)	S	S	S	+	+	+	−	D
9	Interior climate comfort	S	S	S	S	S	+	−	D
10	Weight	−	+	S	−	S	S	+	D
11	Security	S	S	S	S	S	S	−	D
12	Emissions	S	+	+	S	+	S	+	D
13	Communication and entertainment	+	S	S	S	S	S	−	D
14	Costs	S	−	S	+	−	S	S	D
15	Customer life cycle	S	S	S	S	S	S	+	D

(Continued)

TABLE 10.2 (CONTINUED)
Pugh Diagram Comparing Three Vehicle Concepts with Four Competitors and the Existing Vehicle as the Datum

Sr. No.	Attribute	Concept #1	Concept #2	Concept #3	Competitor #H	Competitor #T	Competitor #B	Competitor #W	Existing Vehicle (Datum)
16	Product and process complexity	−	−	S	+	+	S	+	D
	Sum of S	8	6	14	6	8	9	4	
	Sum of +	6	8	1	5	3	6	5	
	Sum of −	2	2	1	5	5	1	7	
	Total score = Sum of + minus sum of −	4	6	0	0	−2	5	−2	

Note: + = Better than the datum; − = worse than the datum; D = datum; S = same as the datum.

3. Characteristics of the driver's field of view (visibility) from the vehicle (e.g., command seating position, sitting in a well type feel) by coordinating basic vehicle package parameters such as height of the seating reference point (SgRP) from ground, height of the belt line and top of the instrument panel, and hood height.

4. Entry/egress ease or difficulty is also affected by proper coordination between vehicle package parameters such as height of the SgRP from ground, height of the top of the rocker panel from the ground, lateral location of the SgRP from the outer edge of the rocker panel, and entrance height of the vehicle.

EVALUATION OF VEHICLE CONCEPTS

Vehicle concepts are typically evaluated in market research clinics. The ultimate objective of conducting one or more market research clinics is to determine whether the new vehicle development program should be continued and which vehicle concept or concepts (one or more, depending on the program objectives) should be selected for further development. When multiple concept vehicles are shown with other comparison vehicles, evaluations are also conducted on a number of features included in different concepts and comparison vehicles. The results of these evaluations can be further used to select features and/or changes or modifications that should be made to the selected concept.

The setup and procedures used in the market research clinics will vary depending on the scope of the vehicle program, the number of concepts to be evaluated, the content of the concept vehicles, and the other reference vehicles used for comparison. The topic of market research evaluations is covered in Chapter 11.

USE OF A PUGH DIAGRAM FOR CONCEPT SELECTION AND IMPROVEMENTS

A Pugh diagram is an excellent tool to compare and evaluate a number of vehicle concepts with other existing and benchmarked vehicles (see Chapter 17 for more details). Table 10.2 presents a Pugh diagram illustrating a comparison of three different vehicle concepts with five existing vehicles (four competitors and one current vehicle model as the datum) based on vehicle attributes.

The total score shown in the last row of Table 10.2 shows that Concept#2 was better than all other concepts and competitor vehicles. However, Concept#2 can be further improved by studying the attributes of other vehicles, especially where Concept# 2 received − or S scores.

PLANNING FOR MODELS, PACKAGES, AND OPTIONAL FEATURES

In planning a new vehicle, the variety of existing vehicle models in different market segments produced by the auto company and its competitors, along with data on their sales volumes, lists of features, optional packages, unique brand considerations, and so forth, are discussed in various strategy planning sessions to determine the characteristics of future vehicle models. A number of financial analyses are also conducted to determine the costs and benefits associated with developing different alternative vehicle programs (see Chapter 17).

As the vehicle concepts are being developed, the product planners, market researchers, and design team members also conduct a number of planning meetings to understand the customer needs and to determine the combinations of number of models, number of package options for each model, and number of optional features to be offered to the customers. (This issue is also covered in Chapter 15.)

CONCLUDING REMARKS

The development of a vehicle concept is a very important phase in product development, because it allows automakers to create several concepts before committing to one design that is fully developed and engineered. The phase also allows the design team to iterate the design process several more times, and thus, it enables the team to learn about different design issues and improve their designs. The management personnel also get a better understanding of design issues and challenges as different concepts are reviewed. The organization learns and benefits so that chances of failures, costly fixes in later stages, and risks to the program are reduced. Also, the resulting organizational learning makes succeeding phases more efficient.

REFERENCES

Gupta, A. 2009. Development of a Reconfigurable Electric Vehicle. Published by the College of Engineering and Computer Science, University of Michigan-Dearborn. ISBN: 978-0-933691-13-1. Website: http://umdearborn.edu/cecs/IAVS/books/Reconfigurable_Electric_Vehicle.pdf (Accessed June 26, 2015).

Land Rover USA. 2014. Discovery Vision Concept Interior Design. Website: www.landroverusa.com/vehicles/discovery-sport-compact-crossover/gallery/dual-frame-reveal-youtube-gallery-2.html/293-94215/ (Accessed May 17, 2016).

11 Selecting a Vehicle Concept

INTRODUCTION

Selecting a vehicle concept is probably the most important decision point in the new vehicle development program. The decision point is denoted by the concept selection (CS) gateway (refer to Table 2.1 for gateways). During the early phases of the product development (just before the CS gateway), several alternate vehicle concepts are generally created with different exterior and interior shapes and vehicle configurations. The vehicle concepts are usually shown in the form of (a) realistic-looking pictures or projected images (fully textured, shaded, and colored computer-aided design [CAD] models that look like real vehicles with realistic background scenes such as roads or show rooms) on large screens or (b) three-dimensional physical properties such as vehicle bucks or concept vehicles (drivable or nondrivable).

The vehicle concepts are first shown to company management personnel, and then they are shown to representative groups of customers to help select a concept for further development. The exact process used to select a vehicle concept can vary considerably between different vehicle manufacturers and different vehicle programs. However, most manufacturers stage one or more market research clinics to conduct systematic evaluations of the alternate concepts using representative customers. The results of the market research clinics are reviewed with the vehicle design team and various levels of company management, and the final selection of the concept is made with the concurrence of the highest level of company management. In addition, the feedback received in the evaluations and reviews is used to propose changes to the selected vehicle concept.

Once the vehicle concept is selected, everyone involved in the product development is informed about the decision and asked to begin the next phase of detailed design and engineering of the vehicle.

MARKET RESEARCH CLINICS

WHAT IS MARKET RESEARCH?

Market research during product development primarily involves inviting customers to come to a preselected location and participate in one or more interview sessions to provide judgments and/or feedback on the different vehicle concepts. The invited participants represent a carefully selected sample from a population of customers who own or are principal users of certain preselected vehicle models. The preselected vehicle models are generally selected from the same market segment as the vehicle concept, and they typically include the latest models of competitors' products and the vehicle manufacturer's existing model. The participants, who are the owners or principal users (or drivers) of the preselected vehicles, are thus considered to represent the potential future customers of the vehicle concept.

The participants are led by an interviewer through a series of sessions. In each session, the participant is shown one or more vehicles and/or vehicle features and asked to provide responses to a preselected set of questions. The responses to the questions are analyzed using statistical methods, and the results are provided to the team members of the vehicle program and management personnel. The sessions typically include evaluations of exterior characteristics (e.g., size, proportions, shape, and styling from different viewing angles), interior characteristics (e.g., styling and layout of the instrument panel, doors, consoles, and seats), and other new features (e.g., controls, displays, storage spaces, materials, and adoption of certain technologies). The market research clinic sessions typically last for about one and a half hours.

New Concept Vehicle

An auto manufacturer may create a totally new vehicle concept and present it in an auto show and/or conduct one or more special market research clinics to assess the acceptability of the overall vehicle concept. Generally, a full-size concept vehicle with fully developed exterior and interior, along with details of the powertrain and unique features, is shown to customers. This approach is very common when illustrating high-end sports or luxury cars and providing the public, auto critics, and the media with the opportunity to take an initial look.

On many occasions, an upcoming future model of a vehicle is also revealed about a year before its official introduction into the market. For example, during the 2015 Shanghai auto show, several vehicle manufacturers showed new large four-door executive vehicles, which are typically chauffeur driven with special rear seat features (e.g., reclining rear seats, screens, and computer interfaces for communication). The purpose of such a display is generally to gauge interest among prospective buyers and the public regarding the suitability of the vehicle. The questions that were to be addressed in this particular case were "Would such a vehicle concept attract busy executives in meeting their needs and lifestyle? And would the vehicle project their high social status?"

Specific Evaluation Issues

The management personnel of the auto companies generally conduct deep-dive or in-depth market research to obtain customer feedback on a number of details. Some of the issues addressed in the market research clinics are described in the following subsections.

Evaluation Issues for Exterior Clinics

1. Overall size: Determine whether the vehicle being developed has the size that will be preferred by its customers. For example, the question to be researched here is "Is the vehicle too large, about right, or too small for the market?"

2. Exterior styling and proportions: Determine whether the exterior styling and appearance of the vehicle will be liked by its customers when viewed from different directions (e.g., views from front, rear, side, and other angles).
3. Styling themes: Determine the best styling theme (or alternate) among several leading proposed themes or product concepts shown to the participants.
4. Comparisons with other existing vehicles: Determine whether the exterior design of the concept vehicle is better than other leading (or competitive) designs (e.g., better or worse, similar or dissimilar to other design[s]); or recognized as a design belonging to a certain brand (e.g., it looks like a BMW).
5. Styling and appearance of exterior features such as tail lamps, headlamps, bumpers, grill, mirrors, and fenders.
6. Overall preference and image: Determine whether the concept will be liked or disliked and will be categorized as futuristic, contemporary, luxury, legacy, outdated, aerodynamic, masculine, tough, and so forth.

Issues for Interior Clinics

1. Interior package: Determine whether the interior package shown in the concept vehicle or interior buck will meet the customer expectations in terms of locations of major items (e.g., pedals, steering wheel, arm rests, window openings, and storage areas).
2. Interior design: Determine whether the trade-offs and compromises made in the early design phases are reasonable (e.g., interior roominess vs. closely located instrument panel for ease in reaching controls).
3. Interior design acceptance: Determine whether the interior design of the vehicle will be accepted/liked by its customers in terms of shape, proportions, materials color, texture, and so forth.

The data collected from the clinics are summarized and used to support "sign-off" decisions at gateways; for example, package evaluation is conducted, results are presented to the program and engineering management, and sign-offs from all key product engineering offices are accomplished.

PROS AND CONS OF MARKET RESEARCH

Planning and conducting a market research clinic is expensive and time consuming. Further, the validity of the information provided by the participants in the market research clinics can be questioned by many. The participants may respond that they like the concept very much, but whether they will actually buy the vehicle when it is released in the market is difficult to predict. Thus, some management personnel will question the usefulness of the clinics, because they believe that the invited participants may not be able to provide reliable information to predict the future success of the concepts shown to them. The invited participants may not be familiar with design and technological trends. The advantage of conducting market research is that it provides a "heads-up" (early) response from the customers before the final product is made and available to them.

Thus, considering the pros and cons, the management must discuss all issues, along with the usefulness of the market research results, to make their final decision on whether to proceed with the concept, modify it based on the customer response, or cancel the vehicle program.

If the management does not see value in market research clinics, then they are not conducted, and the decision is usually made through internal evaluations (e.g., using company employees, relatives, and friends of employees or visitors) and/or management reviews.

MARKET RESEARCH METHODS USED IN PRODUCT DEVELOPMENT

METHODS TO OBTAIN DATA

Personal interview is the most useful and most commonly employed technique used during concept selection market research clinics. Mail or web-based surveys and telephone surveys are not used during concept selection. Focus group sessions are generally conducted once the personal interview results are available and the program management wants more feedback information from certain groups of individuals to understand the reasons associated with their earlier responses on certain issues. Thus, focus group sessions can be held to get more information from specific groups of individuals (e.g., those who liked or disliked a particular vehicle concept very much) to gain additional insights into the reasons for their response and identification of some particular details (e.g., front-end or rear-end appearance, instrument panel layout) of some of the concepts.

Personal Interview

Personal interview involves a trained interviewer asking each selected participant a number of questions. It is a 1-on-1 interview, that is, one participant interviewed by one interviewer using selected questions on the vehicle concept. The participant is carefully selected about 1–2 weeks prior to the actual interview. The selection process typically involves a market researcher randomly selecting owners of a vehicle of a certain make, model, and model year (MY) residing in selected counties (e.g., owners of 2016 MY Ford Mustangs purchased and residing in Franklin and Marion counties in Ohio) from a list of registered owners obtained from the state agency that registers motor vehicles at the time of purchase. The selection procedure involves the market researcher making a telephone call to each owner to first verify that the participant indeed owns the vehicle and is the principal driver of the vehicle. The verified owner is then invited to participate in the clinic. The offer involves a certain incentive for participation (e.g., $200 for 2 h of participation). The level of the incentive will depend on the type of market segment (e.g., owners of a large luxury vehicle would be given a larger incentive than owners of a small economy vehicle), the amount of time involved, the complexity of the interview procedure, and the location and time of the clinic.

When the participant arrives at the clinic location at the scheduled time, the market researcher first checks their identity and ownership of the vehicle by asking them to show their driver's license and the vehicle registration papers. The participant is then provided with general instructions on the clinic and asked to complete a demographic form to provide information such as name, address, age, gender, profession, education level, income range, characteristics of trips taken with the vehicle (purpose, frequencies, and distances), and vehicles owned in the past few years. In some clinics, a few anthropometric measurements of the participant are also made, such as stature, weight, sitting height, shoulder width, and buttock-to-knee length. The participant is then led to the evaluation area by an interviewer, who provides instructions on how the vehicle concepts are to be evaluated, and the interview is conducted using a predeveloped procedure and a questionnaire. The responses provided by the participant are recorded using a printed form, a laptop, or a tablet computer.

Focus Group Sessions

In focus group sessions, discussions are held within a group of 8–10 participants with a moderator. The participants in each group are selected by the market researcher based on a certain set of characteristics (e.g., make and model of vehicle currently owned, age, gender, educational background, and profession) to discuss product issues and concepts. The participants in each group have similar characteristics (for example, older males who currently own a 2015 MY Toyota Camry and disliked the styling of the concept vehicle in the prior personal interview). Inviting participants with similar characteristics has the advantages that none of the participants will dominate the discussions (i.e., defend their judgments), and they can collectively think about the issues raised by the moderator and contribute to the discussions (e.g., provide reasons for liking or disliking a certain feature of the concept vehicle). Several groups with different characteristics can be invited separately to discuss the product issues and thus provide feedback from a broader range of individuals.

The goal here is to understand the desires, concerns, and reactions of certain types of individuals regarding preselected topics related to the product concept. For example, the manufacturer may want to know the reasons behind young males disliking a product concept. Thus, participants who especially disliked the product concept in the personal interview (conducted earlier) can be reinvited to further discuss their concerns related to disliking the product. The moderator can systematically lead them through different areas of each issue and create discussions among the group to probe for possible reasons for their dislike in greater detail.

Since the total number of participants included in the focus group sessions is usually small, the gathered information is generally not subjected to any statistical analysis. But the information can be used by the design team to get insights into various thoughts and reasons provided to support their evaluations.

Mail, Web-Based, and Telephone Surveys

Mail, web-based (Internet), and telephone surveys are generally not used for vehicle concept evaluation, primarily because the researcher does not have control over the participant selection (i.e., who is responding, when, and in what situation), and it is not possible for the respondent to see and interact with the physical product (e.g., a vehicle buck) before responding. The response rate (i.e., the percentage of people who will respond) is also very poor in these response-gathering methods.

MARKET RESEARCH CLINICS

Marketing research departments usually design and organize the exterior and interior package evaluation clinics. The team members involved in the vehicle program work with the market researchers to develop questions to be included in the evaluations. The clinics are held in cities representing major target markets. For example, since pickup truck usage is higher in Texas and western states in the United States, a new pickup truck concept can be evaluated by staging a market research clinic in Dallas, TX. Participants who "currently own" certain vehicles are selected and scheduled to arrive at the clinic site at predetermined times. Typically, 75–300 participants may be invited by a random selection procedure from the vehicle registration data in selected counties. The total number of participants is generally selected by statistical calculations to determine the smallest significant differences in percentages of responses (or rating values) to be obtained on different alternative concept and reference vehicles shown in the clinics. For example, a sample size of 512 is required to determine a difference of $\pm 4\%$ at 95% confidence level in a base percentage value of 75. This means that, if one vehicle concept is liked by 75% of the participants, the second vehicle concept must be liked by below 71% or above 79% of the participants to be considered to be different from the first concept with 95% confidence. The formula for computation of the sample sizes is

$$n = \left[z^2 p(1-p) \right] / A^2$$

where:

n = number of participants required

p = mean percentage of responses (in decimal, e.g., $p = 0.75$)

A = absolute accuracy required (in decimal, e.g., $A = 0.04$ for 4%)

z = number of standard deviations (of standardized normal variable) for confidence level desired (e.g., $z = \pm 1.96$ includes 95% of the area under the probability density function of a standardized normal variable with mean equal to 0.0 and standard deviation equal to 1.0)

SOME EXAMPLES OF VEHICLE CHARACTERISTICS EVALUATED IN MARKET RESEARCH CLINICS

Market research clinics are conducted to evaluate a number of issues related to new vehicle concepts. Some examples of types of issues covered in the clinics are

1. Vehicle size: A pickup truck manufacturer wants to find out whether a new pickup truck with a size in between the standard-size pickup truck (e.g., Chevy Silverado, Dodge Ram, Ford F-150) and the small pickup truck (e.g., Chevy Colorado, Dodge Dakota, Ford Ranger) can be designed and marketed in the U.S. market.
2. Exterior styling: A pickup truck manufacturer wants to determine the acceptability of the exterior styling of its new pickup truck in comparison with its existing product and other benchmarked products.

3. Minivan seat height: The vertical height (measured from the ground) of the driver's and other seats is one of the key variables related to success of the minivan segment. Too low a seat height is perceived to cause difficulty in entering and egressing from the vehicle, and it does not provide the command seating position (with better view from higher eye height). On the other hand, a very high seat height causes difficulty in climbing up or down, especially for female drivers. Thus, market research and human factors evaluations are conducted to determine the range of the most preferred seat height.

4. Interior package of a mid-size entry-luxury sedan: An entry-luxury mid-size sedan must be perceived to be better (more luxurious) than an economy mid-size sedan. Market research is conducted to determine whether extra features and craftsmanship enhancements (e.g., higher-quality materials, better surface finish, and smaller gaps between adjoining components) incorporated into the vehicle interior are perceived by the participants to be improvements over the economy vehicles.

5. Understandability of complex control units: The layout, type of controls, and labeling of complex control units such as center stack–mounted climate control, radio, and other components need to be evaluated for ease of understanding, that is, how well drivers understand the functions of the controls and whether they can operate them correctly without errors.

6. Interior spaciousness of a sedan: The vehicle interior package space for the front and rear occupants must provide a feeling of spaciousness (i.e., the space is plentiful). The feeling of spaciousness is a complex function of a number of vehicle package parameters, such as lateral distance between left and right occupant locations in the same seating row, longitudinal distance between seating reference points of seating rows, shoulder room and headroom, and minimum distances of the pillars, header, and instrument panel from the driver's eyes. The market research clinic must be designed to ensure that all relevant occupant package parameters related to the perception of interior spaciousness are identified and evaluated by the participants.

COMMONLY EVALUATED VEHICLE CHARACTERISTICS COVERED IN MARKET RESEARCH CLINICS

This section provides a list of items that can be included in market research clinics for exterior and interior evaluations. Typical questions that can be asked to obtain ratings for the evaluations are provided in parentheses following each item. Three-point direction magnitude scales (which provide three choices related to a given vehicle dimension) and 10-point rating scales are used to obtain participants' responses; these are also included in the questions below.

Exterior Evaluation Characteristics

1. Overall vehicle size (Is the overall size of the vehicle too large, about right, or too small?)

2. Overall vehicle length (Is the overall length of the vehicle too long, about right, or too short?)
3. Overall vehicle width (Is the overall width of the vehicle too wide, about right, or too short?)
4. Overall vehicle height (Is the overall height of the vehicle too tall, about right, or too short?)
5. Wheelbase (Is the wheelbase, i.e., the distance between front and rear wheels, too long, about right, or too short?)
6. Front overhang (Is the front overhang, i.e., the distance from the front bumper to the center of the front wheels [when viewed from the vehicle side], too large, about right, or too small?)
7. Rear overhang (Is the rear overhang, i.e., the distance from the rear bumper to the center of the rear wheels, too large, about right, or too small?)
8. Ground clearance (Is the ground clearance too large, about right, or too small?)
9. Overall vehicle appearance (Provide rating on the overall appearance of the vehicle using the following 10-point scale, where 10 equals liked very much and 1 equals disliked very much)
10. Vehicle appearance (styling) in side view (Provide rating on the vehicle appearance in side view using the following 10-point scale, where 10 equals liked very much and 1 equals disliked very much)
11. Vehicle appearance in front view (Provide rating on the vehicle appearance in front view using the following 10-point scale, where 10 equals liked very much and 1 equals disliked very much)
12. Vehicle appearance in rear view (Provide rating on the vehicle appearance in the rear view using the following 10-point scale, where 10 equals liked very much and 1 equals disliked very much)
13. Hood length (Is the length of the vehicle hood too long, about right, or too short?)
14. Trunk length (Is the length of the vehicle trunk too long, about right, or too short?)
15. Cargo area length (Is the length of the cargo area too large, about right, or too small?)
16. Adjectives used to describe the overall vehicle styling (Select the objectives from the following list that apply to this vehicle: Modern, Traditional, Retro, Tough, Sporty, Masculine, New, Old, Sharp, and so forth)

Interior Evaluation Characteristics

1. Overall interior space (Is the overall space inside the occupant compartment too large, about right, or too small?)
2. Height of the driver's seat (Is the height of the driver's seat too high, about right, or too low?)
3. Height of the vehicle roof (Is the height of the vehicle roof, while seated in the driver's seat, too high, about right, or too low?)
4. Height of the vehicle floor (Is the height of the vehicle floor from the ground too high, about right, or too low?)

5. Legroom (Is the fore-aft leg space, while seated in the driver's seat, too large, about right, or too small?)
6. Headroom (Is the space above your head, while seated in the driver's seat, too generous, about right, or too small?)
7. Shoulder room (Is the space for your shoulders, while seated in the driver's seat, too generous, about right, or too small?)
8. Eye height from the driver's seat (Is your eye height, while seated in the driver's seat, too high, about right, or too low?)
9. Space in front of the head (Is the space in front of your head, while seated in the driver's seat, too generous, about right, or too small?)
10. Space to the left (outboard) side of the head (Is the space to the left side of your head, while seated in the driver's seat, too generous, about right, or too small?)
11. Longitudinal fore/aft location of the gas pedal (Is the gas pedal located too far, about right, or too close to you?)
12. Width of the vehicle at shoulder height (Is the inside width of the vehicle at shoulder height too wide, about right, or too short?)
13. Visibility to the front of the vehicle (Is the visibility to the front of the vehicle from the driver's seat excellent, adequate, or insufficient?)
14. Visibility to the left side of the vehicle (Is the visibility to the left side of the vehicle from the driver's seat excellent, adequate, or insufficient?)
15. Visibility to the right of the vehicle (Is the visibility to the right side of the vehicle, while seated in the driver's seat, excellent, adequate, or insufficient?)
16. Visibility to the rear of the vehicle (Is the visibility to the rear of the vehicle, while seated in the driver's seat, excellent, adequate, or insufficient?)
17. Ease in getting into the driver's seat (Provide rating on the ease in getting into the driver's seat using the following 10-point scale, where 10 equals liked very much and 1 equals disliked very much)
18. Ease in getting out of the vehicle (Provide rating on ease in getting out of the driver's seat using the following 10-point scale, where 10 equals liked very much and 1 equals disliked very much)

Chapter 21 presents additional information on the development of questionnaires and data analysis methods for evaluations conducted in interior and exterior market research clinics. Chapter 22 also provides an example of data from interior package evaluations (see Table 22.5).

Exterior Buck Preparation and Evaluation Setup

The exterior bucks of a concept vehicle are generally shown to customers along with other existing vehicles. For example, if a manufacturer wants to evaluate a new concept vehicle (e.g., as a 2021 MY vehicle) to replace its existing mid-size sedan, the existing model (2016 MY) and two other competitors' 2016 MY vehicles can be shown and evaluated to serve as references in the market research clinic. The exterior buck created to represent the concept vehicle should have a fit and finish quality as good as that of any other production vehicle. The exterior buck should be made to true scale (full size). The buck is generally made of materials such as fiberglass, wood, metals, plastics, and glass and made to look like a real vehicle from the exterior. But, most likely, it will not be a working vehicle.

To control the exposure of the vehicles in the market research clinic from certain views, such as side, front, rear, or angled views, the setup of the vehicles should be carefully planned. Figures 11.1 and 11.2 present a setup to view the side and front views, respectively, of four vehicles in a market research clinic.

In Figures 11.1 and 11.2, Vehicle S is the concept vehicle, which is compared with three other vehicles: Vehicle M (the manufacturer's existing vehicle) and Vehicles W and P (the leading competitors' vehicles, also used for benchmarking). Figure 11.1 shows that the participant (labeled as the subject) in the middle of the diagram is asked to stand inside the square and look at the four vehicles positioned such that the side views of the four vehicles are geometrically similar when viewed from the middle square. The subject is asked to answer a number of questions while standing in the square about the side views of the four vehicles. For convenience and quick change of vehicle views, the four vehicles are placed on turntables, which are synchronized and can be indexed (rotated) at any angular position to allow comparisons. Figure 11.2 shows that the vehicles are indexed so as to allow comparisons of the front views of the four vehicles.

Chapter 22 presents an example of the results of a market research clinic for concept selection (see Table 22.6).

INTERIOR BUCK PREPARATION FOR PACKAGE SURVEYS

Interior package bucks are built to full scale using wood and/or aluminum armature (scaffold), and the interior surfaces of the instrument panel, roof, doors, center console, and other trim parts are generally modeled using fiberglass material and

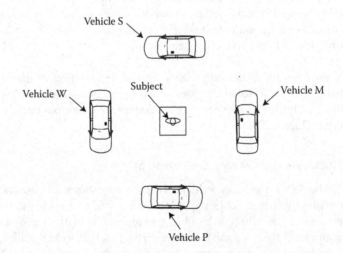

FIGURE 11.1 Plan view of the setup for comparative evaluation of side views of four vehicles.

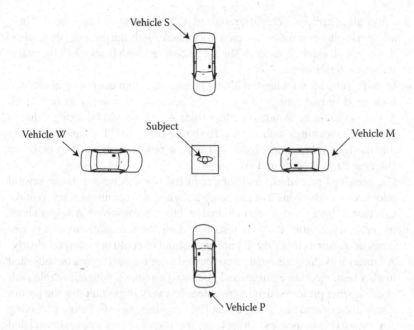

FIGURE 11.2 Plan view of the setup for comparative evaluation of front views of four vehicles.

mounted in their design positions. Seats with seat tracks, steering column, pedals, and shifter are also mounted. Seats, steering column, and pedal movements/adjustments can be made corresponding to design (nominal) values and adjustment ranges. The participants (one at a time) are asked to sit in the seats, adjust to their preferred driving position, and evaluate various items and features of the buck using a predeveloped questionnaire.

Other comparison vehicles can be also included for the interior evaluations. To avoid any biases caused by the exterior of the vehicles during these interior evaluations, the exteriors of the vehicles and the bucks should be covered with black cloth or other similar envelopes. The interior materials used in the four vehicles should have the same material and same neutral color (e.g., gray) to avoid biases due to differences in interior materials and colors.

PRECAUTIONS FOR CLINICS TO AVOID BIASES

1. For exterior clinics, all vehicles should be presented in identical views and with the same exterior surface characteristics. For example, the concept vehicle and the three reference vehicles shown in Figures 11.1 and 11.2 should be painted with the same neutral exterior color (e.g., silver or gray) to avoid introducing biases in participants' perceptions of exterior characteristics of the vehicles.
2. Remove or mask all brand identity (e.g., logos, vehicle name badges, or labels).

3. Cover all nonrelevant characteristics (with black cloth or black tape). Thus, all interior items must be covered with black cloth during exterior evaluations, and all exterior surfaces should be covered with black cloth for evaluation of interior items.

4. Identify vehicles with neutral letters or numerals that cannot be associated with good or bad ratings. For example, consecutive numbering (e.g., 1, 2, 3, 4, ...) should be avoided. Letters from A to F should be avoided due to associated meanings such as "A is the best rating" and "F means fail." Note that the four vehicles in Figure 11.1 are purposely designated to bias-free letters such as S, M, P, and W.

5. Use pretested procedures and subject instructions. Always conduct several pilot tests (or "dry runs") of the procedures and subject instructions to make sure that all biases and interpretational problems are resolved. Always debrief the participants after the pilot tests by asking them to state steps in procedures or questions that they did not understand or could not interpret clearly.

6. All interior evaluations must be conducted after the participant has adjusted his/her seat, steering column, and pedals (if equipped with adjustable pedals) to his/her preferred driving position. It is very important that the participants are explicitly asked to adjust their driving position before providing any evaluation responses. Otherwise, the responses on evaluation related to the locations of all interior items (e.g., reach distance to controls, viewing distances related to visibility of displays, field of view available to the driver) will not be correct. In static market research clinics, where the participants are not allowed to drive the vehicle, the participants generally do not adjust the seat and primary driving controls to their preferred driving position (unless they are specifically instructed to do so). They may leave the seat in the position selected by the previous participant or leave the seat at a further rearward and more reclined position.

SOURCES OF ERRORS

Errors can be introduced during the entire process of collecting responses from subjects invited to participate in market research. A number of different types of error can affect the data gathered during the surveys. The types of errors are

1. Participant selection error: Participants need to be carefully selected to ensure that they represent the "target" customer population. Customers who meet the characteristics of the target must be selected based on a truly random process. For example, the following three situations will not allow random selection of participants: (a) use of restricted databases (i.e., only certain characteristics of individuals are included in the database), (b) unavailability of certain participants during the selected interviewing times (e.g., only retirees and unemployed participants can attend afternoon market research clinic sessions on weekdays), and (c) participants in financial need (i.e., participants who are willing to attend the market research clinic only because of the incentives [e.g., money] offered to attend).

2. Interviewer error: An interviewer can influence the respondent's answers by his/her appearance, approach, and questioning attitude/behavior. Some respondents may purposely skew their response to impress or ignore the interviewer. The participant's response to the question can be affected by the wording used in the question or the interviewer's tone in asking the question. Thus, it is important to carefully select interviewers and train them extensively to ensure consistency in following the predeveloped procedure and subject instructions and avoid interviewer biases. The interviewers should be asked to read preapproved instructions precisely and not state or abbreviate the instructions using their own words. Otherwise, different participants may interpret each question differently. The interviewers should be constantly monitored to ensure that they are preforming the required tasks precisely. Cheating by interviewers (e.g., not recording or inaccurate recording of responses) is always a problem, especially when a large number of participants are to be interviewed.

3. Respondent error: In designing and administering the questionnaire, precautions must be taken to ensure that the following errors do not occur: (a) no response error (participant fails to provide response); (b) response bias (participant, due to his/her own tendency or affiliation, may deliberately falsify his/her response or misinterpret the instructions). The interviewers must be trained to identify such behaviors and notify supervisors to take appropriate action (e.g., disqualify the participant).

Additional information on survey planning and errors in surveys can be found in Zikmund and Babin (2009).

Types of Survey Questions and Data Analyses

The questions included in market research clinics typically require the participant to provide (a) demographic data (e.g., age, income, and education level), (b) ratings using predeveloped scales (e.g., acceptability ratings using a scale with numbers and/or adjectives/descriptors), (c) ratings using direction magnitude scales (e.g., steering wheel is located too far, about right, or too close from me), (d) categorization of the concept vehicle or its characteristics, such as like or dislike, economy or luxury, or outdated or modern.

For example, to obtain ratings on the exterior appearance of each vehicle, the interviewer will provide the following instructions to the participant:

Please walk to the center of the display area [see Figures 11.1 and 11.2], stand in the square marked on the floor, and look around to view all the four vehicles. Please rate the following using the 10-point scale below (Enter the number that best describes your feelings in the form [shown in Table 11.1].)

Table 11.1 presents a data recording form for exterior evaluations of the four vehicles S, M, P, and W. The table shows that the four vehicles are evaluated (i.e., rated on the 10-point scale) by indexing the vehicles such that they can all be viewed from

TABLE 11.1

An Example Data Recording Form for Simultaneous Comparisons of Exterior Appearance of Four Vehicles

Exterior Evaluation Characteristics	Like Very Much		Like Somewhat		Neither Like nor Dislike		Dislike Somewhat		Dislike Very Much	
Rating -->	10	9	8	7	6	5	4	3	2	1

Side View Evaluation (Figure 11.1)

Side view appearance of
  Vehicle S

Side view appearance of
  Vehicle M

Side view appearance of
  Vehicle P

Side view appearance of
  Vehicle W

Overall length of
  Vehicle S

Overall length of
  Vehicle M

Overall length of
  Vehicle P

Overall length of
  Vehicle W

Overall height of
  Vehicle S

Overall height of
  Vehicle M

Overall height of
  Vehicle P

Overall height of
  Vehicle W

Front View Evaluation (Figure 11.2)

Front view appearance of
  Vehicle S

Front view appearance of
  Vehicle M

Front view appearance of
  Vehicle P

Front view appearance of
  Vehicle W

Overall width of
  Vehicle S

Overall width of
  Vehicle M

Overall width of
Vehicle W

Overall width of
Vehicle P

Overall height of
Vehicle S

Overall height of
Vehicle M

Overall height of
Vehicle P

Overall height of
Vehicle W

Rear View Evaluation

Rear view appearance of
Vehicle S

Rear view appearance of
Vehicle M

Rear view appearance of
Vehicle P

Rear view appearance of
Vehicle W

Overall width of
Vehicle S

Overall width of
Vehicle M

Overall width of
Vehicle W

Overall width of
Vehicle P

Overall height of
Vehicle S

Overall height of
Vehicle M

Overall height of
Vehicle P

Overall height of
Vehicle W

Overall Exterior Appearance (All Views)

Overall exterior
appearance of Vehicle S

Overall exterior
appearance of
Vehicle M

Overall exterior
appearance of Vehicle P

Overall exterior
appearance of
Vehicle W

the same angles and compared while viewing from (a) side view, (b) front view, and (c) rear view. It should be noted that all the vehicles can also be simultaneously and dynamically compared by rotating them at the same constant slow speed (e.g., one revolution per minute) while maintaining the same synchronized viewing angles to obtain ratings on the overall vehicle exterior appearance comparisons shown in the last four rows of Table 11.1. In addition, the vehicles can be compared and evaluated from other viewing angles, such as front quarter view (viewing toward the front from 45° to the x-axis of the vehicle) and rear quarter view (viewing toward the rear from 45° to the x-axis of the vehicle).

The responses provided by each participant to each question are recorded by the interviewer during each session of the market research clinic. The data obtained for each question are summarized both over all the participants and by a combination of characteristics of participants (e.g., gender, age group, education level, brand of vehicle owned by the participant, stature, and weight) for each vehicle. The data summaries typically involve determining the distribution of ratings, mean values, percentages of responses above or below a specified value, number of observations of certain events (e.g., participant bumped head while entering the vehicle), and a summary of comments made by the participants.

Statistical analyses are performed using data for each question to determine whether the responses showed differences due to differences in vehicles (e.g., whether the ratings of the concept vehicle were higher or lower than the ratings of each of the reference vehicles) and whether demographic and anthropometric characteristics of the participants affected their ratings or preferences for the vehicles.

Chapter 22 presents an example of the results of a market research clinic for concept selection (see Tables 22.4 through 22.6).

TYPES OF MARKET RESEARCH CLINICS

Market research clinics can involve evaluations under static or dynamic test conditions, whole vehicle or part vehicle (e.g., systems or subsystems evaluations), or some combination of hardware or software features. It is always preferable to conduct evaluations using the whole vehicle under the full set of actual usage conditions (both static, i.e., when the vehicle is parked, and dynamic, i.e., when the vehicle is being driven on public roads under natural unrestricted traffic situations). However, a working property or prototype of the whole vehicle may not be available in the early stages of the vehicle development. The cost of producing a working prototype is also very high. Thus, creative approaches are developed to obtain the necessary information by using a combination of test properties and evaluation methods.

Static versus Dynamic Clinics

When drivable vehicle properties (e.g., a prototype vehicle) are available, dynamic driving tests are always preferable to static clinics, because during the drives, the participant can experience a number of vehicle characteristics, such as acceleration, braking, steering feel, gear shifting, wind noise, engine sound, visibility of the roadway, and operation of controls and displays. The dynamic clinics are generally performed during the vehicle validation phase just prior to Job#1 (see Chapter 21).

Driving simulators can also be used to evaluate dynamic aspects of the vehicle if a high-fidelity driving simulator is available. Many in-vehicle devices that involve controls and displays, such as audio systems and navigation systems, can be evaluated to study the driver workload and the driver's ability to perform a number of in-vehicle tasks safely without large lane deviations (see Bhise [2012] for more information).

In the absence of drivable prototypes and driving simulators, many vehicle systems can also be evaluated by installing them in other existing vehicles for dynamic tests. Otherwise, static laboratory tests can be conducted to explore a number of operational issues with limited validity of results (see Bhise [2012] for more information).

CONCLUDING REMARKS

Concept selection should be conducted by using the data obtained during evaluation of alternate vehicle concepts along with other existing vehicles. The data should be obtained by asking representative customers to participate in evaluation studies conducted in carefully planned market research clinics. The results of the evaluation should be used to select and improve a vehicle concept. The process of creating many vehicle concepts, evaluating several concepts, and using the collected data generally results in developing a superior vehicle design.

REFERENCES

Bhise, V. D. 2012. *Ergonomics in the Automotive Design Process*. Boca Raton, FL: The CRC Press.

Zikmund, W. G. and B. J. Babin. 2009. *Exploring Market Research*. 9th Edition. London: Cengage Learning.

12 Managing Vehicle Development Programs

INTRODUCTION

Vehicle development programs are large and very complex. The management of such complex programs requires (a) a high level of understanding of the vehicle development process, (b) information about the vehicle to be developed and its competitors and customers, (c) corporate needs and resources, organizational structure and people, and processes for securing and managing resources, including teams of professionals and suppliers, and (d) tight control over the program phases in terms of timings, cost, and trade-offs between product characteristics and performance.

The success of a vehicle program, just like the performance of a very large orchestra, requires discipline and coordinated actions of many individuals in the program teams. The number of an auto manufacturer's employees involved in a vehicle program varies greatly between vehicle manufacturers and sizes of vehicle program. However, in a typical new vehicle development, the involvement of 400–800 technical people in core engineering and design functions is not unusual. Further, the number of suppliers and their technical personnel assigned to developing their respective supplied entities varies greatly depending on the size of the program and the outsourcing policies of the auto manufacturer.

The selection of a program manager who will lead the vehicle program is probably the most critical decision faced by the senior management of the auto manufacturer.

PROGRAM MANAGER

Vehicle program manager (also called *chief program manager* or *chief program engineer* in some automotive companies, or *Shusa* in Toyota) is a high-level management position (Womack et al., 1990). The chief program manager generally has the total responsibility for making all decisions related to the vehicle. In many automotive companies, the chief program manager reports to an executive director of a specified vehicle division or a vice president of product development. All design, engineering, and marketing managers assigned to the vehicle program directly report to the chief program manager. All communications, product reviews, and decisions on the program are generally made in the program steering team meetings. In the steering team meetings, chaired by the chief program manager, all the program activities are reviewed by their respective chief engineers, the chief designer, and the marketing manager responsible for the program. Each chief engineer (or manager) similarly manages activities through various teams and cross-functional teams involved in developing various levels of entities (i.e., vehicle, system, subsystem, and component level) (see Figure 1.3 for team structure).

The chief program manager is responsible for formulating, managing, and carrying the vehicle program to completion till the launch of the new products. The program manager must have working knowledge and experience in project planning, scheduling, and budgeting. He/she must be a team builder, a coach, a motivator, and an excellent communicator. He/she must be a big-picture thinker, the integrator of a lot of technical information, and a quick and forceful decision maker. And he/she should have outstanding skills as a manager and controller of program timings and costs.

The basic job duties of the program manager can be described as to

1. Formulate multifunctional product teams with specific roles and responsibilities to the program
2. Plan and manage projects/programs by coordinating product teams and their activities to meet established program goals, schedules, and budgets at program gateways
3. Lead program finance and budgeting activities related to product development, purchasing services and materials, capital equipment and facilities expenditures, and so forth
4. Coordinate development, specification, and procurement of first-generation equipment
5. Manage coordination of supplier base (creating statements of work, quoting, procurement and delivery coordination, evaluation/rating)
6. Provide product-specific support to marketing (through presentations, datasheets, customer visits, market research data/customer demographics, formulation of product business and launch plan, and product pricing)

Thus, the program manager must possess the needed skills and experience:

1. Demonstrated success in managing product launches from conception to manufacturing hand-off
2. Extensive experience in formulating and managing project/program schedules
3. High proficiency in project management issues and activities
4. Ability to identify and focus team members on tasks critical toward successful project completion
5. Demonstrated understanding and utilization of the basic program management disciplines, including critical path, design-to-cost, value analysis, concurrent engineering, and risk management
6. Experience of and exposure to working in automotive product development and manufacturing operations
7. Experience in implementing quality management systems and risk management in product programs
8. Working-level understanding of basic business administration disciplines such as finance, statistics, economics, marketing, and business strategy

PROGRAM VERSUS PROJECT MANAGEMENT

The life cycle of a complex product can be managed as a program. The program will involve the prime responsibilities of designing the right product, producing it, servicing it during its operating life, and finally, closing the production operations and disposing of or recycling the products. The entire program is generally divided into a number of manageable projects, such as (a) developing the product, (b) building the needed tools and production equipment, (c) building plants and installing equipment to get ready for production, (d) recruiting and training people to operate the plant, (e) generating a marketing plan and training dealers to sell and service the product, and (f) producing vehicles at the required rate.

Thus, a program usually contains many projects. The outputs of projects are used to create the program outcomes. Thus, a program can be either a large project or a group of projects. Each project can have a project manager. The project manager's job is to ensure that his/her project succeeds. The program manager, on the other hand, may not spend much effort on the management of individual projects, but is concerned with the aggregate result or the end-state. For example, in an automotive company, a program may include one project to introduce new products to take advantage of rising markets in emerging countries and another project to protect against the downside of falling markets in developed countries. These projects are opposites with respect to their success conditions, but they fit together in the same program.

Program management thus provides a layer above the management of projects and focuses on selecting the best group of projects, defining them in terms of their objectives, and providing an environment in which projects can be run successfully. Program management also emphasizes the coordinating and prioritizing of resources across projects, managing interfaces between the projects, and the overall costs and risks of the program. The program manager should avoid micro-management of the projects; he should leave the project management to the project managers and concentrate on the success of the overall program. Brown (2008) provides additional information on this topic.

PROGRAM MANAGEMENT FUNCTIONS

The program management functions typically include the following activities:

1. Projects management: (a) coordinating projects through management of a master plan, (b) status reporting, (c) issues management, and (d) resource management
2. Performance management: (a) cost measurement, (b) benefits measurement, and (c) analysis of business data
3. Change management: (a) change facilitation, (b) change communication, and (c) workforce training and transition
4. Knowledge management: (a) documentation and sharing of lessons learned from past projects and programs, (b) management of standards and best practices, (c) outputs of product and process benchmarking, and (d) customer complaints and feedback data gathering

The program manager works with other departments to ensure that professionals with the right combinations of expertise are available to perform the design responsibilities for the systems assigned to the departments. The program manager also needs to understand the matrix management organizational structure and work with various functional (or core) departments to ensure that the necessary expertise is available to professionals assigned to his/her vehicle program.

DEVELOPMENT OF DETAILED PROJECT PLAN

The project development activity requires many inputs from a number of stakeholders and activities associated with the project and other issues that can affect the activities of the project. The information gathered is used to develop a project plan. The key project development activities include

1. Collecting inputs from all stakeholders
2. Creating a common understanding of all the projects
3. Preparing documentation of technical plan, management plan, and systems engineering management plan (SEMP) (covered in a later section of this chapter; see Figure 12.3) for each project in the program
4. Supporting the implementation and management of the systems engineering (SE) process, involving development of requirements; functional analysis and allocation of requirements to systems; interface analysis; balanced product design; detailed design; designing and building tools and manufacturing facilities; conduction verification and validation tests; sales, marketing and service; and finally, retirement of the product and disposal of facilities.

PROJECT MANAGEMENT

Project management is the discipline of planning, organizing, securing, and managing resources to bring about the successful completion of the specific project goals and objectives.

The traditional phased approach involves a sequence of the following six phases to be completed:

1. Project proposal and pre-project preparations
2. Project initiation
3. Project planning and design
4. Project execution and construction
5. Project monitoring and controlling systems
6. Project closing/termination

Figure 12.1 presents a flow chart of these activities in relation to a project. The project involved a series of tasks to be performed. It is important that the project work must be clearly understood, with details of all the tasks to be performed, their sequence, the resources (people, equipment, and funds) needed, and the time

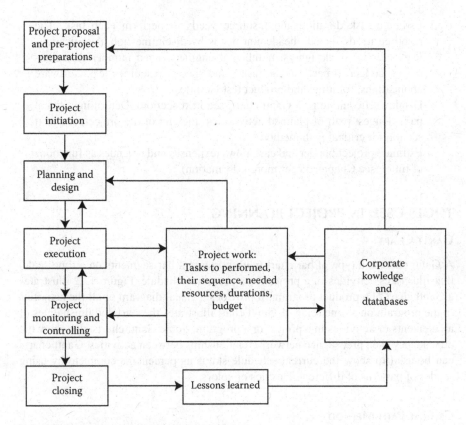

FIGURE 12.1   Project management activities.

required. The responsibility for performing each task is generally assigned to one or more engineering departments depending on the specialized functions that need to be performed in each task. The corporate knowledge and databases are generally maintained by various functional departments of the organization.

Not all projects will go through every phase, as some projects may be terminated before they reach completion. Some projects may not follow structured planning and/or monitoring stages. Some projects will go through Phases 3–5 multiple times. Many industries use variations of these project phases.

STEPS IN PROJECT PLANNING

The basic steps involved in planning a project include

1. Develop a work breakdown structure (WBS) (see next section) of all activities by listing each task in each of the activities. Each task is defined as a group of all the steps or actions to be completed to accomplish the task.
2. Identify task inputs, outputs, and deliverables.
3. Establish task precedence relationships.
4. Determine start and finish time for each task.

5. Estimate task duration and resource needs to perform each task. The resource needs include headcount needs by disciplines/job classifications (e.g., number of designers, number of engineers, and number of technicians), budget to perform the tasks, and special resources (e.g., software applications, training, and product test facilities).
6. Display schedule (e.g., a Gantt chart; see next section). Determine critical path (longest path of planned activities to the end of the project; see next section for critical path method).
7. Estimate project budget and cash flows (expenses and revenues as functions of time) (see Chapter 19 for more information).

TOOLS USED IN PROJECT PLANNING

GANTT CHART

A Gantt chart is a type of bar chart (with horizontal bar segments on a time scale) that illustrates activities in a project or a program schedule. Figure 12.2 illustrates a Gantt chart of a product program. It provides a visual diagram of all the activities in the program on a time scale. A Gantt chart illustrates the start and finish dates of all elements or activities in a project or a program. Some Gantt charts also show the dependency (i.e., precedence network) relationships between activities. Gantt charts can be used to show the current schedule status as percentage complete by using shades of patterns of different densities or colors.

CRITICAL PATH METHOD

The critical path method (CPM) is used for scheduling a set of project activities. The essential technique for using CPM is to construct a model of the project that includes

1. A list of all activities required to complete the project (typically categorized within a work breakdown structure)
2. The time (duration) of each activity
3. The dependencies (or sequence of completions) between the activities
4. Project beginning and end dates

Using these values, CPM calculates the longest path of planned activities to the end of the project and the earliest and latest times at which each activity can start and finish without making the project longer. This process determines which activities are "critical" (i.e., on the longest time path) and which have "total float" (i.e., can be delayed without making the project longer). In project management, a critical path is the sequence of project network activities that add up to the longest overall duration. This determines the shortest time possible to complete the project. Any delay to an activity on the critical path directly impacts the planned project completion date (i.e., there is no float on the critical path). A project can have several parallel near-critical paths. An additional parallel path through the network with the total durations shorter than the critical path is called a *subcritical* or *noncritical* path.

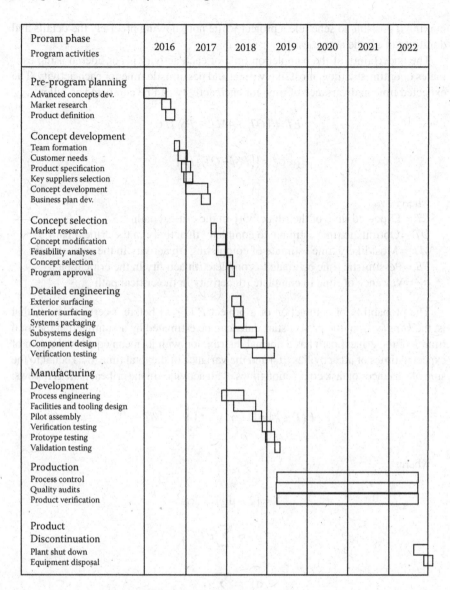

FIGURE 12.2 Gantt chart of a product program.

PROGRAM (OR PROJECT) EVALUATION AND REVIEW TECHNIQUE

The program (or project) evaluation and review technique (PERT) is a model for project management designed to analyze and represent the tasks involved in completing a given project. It is commonly used in conjunction with CPM. PERT is a method for analyzing the tasks involved in completing a given project, especially the time needed to complete each task, and identifying the minimum time needed to complete the total project. PERT was developed primarily to simplify the planning and scheduling of large and complex projects. It is able to incorporate uncertainty by making it possible to schedule a project while not knowing precisely the details and durations of all the activities.

The uncertainty of the completion time of each activity is considered using estimates of optimistic time, most likely time, and pessimistic time for each activity. The expected time and variance of time for each activity can be computed as follows:

$$ET_i = (OT_i + 4MT_i + PS_i)/6$$

$$\sigma_i^2 = ((PS_i - OT_i)/6)^2$$

where:
ET_i = Expected time of the ith activity in the critical path
OT_i = Optimistic time estimate to complete ith activity in the critical path
MT_i = Most likely time estimate of completing ith activity in the critical path
PS_i = Pessimistic time estimate to complete ith activity in the critical path
σ_i^2 = Variance of time to complete ith activity in the critical path

The probability of completion of a project P $(T \leq k)$ before a certain date k, that is, the kth day from the project start date, can be estimated by assuming that the total time T of the critical path has a normal distribution with its mean equal to the sum of expected times of all activities (μ_T) and the variance of the total time (σ_T^2) equal to the sum of variances of task completion times of all activities in the critical path as follows:

$$P(T \leq k) = (1/\sigma_T \sqrt{2\pi}) \int_{-\infty}^{k} e^{-Y} dT$$

where:

$$Y = (T - \mu_T)^2 / 2\sigma_T^2$$

$$\mu_T = \sum_i ET_i$$

$$\sigma_T^2 = \sum_i \sigma_i^2$$

If the project has more than one critical path, then the probabilities of completion of each of the paths before a certain date can be computed. The probability of completion of the project can be computed by multiplying the probabilities of completing all the critical paths before a certain date (if the paths are independent of each other).

PERT is event oriented rather than start and completion oriented, and is used more in projects in which time, rather than cost, is the major factor. It is applied to very large-scale, one-time, complex, nonroutine infrastructure and research and development projects.

WORK BREAKDOWN STRUCTURE

WBS is a tool used to define and group discrete work elements of a project in a way that helps organize and define the total work scope of the project. A WBS element may involve a task (or a function to be performed) related to the design or production of a product, processing of data, providing a service, or any combination of tasks. A WBS also provides the necessary framework for detailed cost estimation and control along with providing guidance for schedule development and control. Additionally, WBS is a dynamic tool and can be revised and updated as needed by the project manager.

The outputs of the WBS are generally shown in a series of block diagrams using flow charts and tree structures (e.g., with hierarchical levels similar to the decomposition tree shown in Figure 2.5). Each block represents a task and provides many task details and parameters (e.g., time required, dates, costs, assigned to). The WBS typically displays (a) various elements of the project, (b) distribution (or number) of work elements of the project in different tasks, (c) distribution of the costs or budgeted amounts between the elements of the project, and (d) subdivision of larger work elements into smaller elements. Some versions of the WBS may not consider timings or order of execution of the tasks. (However, many project management software applications used in WBS analysis can create Gantt charts and conduct CPM and PERT analyses.)

PROJECT MANAGEMENT SOFTWARE

Several project management software systems are currently available (e.g., Oracle, Microsoft Project, and Project Standard 2010, developed and sold by the Microsoft Corporation [2012]). The software programs are designed to assist a project manager in developing a plan, assigning resources to tasks, tracking progress, managing the budget, and analyzing workloads. Microsoft Project allows the creation of color-enhanced time charts with milestones, tasks, phases, people, and so on, and can share databases with other applications (e.g., Microsoft Excel). Many of the software packages allow online sharing of data between project managers, program managers, and team leaders. Thus, all team members have instant access to the project data and many features such as input changes, assign tasks, create personalized dashboards of projects, view calendars, prepare reports, track project issues, create customized charts and graphs, and assign tasks.

OTHER TOOLS

Many other tools are available for specialized analyses such as investment analysis, cost–benefits analyses, expert surveys, simulation models and predictions, risk profile analyses, surcharge calculations, milestone trend analysis, cost trend analysis, target versus actual comparisons of dates, time used, and costs incurred and head count. These analyses can facilitate communication of project status and improve the efficiency and capabilities of project and program managers, especially when these tools are available online and accessible with extensive databases on existing, past, and other similar projects for comparison purposes. The tools also allow managers to create different types of project timings, budget and progress reports for communications and control of project schedules, cash flows, and problems involving different types of risks.

SYSTEMS ENGINEERING MANAGEMENT PLAN (SEMP)

A SEMP is a higher-level plan (not very detailed) for managing the SE effort to produce a final operational product (or a system) from its initial requirements. Just as a project plan defines how the overall project will be executed, the SEMP defines how the engineering portion of the project will be executed and controlled. The SEMP describes how the efforts of system designers, test engineers, and other engineering and technical disciplines will be integrated, monitored, and controlled during the complete life cycle of the product. In other words, the SEMP describes what each team (or department) needs to do and when to achieve the vehicle program objectives.

Figure 12.3 presents a flow chart illustrating the relationship of the SEMP to project work and project management. The SEMP thus uses the information on project work and specifies all major engineering activities in terms of details such as what needs to be performed, how, and when.

For a small project, the SEMP might be included as part of the project plan document, but for any project or program of greater size or complexity, a separate document is recommended. The SEMP provides the communication bridge between the project management team and the technical teams. It also helps coordinate work between and within the different technical teams. It establishes the framework to realize the appropriate work (or tasks to be performed) that meet the entry and success criteria of the applicable project phases. The SEMP provides management with the necessary information for making SE decisions. It focuses on requirements, cascading of the product-level requirements down to lower-level entities, design, development (detailed engineering), test, and evaluation. Thus, it addresses the traceability of stakeholder requirements and provides a plan to ensure that the right product (or system) will be developed during the entire project.

CONTENTS OF SEMP

The purpose of this section is to describe the activities and plans that will act as controls on the project's SE activities. For instance, this section identifies the outputs of each SE activity, such as documentation, meetings, and reviews. This list of required outputs will control the activities of the team and thus, will ensure the satisfactory completion of the activities. Some of these plans may be completely defined in the SEMP (in the framework or the complete version). For other plans, the SEMP may only define the requirements for a particular plan. The plan itself is to be prepared as one of the subsequent SE activities, such as may be the case with a verification plan or a validation plan. Almost any of the plans described in this section may fall into either category. It all depends on the complexity of the particular project (or

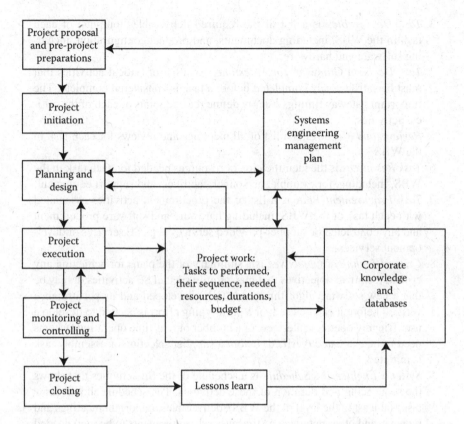

FIGURE 12.3 Relationship of the systems engineering management plan (SEMP) to project work and project management.

program) and the amount of upfront SE that can be done at the time the SEMP is prepared.

The first set of required activities relates primarily to the successful management of the project. These activities are likely to have already been included in the project/program plan, but they may need to be expanded in the SEMP (USDOT Federal Highway Administration, 2007). Generally, they are incorporated into the SEMP, but on occasion, they may be developed as separate documents. The items that can be included in the SEMP are shown in the following list. The items and their descriptions, provided by the U.S. Department of Transportation (USDOT Federal Highway Administration, 2007), were modified to meet the needs of complex product development.

1. *WBS* consists of a list of all tasks to be performed on a project, usually broken down to the level of individually budgeted items.
2. *Task Inputs* is a list of all inputs required for each task in the WBS, such as source requirements documents, drawings, interface descriptions, and standards.

3. *Task Deliverables* is a list of the required deliverables (outputs) of each task in the WBS, including documents, and product configuration, including software and hardware.

4. *Task Decision Gateways (or Milestones)* is a list of critical activities that must be satisfactorily completed before a task is considered complete. The important gateway timings usually define the endpoints of each of the critical activities.

5. *Reviews and Meetings* is a list of all meetings and reviews for each task in the WBS.

6. *Task Resources* is the identification of resources needed for each task in the WBS, including, for example, personnel, facilities, and support equipment.

7. *Task Procurement Plan* is a list of the procurement activities associated with each task of the WBS, including hardware and software procurement and any contracted or supplier-provided services (e.g., SE services or development services).

8. *Critical Technical Objectives* is a summary of the plans for achieving any critical technical objectives that may require special SE activities. It may be that a new software algorithm needs to be developed and its performance verified before it can be used; or a prototyping effort is needed to develop a user-friendly operator interface; or a number of real-time operating systems need to be evaluated (verified) before a supplier selection or assembly task is initiated.

9. *Systems Engineering Schedule* is a schedule of the SE activities that shows the sequencing and duration of these activities. The schedule should show tasks (at least to the level of the WBS), deliverables, important meetings and reviews, and other details (e.g., timings and requirements to be met) needed to control and direct the project. The SE schedule is an important management tool. It is used to measure the progress of the various teams working on the project and to highlight work areas that need management intervention.

10. *Configuration Management Plan* describes the development team's approach and methods to manage the configuration of the systems within the products and processes. It will also describe the change control procedures and management of the system's baselines as they evolve.

11. *Data Management Plan* describes how and which data will be controlled, the methods of documentation, and where the responsibilities for these processes reside. The data should include product design (e.g., computer-aided design [CAD] models or data), schedules of different events (e.g., reviews, tests), results of tests, costs, communications, and so forth.

12. *Verification Plan* is always required. This plan is written along with the requirements specifications. However, the part related to tests to be conducted can be written earlier (e.g., included in the systems design standards). The verification procedures are generally developed by the core engineering experts, and they define the step-by-step procedures for conducting verification tests.

13. *Validation Plan* is required. It ensures that the product being designed is the right product and will meet all the customer needs.

The second set of plans can be designed to address specific areas of the SE activities. They may be included entirely in the SEMP, or the SEMP may give guidance for their preparation as separate documents. The plans included in the first set listed are generally universally applicable to any project. On the other hand, some of the plans included in this second set are required on an as-needed basis. The unique characteristics of a project will dictate their need. For a complex product such an automobile, many of these second-set plans are required items. These items are described in the following list. The items and their descriptions provided in USDOT Federal Highway Administration (2007) were modified to meet the needs of complex product development.

1. *Software Development Plan* describes the organizational structure, facilities, tools, and processes to be used to produce the project's software. It also describes the plan to produce custom software and procure commercial software products.

2. *Hardware Development Plan* describes the organizational structure, facilities, tools, and processes to be used to produce the project's hardware. It describes the plan to produce custom hardware (if any) and to procure commercial hardware products.

3. *Technology Plan* (if needed) describes the technical and management process for applying new or untried technology. Generally, it addresses performance criteria, assessment of multiple technology solutions, and fallback options to existing technology.

4. *Interface Control Plan* identifies all important interfaces within and between systems (within the product and external to the product) and identifies the responsibilities of the organizations on both sides of the interfaces.

5. *Technical Review Plan* identifies the purpose, timing, place, presenters and attendees, topics, entrance criteria, and exit criteria (resolution of all action items) for each technical review to be held for the project/program.

6. *System Integration Plan* defines the sequence of activities that will integrate various product chunks involving components (software and hardware), subsystems, and systems of the product. This plan is especially important if there are many subsystems and systems are designed and/or produced by different development teams from different organizations (e.g., suppliers).

7. *Installation Plan* or *Deployment Plan* describes the sequence in which the parts of the product are installed (deployed). This plan is especially important if there are multiple different installations at multiple sites. A critical part of the deployment strategy is to create and maintain a viable operational capability at each site as the deployment progresses.

8. *Operations and Maintenance Plan* defines the actions to be taken to ensure that the product remains operational for its expected lifetime. It defines the maintenance organization and the role of each participant. This plan must cover both hardware and software maintenance.

9. *Training Plan* describes the training to be provided for both maintenance and operation.

10. *Risk Management Plan* addresses the processes for identifying, assessing, mitigating, and monitoring the risks expected or encountered during a project's life cycle. It identifies the roles and responsibilities of all participating organizations for risk management.
11. *Other plans* that might be included are, for example, a safety plan, a security plan, and a resource management plan.

This second list is extensive and by no means exhaustive. These plans should be prepared when they are clearly needed. In general, the need for these plans become more important as the number of stakeholders and systems involved in the project increases.

The SEMP must be written in close synchronization with the project plan. Unnecessary duplication between the project plan and the SEMP should be avoided. However, it is often necessary to put further expansion of the SE efforts into the SEMP, even if they are already described at a higher level in the project plan.

CHECKLIST FOR CRITICAL INFORMATION

The USDOT Federal Highway Administration (2007) guide also provides a checklist to ensure that the SEMP includes

1. Technical challenges of the project
2. Description of the processes needed for requirements analysis
3. Description of the design processes and the design analysis steps required for an optimum design
4. Identification and documentation of any necessary supporting technical plans, such as a verification, an integration, and a validation plan
5. Description of stakeholder involvement when it is necessary
6. Identification of all the required technical staff and development teams, and the technical roles to be performed by the system's owner, project staff, stakeholders, and development teams
7. Description of the interfaces (or interactions) between the various development teams

ROLE OF SYSTEMS ENGINEERS

The role of the systems engineers assigned to the program is essentially to do what is needed to implement the SE process. A carefully developed SEMP will provide a clear roadmap for the systems engineers. They should work closely with all other team members, technical and program planning, to ensure that all basic SE steps are followed (see Figures 2.1 and 2.3).

The systems engineers will usually play the key role in leading the development of the product and/or system architecture, defining and allocating requirements, evaluating design trade-offs, balancing technical risk between systems, defining and assessing interfaces, and providing oversight of verification and validation activities. The systems engineers will usually have the prime responsibility for developing many of the project documents, including the SEMP, requirements/specification documents, verification and validation documents, certification packages, and other technical documentation (NASA, 2007).

SE is about trade-offs and compromises, about generalists rather than specialists. It is about looking at the "big picture" and ensuring not only that they get the design right (meet requirements) but that they get the right design. Thus, a system engineer needs to perform the following tasks:

1. Understand customer and program needs
2. Obtain required data
3. Develop SEMP
4. Communicate the SEMP to program teams
5. Provide recommendations to program teams on SE tasks
6. Assist teams in conducting necessary trade-off analyses
7. Continuously communicate with program teams to perform above tasks

VALUE OF SYSTEMS ENGINEERING MANAGEMENT PLAN

A carefully developed and well-executed SEMP will enable proper implementation of SE during the program; that is, all the SE steps, from obtaining customer needs, to product validation in product development and subsequent steps during the product operations and disposal stages, are completed by the program teams in a timely manner.

The value of the SEMP can be summarized as follows:

1. It will facilitate reducing the risk of schedule and cost overruns and will increase the likelihood that the SE implementation will meet the user's needs.
2. It will engage the right specialists at the right (needed) time (because they will know what needs to be done) and make sure that the design team members perform the right tasks (e.g., analyses or tests), thus resulting in improved stakeholder participation.
3. The product team will be more adaptable, and the developed products and systems will be resilient and meet customer needs.
4. All entities within the product will be verified for functionality, and thus, the product should have fewer defects.
5. The experience gained and lessons learned during the implementation of the SEMP can be used to create improved SEMP documentation for the next program.

EXAMPLE OF A SYSTEMS ENGINEERING MANAGEMENT PLAN

Table 12.1 presents a high-level SEMP for an automotive product program. The SEMP plan is organized in 19 steps. The second column of the table presents a brief description of each step. The third and fourth columns present the start and end times from Job #1 for each step. The fifth column presents the analyses to be performed and

TABLE 12.1

Example of a Systems Engineering Management Plan for an Automotive Product

		Months from Job #1			
	Steps Involved				
Step No.	Description	Start	End	Analyses, Tools, and Methods Used	Teams and/or Departments Responsible for Analyses and Applications of Tools
				Program Kick-off	
1	*Benchmarking, Market Research, and Customer Surveys:* Existing leading competitors' vehicles are benchmarked to study their characteristics and features. Market research is conducted using customers of leading vehicles in the market segment. Business needs and government requirements are studied.	−45	−40	Benchmarking analyses using, measurements, expert reviews, and photographs	Design and engineering teams, product planning and vehicle attribute engineers
				Customer interviews to determine customer needs	Market research department and design team
				Pugh charts comparing early concepts with the manufacturer's existing and competitors' vehicles	Product planning and engineering
				QFD studies translating customer needs	Design and engineering teams
				Review automotive magazines and study databases on customer complaints and warranty data	Design and engineering teams
				Study business needs and government requirements	Product planning and engineering

2	*Vehicle Specifications, Business Plan, Technology Plan, and Concept Design:* Vehicle specifications and attribute requirements are developed. Concept development and finalization are carried out along with a business plan to determine vehicle program confirmation to proceed with the next phases.	−39	−36	Vehicle specifications and vehicle attribute requirements analysis	Design and engineering teams. Vehicle attribute engineering
				Vehicle concept drawings, CAD models of alternate vehicle concepts	Design and engineering teams
				Cost flow analyses over program life cycle	Finance, planning, and engineering teams
				Business plan	Program planning, engineering, and program management. Senior company management
				Technology plan	All engineering functional departments and vehicle design team
				Manufacturing feasibility and cost analysis	Manufacturing, engineering, product planning, and engineering departments
3	*Supplier Selection and Team Formation:* Supplier capabilities are evaluated, and suppliers for various vehicle entities are selected. Supplier personnel are integrated with various vehicle engineering and design teams.	−40	−37	Supplier evaluations	Supplier selection and purchasing teams
				Supplier selection	Supplier selection and purchasing teams
				Supplier integration	Supplier engineers and vehicle design and engineering teams

(Continued)

TABLE 12.1 (CONTINUED)

Example of a Systems Engineering Management Plan for an Automotive Product

Step No.	Steps Involved		Months from Job #1		Analyses, Tools, and Methods Used	Teams and/or Departments Responsible for Analyses and Applications of Tools
	Description		Start	End		
					Detail Design Phase—Systems Level	
4	*Exterior and Interior Surface Development and Body Design:* The exterior and interior surfaces are designed to meet all application vehicle attribute requirements.		−39	−29	Market research clinics for exterior and interior design approval	Marketing research department. Design and engineering teams
					CAD model development. Occupant package design using packaging tools (SAE occupant packaging standards)	Packaging and ergonomics attribute engineering. Design and engineering teams
					Full-scale clay models development	Design and engineering team. Aerodynamics attribute engineering
					CFD analyses and wind tunnel testing	Aerodynamics engineering department.
					Interior and exterior bucks prepared and used for management and customer reviews	Buck building/fabrication department. Design and engineering teams
5	*Chassis and Suspension Design, Verification, and Sign-off:* The frame or unitized body of the vehicle designed simultaneously with the suspension of the vehicle.		−32	−23	CAE, DFMA, and FMEA analyses	Body and chassis engineering. Design and engineering teams.

#	Task			Methods/Tools	Responsible Organization
6	*Occupant Packaging, Ergonomics and Interior Design and Sign-off:* Occupant packaging is carried out to accommodate the users comfortably inside the vehicle. Ergonomic analyses and evaluations are conducted.	−30	−26	Body and chassis testing equipment (e.g., measurement of strength, stiffness, crush/deflections, vibrations)	Body and chassis engineering.
				Occupant packaging tools (e.g., SAE standards). Ergonomics design guidelines and tools. SAE H-point machine. Field of view analysis. Entry/exit evaluations	Interior packaging/ergonomics team
				Packaging bucks. Programmable vehicle model (PVM). Package evaluation surveys.	Interior packaging/ergonomics team.

Detail Design Phase—Subsystems and Component Level

#	Task			Methods/Tools	Responsible Organization
7	*Powertrain Optimization and Integration:* The powertrain for propelling the vehicle is designed and packaged into the available space in the vehicle body.	−36	−26	CAD analysis. Powertrain packaging bucks	Vehicle package engineering, body and chassis, and powertrain design team
				CAFE rules and EPA standards	Powertrain team. Government standards compliance department
				Engine dynamometer, emissions testing laboratory, and drive evaluations	Emission and fuel economy attribute engineering. Powertrain engineering
8	*Electrical Systems Architecture and Design:* The electrical systems for the various vehicles are designed and packaged.	−37	−26	CAD analysis	Electrical engineering. Body engineering

(Continued)

TABLE 12.1 (CONTINUED)

Example of a Systems Engineering Management Plan for an Automotive Product

Step No.	Steps Involved — Description	Months from Job #1 Start	End	Analyses, Tools, and Methods Used	Teams and/or Departments Responsible for Analyses and Applications of Tools
9	*Climate Control System Design:* Thermal loads generated by external environment, powertrain, and hybrid battery pack are analyzed to design cooling, heating, and ventilating systems.	−38	−26	CAE analysis of electrical and electronics systems. Simulation software (e.g., MATlab and Simulink)	Electrical and electronics engineering. All vehicle attribute engineering departments
				CAD, CAE, FMEA, and DFMA analyses	Thermal, aerodynamics, and climate control systems engineering. Body, electrical and package engineering. Vehicle attribute engineering
				Climatic wind-tunnel tests. Drive evaluations using mechanical and early prototypes	Climate control engineering. Vehicle integration engineering
Design Verification and Vehicle-level Validation and Reviews					
10	*Subsystems Assembly, Verification, Validation, and Sign-off:* The different subsystems are assembled together and tested to verify their functionality. Vehicle-level validation tests are conducted to ensure that the vehicle with the right characteristics is being developed.	−35	−28	FMEA and Interface analyses. Component-, subsystem-, system-, and vehicle-level assemblies are tested in laboratory and road tests (e.g., vibration, durability, performance) to ensure that they meet all applicable requirements.	Vehicle attribute engineering. Vehicle integration engineering.

11	*Tooling Design, Optimization, and Validation:* Tools are developed to ensure meeting all applicable system design and manufacturing requirements.	−32	−25	CAD and CAE analyses. DFMA and FMEA	Manufacturing process, tooling, and assembly engineering. Vehicle integration and attribute engineering. Quality engineering

Pilot Manufacturing and Assembly and Prototype Evaluations

12	*Prototype Build and Testing:* The early prototypes are tested to verify their performance in real-world conditions.	−25	−16	Proving grounds Wind tunnel—aero, thermal, and climatic evaluations (e.g., drag, air, heat, water leakage) Vibration testing machines NVH and anechoic chamber test equipment Crash tests (e.g., frontal/offset, side, rear, and roof crush; fuel spill/integrity)	All engineering teams
13	*Pilot Production:* A pilot production of the final prototype is carried out in small batches to check tooling and ease of assembly.	−18	−10	Work station design and ergonomics guidelines and evaluations. Production improvement techniques, for example, JIT, Jidoka, balancing assembly lines	Manufacturing process, assembly, industrial, and plant engineering
14	*Assembly Plant Modifications, Tooling, and Optimization:* The final assembly plant is modified based on the optimization of the pilot plant, tooling, work flow, and so forth.	−12	−4	Work station design and ergonomics guidelines and evaluations. Production improvement techniques, for example, JIT, Jidoka, balancing assembly lines	Manufacturing process, assembly, industrial, and plant engineering

(Continued)

TABLE 12.1 (CONTINUED)
Example of a Systems Engineering Management Plan for an Automotive Product

Step No.	Steps Involved Description	Months from Job #1 Start	End	Analyses, Tools, and Methods Used	Teams and/or Departments Responsible for Analyses and Applications of Tools
15	*Vehicle Validation and Sign-off:* The final build vehicles are evaluated by customers and management for validation of the vehicle. Final regulatory requirements compliance tests and regulatory sign-off.	−10	−5	Proving grounds. Market research and drive evaluations	Market research department. Engineering and program management
Product Launch, Production, Refinements/Updating, and Retirement					
16	*Marketing and Product Launch:* Marketing, sales, dealership training, and production promotional activities implemented and monitored.	−3	0	Advertisements: brochures, websites, auto shows, test drives	Marketing and sales departments. Quality engineering. Financial, engineering and program management
17	*Production Ramp-up/down:* The final production is ramped up or slowed down based on the supply/demand projections.	0	56	Tracking tools for vehicle sales, customer and dealership feedback, warranty rate, and so forth.	Marketing and sales. Quality engineering
18	*Model Changes, Refinements, and Facelifts:—* The new vehicle is refreshed on as-needed basis to keep up with the competition.	12	24	Design and technology trends, customer reviews, benchmarking competitors' products, sales, and forecasts	Marketing, sales, and product planning
19	*Product Retirement:* The product is finally retired as a last resort if the sales decline drastically.	56	60	Tracking tools for vehicle sales, customer and dealership feedback, warranty rate, and so forth	Marketing, sales, and product planning

Note: CAE, computer-aided engineering; CAFE, corporate average fuel economy; CFD, computational fluid dynamics; DFMA, design for manufacturing and assembly; EPA, Environmental Protection Agency; FMEA, failure modes and effects analysis; JIT, just in time; NVH, noise, vibrations, and harshness; PVM, programmable vehicle model; SAE, Society of Automobile Engineers.

tools and methods to be used in each step, and the last column presents the teams and departments responsible for conducting the work planned in the steps. The SEMP thus identifies details such as what should be conducted, when, using what analyses and methods, and the organizations responsible for performing the steps. It is assumed that the professionals responsible for performing each step are experts and more knowledgeable than the systems engineers preparing the SEMP. The SEMP aids in performing the important tasks of making sure that the right experts are conducting the right analyses using the right tools and methods at the right time in the program in a coordinated manner. This coordination of the right combination of experts at right time is very important; otherwise, each organization may just perform its analyses (i.e., analyzing and designing the systems for which they are responsible) without coordinating and interfacing design activities related to different interfacing systems. The coordination typically occurs through many informal and formal meetings between different design teams to review progress and to resolve problems (e.g., trade-offs between different attributes and system functions and interferences between different vehicle systems) and to review designs and seek the approval of higher levels of management at certain preselected events and gateways in the vehicle program schedule.

COMPLEXITY IN PROGRAM MANAGEMENT

Programs that require separate program management functions, processes, and people for their management are inherently more complex than simple programs that are generally managed by technical persons responsible for product development. Simple programs do not require additional processes or people to manage the program (the number of management tasks are generally small, and the responsibilities in small teams are shared).

Thus, for the management of complex products, the program management should undertake the following:

1. Divide the complex product into a manageable number of smaller "chunks" (note that a chunk can include one or more systems of the product)
2. Create an organization structure with multiple teams (for different systems or chunks of the product) to manage the complex product program
3. Select team members based on their expertise and capabilities to understand the "big picture," that is, the technical issues related to the functioning of the entire product and the interfaces between and within their assigned chunk(s)
4. Train team members to select and apply tools (covered in this book, and other tools in various specialized disciplines)
5. Require each team to create requirements for their assigned chunk based on the customer needs and customer attributes created by the product planning activity
6. Require each team to provide information to the program control team on the status of their deliverables according to their WBS
7. Require each team to select and apply necessary tools (covered in Chapters 12 through 16) during the product development phases and report results to their parent team during design reviews and program management meetings

TIMINGS IN PROJECT MANAGEMENT

To facilitate timely completion of planned activities, the program management should include

1. Gateways/milestones (timely targeted decision points) in the program schedule (see Figure 2.4 and Table 2.1)
2. Reviews by different specialized areas (by attributes, specialized design and user groups [e.g., technical experts, users, service personnel, and maintenance personnel], peer reviews, subsystems reviews, system reviews, and so forth)
3. Definition of work to be completed at each milestone
4. Formal approval plan to proceed to the next phases
5. Plan to handle disapprovals or open issues that will involve rework, delays, and workload balancing problems (overtime costs and/or program slippage)
6. Good communication on the status of program timings (ahead of schedule, on schedule, or behind schedule) and unresolved problems

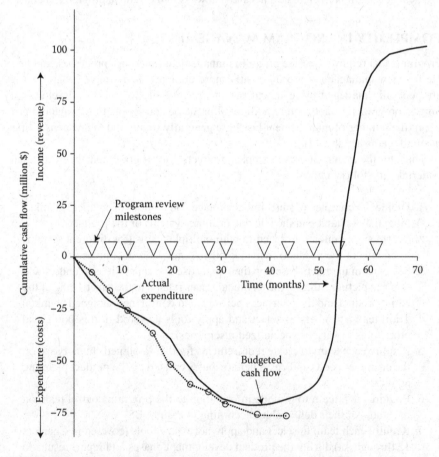

FIGURE 12.4   Comparison of actual expenditure with budgeted cost flow.

COST MANAGEMENT

The program management should prepare cost and timing charts to communicate and control costs and timings. Various types and formats of charts can be used to control and communicate information on budget levels and comparisons between budgeted costs and costs incurred and projected costs as functions of time (especially, cost overruns). Figure 12.4 presents an example of a time chart comparing cumulative budgeted cash flow with actual expenditure.

CHALLENGES IN PROJECT MANAGEMENT

A hectic pace and staying on top of all relevant internal and external issues or problems that can affect the project or program pose constant challenges to the program management. Examples of internal factors are failures in meeting verification test requirements, breakdown of critical test equipment, changes in personnel, bad weather, power outages, and so forth. Examples of external factors are delays caused by supplier problems, changes in the state of the economy, changes in budgets, new technological developments that can change program objectives, political problems in countries affected by the program, and so on.

Thus, program management personnel must be able to handle multiple problems simultaneously, constantly maintain communication with the lower and higher levels of team organization, anticipate problems, and be prepared for various possibilities. Generally, program managers with a technical background and familiarity with the technical aspects and issues will be able to handle and foresee possible developing problems more quickly than nontechnically oriented program managers.

Component-level problems during the design stage and failures in some verification tests can affect the progress of work on the higher product levels (i.e., risk of not meeting deliveries of subsystems, systems, and product for verifications and validations, effect on costs due to redesign, rework, retests, etc.). Therefore, it is essential that technical problems and the resulting timing and cost problems need to be tracked and communicated immediately through the higher levels so that the necessary corrective or precautionary actions can be taken to minimize the program risks. Progress charts on technical issues, timing, and costs need to be kept up to date and reviewed through an appropriate reporting structure.

CONCLUDING REMARKS

Project and program management involve a lot of challenges, even in situations where the path to the desired deliverable seems obvious. The difficulties arise due to frequent changes in people, technology, and competition. Some examples of sudden changes are (a) the project may be progressing fine till a key team member suddenly resigns, (b) a revolutionary new product is about to be introduced into the market, or (c) a major competitor launches a product almost identical to the manufacturer's new product. These situations would force changes in the program. Program complexity will only increase over time, as the rates of technological innovation are increasing rapidly. Further, changes in the organizational culture, working environment, and economic situation and scarcity of resources can also add substantial challenges in controlling and successfully completing programs.

REFERENCES

Brown, J. T. 2008. *The Handbook of Program Management: How to Facilitate Project Success with Optimal Program Management*. New York, NY: McGraw-Hill.

Microsoft Corporation. 2012. Project 2010. Website: www.microsoft.com/project/en-us/demos.aspx (Accessed: May 19, 2016).

NASA (National Aeronautics and Space Administration). 2007. *NASA Systems Engineering Handbook*. Website: http://www.acq.osd.mil/se/docs/NASA-SP-2007-6105-Rev-1-Final-31Dec2007.pdf (Accessed: November 18, 2016).

USDOT Federal Highway Administration. 2007. *Systems Engineering for Intelligent Transportation Systems*. Report no. FHWA-HOP-07-069.

Womack, J. P., D. T. Jones, and D. Roos. 1990. *The Machine That Changed the World*. New York, NY: Macmillan.

13 Computer-Aided Technologies

INTRODUCTION

Computers have made profound changes and improvements to almost every process used in the automobile industry. In this chapter, we will review how computerization has affected automotive product development through applications in areas such as three-dimensional (3-D) product design, simulation, collaboration, visualization, digital prototyping, databases on decomposition of vehicles to component levels, requirements, and preparation of forms and electronic reports. These changes have helped in designing better products by reducing errors, development time, and development costs. Some areas of applications include (a) 3-D computer-aided design (CAD) in modeling and visualization, (b) dimensional measurements, (c) scanning and milling to create vehicle models, (d) computer-aided engineering (CAE) in engineering analyses, and (e) robots and computer-controlled machinery in various manufacturing and assembly processes. Program management has also benefited greatly from reduction in the time taken to prepare project and program schedules, budgets, communications between and within teams, and so on.

COMPUTER-AIDED TECHNOLOGIES

Computer-aided technologies are used to conduct a variety of product planning, design, engineering, manufacturing, and data management tasks. The technologies and the acronyms used to the industry are listed here:

1. Computer-aided design (CAD)
2. Computer-aided architectural design (CAAD)
3. Computer-aided design and drafting (CADD)
4. Computer-aided engineering (CAE)
5. Computer-aided manufacturing (CAM)
6. Computer-aided process planning (CAPP)
7. Computer-aided quality assurance (CAQ)
8. Computer-aided reporting (CAR)
9. Computer-aided requirements capture (CAR)
10. Computer-aided rule definition (CARD)
11. Computer-aided rule execution (CARE)
12. Computer-aided software engineering (CASE)
13. Computer information system (CIS)
14. Computer-integrated manufacturing (CIM)
15. Computer numerical controlled (CNC)

16. Computational fluid dynamics (CFD)
17. Electronic design automation (EDA)
18. Enterprise resource planning (ERP)
19. Finite element analysis (FEA)
20. Knowledge-based engineering (KBE)
21. Manufacturing process management (MPM)
22. Manufacturing process planning (MPP)
23. Material requirements planning (MRP)
24. Manufacturing resource planning (MRP II)
25. Product data management (PDM)
26. Product life-cycle management (PLM)

CLAIMS: ADVANTAGES AND DISADVANTAGES OF COMPUTER-AIDED TECHNOLOGIES

Computer-aided technologies have steadily gained popularity since the 1970s in the auto industry. In the early 1970s, large X-Y plotters were used with mainframe computers to draw full-size drawings of vehicles. Now, 3-D models of vehicles are generated by designers and package engineers during the early vehicle conceptualization phases. Magnified (full-size) images of these models are projected on large screens for design review during conferences and meetings. These models and their data are electronically transmitted and shared with all design teams at many locations within an auto company and its suppliers. The teams use these models and common databases to perform various engineering analyses and enrich the models and databases by adding detail designs of vehicle systems as they are completed. Computer-aided technologies have greatly improved the productivity, quality (e.g., reduced data transfer errors), costs, and timings of vehicle development programs. The claimed benefits of computerization are

1. Modeling complex shapes easily and quickly (e.g., freeform and parametric-shaped objects; using a large number of functions and subroutines such as copy, paste, reflect, and extrude to draw complex objects (Cozzens, 2007)
2. Scanning, digitizing, and modeling of complex 3-D surfaces
3. Precisely manipulating geometry (e.g., by changing parametric data with different inputs and databases)
4. Reusing engineering data to create variations in designs (e.g., copy, paste, and modify)
5. Using standard components, subsystems, and systems from libraries in databases to populate assemblies
6. Making product design changes quickly and easily
7. Updating designs to reflect changes automatically
8. Integrating electrical, mechanical, thermal, and aerodynamic design and analyses
9. Creating 3-D prototype parts using 3-D printing machines
10. Creating 3-D wiring layouts by importing electrical schematics
11. Automating cable harness design in the 3-D environment

12. Creating 3-D product design models from schematic drawings/sketches
13. Getting to market faster by reducing rework, reusing data, and facilitating data sharing and collaboration (using common databases)
14. Improving product design quality and customer satisfaction (reducing design and manufacturing defects)
15. Reducing warranty claims (saving repair costs)
16. Optimizing product performance for intended environment
17. Reducing manufacturing costs
18. Creating accurate design documentation
19. Evaluating product manufacturability
20. Reducing cost during product design process
21. Creating innovative products (e.g., using iterative design)
22. Reducing physical prototypes to save money (e.g., using computer simulations and prototyping)
23. Reducing overengineering of products (e.g., reducing safety factors, optimizing materials and weight)
24. Saving time with automated cloud-based simulation
25. Making sustainable and economical products
26. Exploring alternative manufacturing processes to lower costs
27. Optimizing material selection for impact (crashworthiness), weight, and cost
28. Identifying hazardous materials usage (reducing exposure to toxic and carcinogenic materials)

The disadvantages of computerization are few, but they include

1. Added costs of purchasing and maintaining computers and peripheral equipment.
2. Software procurement, development, interfacing (i.e., using outputs of one computer application as inputs to another computer application without human data manipulation), and maintenance costs.
3. Added costs of information technology staff to continually support computer systems (e.g., for software installations, upgrades, troubleshooting, user support, training, project planning, and hardware configuration).
4. Computer system security and protection costs (e.g., avoiding data hacking/theft and system breakdowns).
5. Poorly understood assumptions made in applying software (i.e., improper use of software applications).
6. Limitations of software applications not considered during analyses.
7. Untrained users of the software making incorrect interpretations of results (i.e., need to provide additional training as computer systems and/or software applications are upgraded).
8. Databases and software applications must be continually updated with the latest available information/versions; otherwise, decisions based on outdated, incomplete, and missing information and/or models may lead to uncompetitive designs.

COMPUTER-AIDED DESIGN, ENGINEERING, AND MANUFACTURING

CAD, CAE, and CAM are interdependent industrial computer application technologies that have greatly influenced the chain of processes between the initial design and the final realization of the product. Ongoing refinements in CAD/CAE/CAM systems continue to save auto manufacturers costs, time, and resources over non-computerized or older methods. As a consequence, CAD, CAE, and CAM technologies are responsible for massive gains in both productivity and quality, particularly since the 1980s. The CAD and CAM methods can be used separately or sequentially, and in general, CAD is used more commonly than CAM.

CAD primarily involves creating computer models defined by geometrical parameters. These models typically appear on a computer monitor as 3-D representations of a part or a system of parts, which can be readily altered by changing relevant parameters. CAD systems enable designers to view objects under a wide variety of representations and to test these objects by simulating real-world conditions (e.g., movements of components during operational situations).

CAM picks up where CAD leaves off by using geometrical design data to control automated manufacturing machinery. CAM systems are associated with CNC or direct numerical control (DNC) systems. These systems differ from older forms of numerical control (NC) in that geometrical data is encoded mechanically. Since both CAD and CAM use computer-based methods for encoding geometrical data, it is possible for the design and manufacturing processes to be highly integrated. More information on this topic can be obtained from Encyclopedia of Business (2015) and Groover (2008).

COMPUTER-AIDED ENGINEERING (CAE) METHODS AND VISUALIZATIONS

CAE methods are CAE software applications that are used in solving specialized engineering analyses, such as structural analysis (e.g., modeling stresses, deflections, movements, fractures/failures, optimization of component strength and weight), noise and vibrations analysis, heat transfer/thermal stress analysis, fluid flow and aerodynamics, optical ray tracing for visual reflection and glare analysis, and mapping electromagnetic fields. The outputs of these analyses can be visualized by creating graphic, color-coded, and/or animated views of states of the systems being evaluated. For example, deflections/deformations under loads can be illustrated by physical movements within components, vibrations can be shown in movements in cyclic patterns within the vibrating components, stress levels can be shown by the use of color codes, and aerodynamic/fluid flows can be shown in color-coded flow contours (Peddiraju et al., 2009). The visualization techniques help provide a better understanding of engineering issues and their magnitudes during the design process.

PRODUCT VISUALIZATION TOOLS

In designing an automotive product, it is essential that the designers and engineers have a very good understanding of the 3-D space available to create the product. The magnitudes of different vehicle dimensions, areas, spaces, proportions and relative sizes of different components, clearances between components, and interferences between components that cause assembly and operational problems can be better understood when physical models (or mock-ups) of the vehicle and various systems and components are available.

Most currently used CAD tools (e.g., CATIA) allow the development of 3-D models that can be viewed from various viewing locations (i.e., eye points). Many of the models allow dynamic animation of movements of various vehicle systems (e.g., opening and closing of doors, movements of moving components in vehicle suspensions, and driveline [engine, transmission, drive shafts, and final drive]). Also, many virtual reality simulators, such as computer-assisted virtual reality environments (CAVEs) and 3-D goggles and screens, are used to provide a more realistic space perception of the CAD models, along with computer modeling of environments (e.g., roads, vehicles in traffic, vehicle in a show room) and humans (e.g., 3-D driver or user/occupant manikin interacting with the CAD model of the vehicle). These models can be used for various purposes, such as to analyze designs during engineering, to review designs with company management, and to show vehicle designs to potential customers in market research clinics.

DESIGN TOOLS USED IN SPECIALIZED ENGINEERING ACTIVITIES

CONCEPT DESIGN

A number of different CAD software applications are currently used in the auto industry to create, visualize, and modify vehicle concepts more quickly and reduce the need for physical clay models (e.g., using rapid and virtual clay modeling). These applications also allow (a) better communication and review of ideas and concepts in the early phases of the vehicle design process with developed two-dimensional (2-D) sketches and 3-D virtual models, (b) improved collaboration among the design teams and faster decision making with the ability to visualize and review 3-D designs in real time, (c) improved process efficiency in developing concepts and sketches, design surfaces, and visualizing concepts using 2-D sketches or 3-D animation tools, and (d) reducing or eliminating expensive building of physical properties (e.g., mock-ups, bucks, and prototypes).

Since CAD models are immediately available for design reviews, many errors in designs can be caught as the modeling activities progress. All teams and suppliers can be given access to the CAD model so that they get the latest information immediately and can begin designing and adding designs of entities that they are responsible for developing and interfacing with other entities. Thus, the process of concept generation moves faster from modeling to decision making with efficient surface design, visualization, and 3-D concept development.

Some CAD applications used in the industry are (see Car Body Design, 2015)

1. Alias Automotive (Industrial design tools for conceptual design and surface modeling) (Autodesk, 2015a).
2. AutoCAD (Autodesk, 2015b).

3. CATIA (computer-aided three-dimensional interactive application) is a multi-platform CAD/CAM/CAE commercial software suite developed by Dassault Systemes. The latest releases are CATIA V5 and CATIA V6 (Dassault Systemes, 2015).
4. NX is a CAD/CAM/CAE PLM software suite developed by Siemens PLM Software. NX is a parametric solid/surface feature-based modeler and is based on Parasolid geometric modeling kernel (Siemens, 2015).
5. Pro/ENGINEER is software targeted to 3-D product design that offers an integrated, parametric, 3-D CAD/CAM/CAE solution (PTC creo, 2015).
6. IronCAD (IronCAD, 2015).

CAE VERSUS PHYSICAL TESTS AND PROTOTYPE BUILDS

Building physical products or parts of the products (e.g., systems, subsystems, or components) and subjecting them to actual laboratory or field tests provides results that are generally more valid (e.g., representative of actual product) as compared with conducting one or more CAE evaluations. Advantages of the use of CAE tests are that they reduce time and costs associated with building physical test properties, preparing the test equipment, and conducting the required tests. The trend in automotive engineering tests is to conduct extensive CAE analyses first to narrow down evaluation alternatives (i.e., combinations of test or independent variables involved in defining the product design) and then, to conduct limited physical tests when the prototype components, systems, or vehicles are available for physical tests for validation purposes.

DESIGN REVIEW MEETINGS

Design review meetings should always be conducted with the use of one or more product visualization methods such as sketches, drawings, CAD models, or physical models to ensure that all reviewers understand the product configuration and details about the problem being reviewed. Without a visualization tool, different reviewers may think differently about the product and the problem being reviewed. Thus, the use of product visualization methods during product design review meetings is very helpful.

VERIFICATION TESTS

Verification tests are conducted to verify that a designed entity performs and meets its requirements. Tests conducted using CAE methods do not use actual (physical) entities, but CAE tests can be conducted under many different simulations with combinations of manufacturing variations or conditions to determine how any given entity can perform under different possible variations. The advantage of CAE testing is that a large number of verification tests can be conducted within a short period of time and without the expense of creating test entities, test setups, and test equipment.

VALIDATION TESTS

Validation is usually the last test phase before Job #1. The validation is conducted to ensure that the right product is built for the customers. Thus, the validation testing should be conducted using customers. Further, since the actual product is available (i.e., prototype vehicles are available a few months prior to Job #1), CAD methods are rarely used in this phase.

ADVANTAGES OF CAD

Modeling with CAD systems offers a number of advantages over traditional drafting methods that use rulers, squares, and compasses. Designs can be altered without erasing and redrawing. CAD systems offer "zoom" features, analogous to a camera lens, whereby a designer can magnify certain elements of a model to facilitate inspection. Computer models are typically 3-D and can be rotated about any axis, much as one could rotate an actual 3-D model in one's hand, enabling the designer to gain a fuller sense of the object. CAD systems also lend themselves to modeling cutaway drawings (sectional views), in which the internal shape of a part is revealed, and to illustrating the spatial relationships among a system of parts. CAD models can also provide exploded views showing various components or subsystems involved in an assembly.

To understand CAD, it is important to understand its limitations. CAD systems have no means of comprehending real-world concepts, such as the nature of the object being designed or the function that the object will serve. CAD systems function by their capacity to codify geometrical concepts. Thus, the design process using CAD involves transferring a designer's idea into a formal geometrical model. In this sense, an existing CAD system does not actually design anything, but can provide tools, shortcuts, and a flexible environment for a designer to work with.

Other limitations to CAD are being addressed by research and development in the field of expert systems. This field is derived from research into artificial intelligence. One example of an expert system involves incorporating information about the characteristics of materials, for example, their weight, tensile strength, and flexibility, into the CAD software. When this and other information is included, the CAD system could "know" what an expert engineer knows when that engineer creates a design. The system could then mimic the engineer's thought pattern and actually "create" a design. Expert systems might involve the implementation of more abstract principles, such as the nature of gravity and friction or the function and relation of commonly used parts, such as levers or nuts and bolts. Expert systems might also change the way data is stored and retrieved in CAD/CAM systems, supplanting the hierarchical system with one that offers greater flexibility.

Another key area of development in CAD technologies is the simulation of performance. Among the most common types of simulation are testing for response to stress and modeling the process by which a part might be manufactured or the dynamic relationships among a system of parts. In stress tests, model surfaces are shown by a grid or mesh that distorts as the part comes under simulated physical or thermal stress. Dynamic CAD/CAE tests serve as a complement to or a substitute for building working prototypes. The ease with which the characteristics of a modeled entity can be changed facilitates the development of optimal dynamic efficiencies in terms of both functioning and manufacturing. Simulation is also used in electrical/electronic design tasks by creating simulated current flows through circuits and observing the behavior of the electrical systems.

The processes of design and manufacture are, in some senses, conceptually separable. Yet, the design process must be undertaken with an understanding of the nature of the production process. It is necessary, for example, for a designer to know the properties of the materials with which the part might be built, the various techniques by which the part might be shaped, and the scale of production that is economically viable. The conceptual overlap between design and manufacture is suggestive of the potential benefits of CAD and CAM and the reason why they are generally considered together as a system.

CONCLUDING REMARKS

Because of major advantages of computer-assisted technologies, the older, manually intensive design and physical testing processes have been replaced throughout the industry. To ensure that the new technologies are properly applied, care must be taken to understand their limitations.

REFERENCES

Autodesk. 2015a. Industrial design and class-A surfacing software. Website: www.autodesk.com/products/alias-products/overview (Accessed: July 4, 2015)

Autodesk. 2015b. Autodesk website. Website: www.autodesk.com/suites/product-design-suite/overview (Accessed: July 4, 2015)

Car Body Design. 2015. CAD Software. Website: www.carbodydesign.com/directory/design-software/cad-software/ (Accessed: July 4, 2015)

Chang, T.-C., R. A. Wysk, and H.-P. Wang. 1998. *Computer-Aided Manufacturing*, 2nd edn. Upper Saddle River, NJ: Prentice-Hall. ISBN 0-13-754524-X.

Cozzens, R. 2007. *CATIA V5 Workbook*. Release 17. Mission, KA: Schroff Development Corporation. ISBN: 978-58503-399-7.

Dassault Systemes. (2015). CATIA website. Website: www.3ds.com/products/catia (Accessed: July 4, 2015)

Encyclopedia of Business. 2015. Computer-Aided Design (CAD) and Computer-Aided Manufacturing (CAM), 2nd edn. Website: www.referenceforbusiness.com/encyclopedia/Clo-Con/Computer-Aided-Design-CAD-and-Computer-Aided-Manufacturing-CAM.html#ixzz37yVAcmU8 (Accessed: July 4, 2015)

Groover, M. P. 2008. *Automation, Production Systems, and Computer-Integrated Manufacturing*, 3rd edn. Upper Saddle River, NJ: Prentice Hall. ISBN 0-13-239321-2.

IronCAD. 2015. IronCAD company website. Website: www.ironcad.com/index.php/industry/automotive-a-transportation (Accessed: July 4, 2015)

Peddiraju, P., A. Papadopoulous, and R. Singh. 2009. CAE Frame Work for Aerodynamic Design Development of Automotive Vehicles. 3rd ANSA & μETA International Conference held on September 9–11, 2009 at Olympic Convention Centre, Porto Carras Grand Resort Hotel, Halkidiki, Greece.

PTC creo. 2015. Pro-Engineer website. Website: www.3hti.com/Creo_Greenfield_Video/index.html?gclid=CJ_ymcSEwsYCFQQHaQodplsBFA (Accessed: July 4, 2015)

Siemens. 2015. NX CAD, CAM and CAE. Website: www.plm.automation.siemens.com/en_us/products/nx/index.shtml (Accessed: July 4, 2015)

14 Vehicle Validation

INTRODUCTION

The term *validation plan* describes the evaluation tests that should be conducted to determine whether the vehicle will meet its customer needs and will be accepted by the customers. The validation tests are carried out at the vehicle level; that is, the whole vehicle with all of its systems is tested. Often, the tests are conducted in customer clinics where representative customers are invited and asked to evaluate the vehicle under various different usage conditions (i.e., driving under different situations and operating/using different vehicle features). Some validation tests are also conducted using various technical experts and key management personnel as the evaluators, with the assumption that they represent the most critical customers.

The primary goal of the validation plan is to test the vehicle to assess whether it will meet all the customer needs. Meeting the customer needs and gaining customer acceptance are the key outcomes of every vehicle program, because the customers are the ones who will invest their money and buy the vehicles. If customers find that the vehicle does not meet their needs, they will most likely purchase a vehicle produced by another company (e.g., from a competitor). Thus, the management of the company wants to review the validation test results to ensure that the vehicle produced will be acceptable to the customers. The validation process essentially also judges the effectiveness of all the product development and management teams in creating the right product, and it also provides an opportunity to make minor refinements to the vehicle before the formal production begins.

SCOPE OF VALIDATION TESTING

WHEN IS VALIDATION PERFORMED?

Validation testing is usually conducted a few months prior to Job#1 (note: after Job#1, the assembled production vehicles will be shipped to the dealers for sale). Just prior to this validation phase, all the engineering tests to verify various components, subsystems, and systems within the vehicle (i.e., verification tests at various lower levels of the vehicle entities) are usually completed, and their results show that all the applicable attribute requirements are met. Thus, the verification tests should not need to be repeated during the validation phase unless additional confirmations are needed (e.g., due to late changes).

The underlying goal of the validation tests is to perform the final tests to determine whether the whole vehicle as "assembled from all the verified systems and components" will meet the needs of its customers. The latest prototype vehicles built at the assembly plant are usually used for validation testing. Many of these vehicles are also used for other tests, such as final durability, performance, and inspection tests.

Here, the management wants to ensure that the vehicle will meet all its customer needs and that customers will be highly satisfied before the final approval is given to release the vehicles for shipment to the dealers. Six or more production prototype vehicles are typically made available for validation tests. The results of the validation tests are used to make minor changes to the vehicles (usually at the assembly plant). The results also help marketing and sales personnel to fine-tune their marketing plans and prepare dealer information to check and fix any late-discovered issues with the vehicles prior to customer delivery.

WHOLE-VEHICLE TESTS

The validation testing should include objective and subjective tests. The objective tests are generally performed using physical measurement devices (e.g., accelerometers, timers, distance measuring devices, force transducers, temperature sensors, and sound level meters) and onboard data recorders, as compared with the subjective tests, which involve judgments of evaluators (typically, customers and/or experts) using various psychophysical measurement methods such as rating on a scale, categorization (acceptable/unacceptable, like/dislike), and paired comparisons (e.g., same as a given reference, better or worse than the reference).

The types of objective and subjective tests and the purposes of the tests are

1. *Objective Tests*
   a. Dynamic destructive tests: crash tests to demonstrate compliance to Federal Motor Vehicle Safety Standards (FMVSS) 200 series requirements (front, side, rear impact, and rollover) and FMVSS 300 series (fuel integrity) requirements (NHTSA, 2015)
   b. Dynamic nondestructive tests: (i) powertrain performance (power, acceleration, sound, and thermal evaluations), (ii) fuel economy and emission tests, (iii) noise and vibration evaluation dynamic tests on different roads at different speeds, (iv) braking and handling, (v) aerodynamic drag tests, (vi) climate control functionality, and (vii) driver behavioral observations during driving and operating controls and using displays (e.g., time taken to perform tasks)
   c. Static tests: (i) fuel economy and emission tests of whole vehicle on a dynamometer, (ii) wind tunnel tests for aerodynamic drag and air/water leakage, and (iii) electromagnetic interference
2. *Subjective Tests*
   a. Dynamic tests: in these tests, customers, experts, and management personnel are asked to drive the vehicles and provide their responses (e.g., ratings and judgments) on the following vehicle characteristics: (i) performance (includes perception of acceleration and time to reach a given speed), handling (perception of how the vehicle handles on the roadway during different maneuvers [e.g., double lane changes, curves on a serpentine course], including steering feel), braking (including brake feel), and occupant comfort (i.e., comfort felt by the occupants while sitting in the seats experiencing the ride, and thermal comfort due to the climate

control), (ii) noise, vibrations, harshness (NVH), squeaks, and rattles (i.e., various sounds and vibrations experienced during driving), (iii) package and ergonomics evaluations (judgments of experts and customers on various vehicle package and vehicle usage–related considerations [e.g., ease/difficulty ratings] during driving), (iv) vehicle visibility and lighting (e.g., field of view, visual obstructions, visibility distances, headlamp beam pattern perception studies using customers), and (vi) overall impression on characteristics such as "fun to drive," "much better than competitor X," or "I would recommend it to a friend" (using customers).

b. Static evaluations of customer perception in areas such as (i) styling and appearance (exterior and interior), (ii) craftsmanship—fit, finish, color/texture harmony, sound and tactile feel, perception of interior materials, (iii) functionality of interior components, and (iv) ease in service and maintenance considerations.

In performing the tests, issues such as (a) what is being measured and how it is measured (equipment and procedures), (b) who evaluates (e.g., a selected sample of customers, an average customer, a very demanding customer, or an expert in a specialized field such as vehicle dynamics or ergonomics) are important considerations. Such considerations are usually discussed with the vehicle program manager and/or senior company management, and their approval is sought before the validation test program is initiated. In addition, whether any additional late models of competitors' vehicles should be included for comparisons is also decided by the program manager.

It is also important to realize that computer-aided engineering (CAE) methods should not be considered as part of the validation testing, as they are based on tests using computer models of the vehicle. The CAE methods are used for verification tests because physical hardware and complete vehicles (early prototypes or production vehicles) are not available during the early engineering phases. The actual production vehicles produced just prior to Job# 1should be used for validation testing.

METHODS USED FOR EVALUATION

All the methods that are generally used in the validation evaluations are shown in Table 14.1. The methods to be used depend on the vehicle attribute and its subattributes to be included in the validation plan. Table 14.1 includes all vehicle attributes and their major subattributes. The validation evaluation methods are based on customer responses, measurement of dimensions, expert reviews, field tests, crash tests, engineering requirements (using methods specified in engineering standards of the auto company, FMVSS, Society of Automotive Engineers (SAE) standards), and so forth.

Many of the important methods used in validation evaluations are briefly described and explained in the following subsections.

CUSTOMER RATINGS

Overall, about 80–150 subjects (at each market research test site) are asked to rate the vehicle and compare it with its competitors on a number of preselected vehicle

TABLE 14.1

Type of Methods Used for Validation of Attributes

Vehicle Attribute	Subattribute	Subattribute Requirements/Sources	Evaluation Method(s)
Package	Seating package (driver and passengers)	Accommodation percentiles and interior dimensions. SAE J 1516, J1517, J4004.	Interior coordinate measurements. Customer ratings on package dimensions.
	Entry/exit	Head/torso, knee, thigh, foot space requirements. Distances from SgRP.	Interior coordinate measurements. Customer ratings on entry/exit ease.
	Luggage/cargo package	Luggage volume requirements. Floor height to ground.	Interior coordinate measurements. Customer ratings on storage space.
	Fields of view—visibility	Wiper/defroster zones, mirror fields, pillar obscuration.	Interior coordinate measurements. Customer ratings on direct and indirect fields and obscurations.
	Powertrain package	Engine, transmission, and drivetrain envelopes.	Interior coordinate measurements. Drive tests on clearances.
	Suspensions and tires package	Suspension and tire envelopes.	Interior coordinate measurements. Drive tests on clearances.
	Other mechanical package	Space requirements for fuel tank, electrical, lighting, climate control systems, and so on. FMVSS 108 requirements.	Interior coordinate measurements.
Ergonomics	Locations—layout of controls, displays, handles, service points, and so on	SAE J1138. Ergonomic requirements.	Interior coordinate measurements. Reach and grasp evaluations during operations.
	Hand and foot reach	SAE J287. SAE J1516 and J4004.	Interior coordinate measurements. Customer ratings in drive clinics.
	Visibility and obscurations	FMVSS 111, SAE J1050, J902,903.	Interior coordinate measurements. Customer ratings in drive clinics.

TABLE 14.1 (CONTINUED)
Type of Methods Used for Validation of Attributes

Vehicle Attribute	Subattribute	Subattribute Requirements/Sources	Evaluation Method(s)
	Operability	Ergonomic guidelines, SAE J1139.	Ergonomics scorecard based on ergonomics engineer"s and customer drive clinics.
Safety	Front impact	FMVSS 204, 208, 212, and 219 requirements.	Sled and crash tests with crash dummies.
	Side impact	FMVSS 214 requirements.	Sled and crash tests with crash dummies.
	Rear impact	FMVSS 301 and 303 requirements.	Sled and crash tests with crash dummies.
	Roof crush	Deformation requirements.	Laboratory tests.
	Sensors, belts, and airbags	Anchorage and dummy tests.	Laboratory evaluations and measurements. Customer belt fit comfort evaluations.
	Other safety features	FMVSS 108, SAE lighting standards.	Photometric and durability tests.
Styling/appearance	Exterior—shape, proportions, and so on.	Exterior design guidelines.	Exterior surface measurements. Customer ratings.
	Interior—I/P = Instrument Panel, Console, trim, and so on.	Interior design guidelines.	Interior surface measurements. Customer ratings in market research tests.
	Luggage/cargo/storage	Customer requirements.	Customer ratings.
	Underhood appearance	Design guidelines.	Customer ratings.
	Color/texture mastering	Color and texture masters.	Expert judgments in matching colors and surface finishes. Customer ratings on craftsmanship.
	Craftsmanship	Craftsmanship guidelines.	Expert and customer ratings. Measurements of mating edges, surfaces, and surface finish.
Thermal and aerodynamics	Aerodynamics	Aero forces, coefficient of drag, and noise requirements.	Wind tunnel and field testing.

(*continued*)

TABLE 14.1 (CONTINUED)
Type of Methods Used for Validation of Attributes

Vehicle Attribute	Subattribute	Subattribute Requirements/Sources	Evaluation Method(s)
	Thermal management	Temperature guidelines.	Static and drive tests.
	Water management	Leak test requirements.	Water and air leak tests.
Performance and drivability	Performance feel	0-60 mph time. Engineering requirements.	Experts and customer ratings.
	Fuel economy	EPA/NHTSA requirements.	EPA test procedures.
	Long-range capabilities	Engineering requirements.	Field tests.
	Drivability	Engineering requirements.	Field tests.
	Manual shifting	Engineering requirements.	Experts and customer ratings.
Vehicle Dynamics	Ride	Engineering requirements.	Experts and customer ratings.
	Steering and handling	Engineering requirements.	Experts and customer ratings.
	Braking	FMVSS 105 requirements.	Field tests.
Noise, vibrations, and harshness (NVH)	Road NVH	Engineering requirements.	Sound measurements. Field tests.
	Powertrain NVH	Engineering requirements.	Sound measurements. Field tests.
	Wind noise	Engineering requirements.	Sound measurements. Field tests.
	Electrical/mechanical	Engineering requirements.	Field tests.
	Brake NVH	Engineering requirements.	Field tests using experts.
	Squeaks and rattles	Engineering requirements.	Field tests using experts. Customer ratings.
	Pass by noise	Engineering requirements.	Sound measurements. Field tests.
Interior climate comfort	Heater performance	Engineering requirements.	Field tests. Customer ratings.
	Air-conditioning performance	Engineering requirements.	Field tests. Customer ratings.
	Water ingestion	Engineering requirements.	Field tests.
Weight	Body	Design assumptions.	Weight measurements.

TABLE 14.1 (CONTINUED)
Type of Methods Used for Validation of Attributes

Vehicle Attribute	Subattribute	Subattribute Requirements/Sources	Evaluation Method(s)
	Chassis	Design assumptions.	Weight measurements.
	Powertrain	Design assumptions.	Weight measurements.
	Climate control	Design assumptions.	Weight measurements.
	Electrical	Design assumptions.	Weight measurements.
Security	Vehicle theft	Engineering requirements.	Expert evaluations.
	Contents/component theft	Engineering requirements.	Expert evaluations.
	Personal security	Engineering requirements.	Expert evaluations.
Emissions	Tailpipe emissions	EPA requirements.	Dynamometer and field tests.
	Vapor emissions	EPA requirements.	Dynamometer and field tests.
	Onboard diagnostics	EPA requirements.	Dynamometer and field tests.
Communications and entertainment	Communication with outside sources	Transmission requirements.	Data transmission tests.
	Communications within the vehicle	Transmission requirements.	Data transmission tests.
	Entertainment	Transmission requirements.	Data transmission tests.
Cost	Cost to the customer	Product planning assumptions.	Cost prediction programs.
	Cost to the company	Product planning assumptions.	Cost prediction programs.
Customer life cycle	Purchase and service experience	Marketing assumptions.	Historic data and customer feedback.
	Operating experience	Marketing assumptions.	Customer feedback.
	Life-stage changes	Marketing assumptions.	Customer feedback.
	System upgrading	Marketing assumptions.	Customer feedback.
	Disposal/recyclability	Recycling requirements.	Material tracking.
Product/process compatibility	Reusability	Reusability requirements.	Field data.
	Commonality	Commonality guidelines.	Analysis of component database.
	Carryover	Tooling budget.	Analysis of component database.
	Complexity	Manufacturing budget.	Analysis of component database.
	Tooling/plant life cycle	Manufacturing strategy. and budget.	Analysis of plant tooling database.

Note: I/P, instrument panel.

characteristics associated with the vehicle attributes (see Table 14.1). The vehicle can be rated in static evaluations (e.g., for evaluations of exterior and interior appearance, interior package and spaciousness) and driving field tests. Field tests are typically conducted for ride, comfort, handling, acceleration, braking, and ergonomic evaluation of various systems (e.g., audio, navigation, climate control, and headlamp beam patterns).

EXPERT REVIEWS

Generally, customers are asked to provide judgments on vehicles used for validation evaluations. However, customers are not very good at finding minor faults in the vehicle. Therefore, trained experts can be asked to detect a range of vehicle faults (from "barely detectable" to "very noticeable") that are not easily detected by the customers. The experts are generally very knowledgeable about what the customer wants, and they can also easily detect problems in the vehicles related to their area of expertise. For example, a noise and vibration expert can easily discern vibrations in suspension components that most customers will not notice in a short test drive. Further, customers can be very biased, and may only find faults with the vehicle based on their previous experience, whereas, the experts can cover a range of issues within their area of expertise. Thus, experts are employed to evaluate problems in the following areas:

1. Drivability, that is, acceleration, ride, steering, handling, and braking
2. Gear-shift feel and jerkiness during shifting
3. Noise and vibrations that can occur in various vehicle systems
4. Fuel economy and tailpipe emissions
5. Craftsmanship, that is, fit, finish, harmony or match in color and textures of components

COMPANY EMPLOYEES AND MANAGEMENT PERSONNEL

The company employees and management personnel are also asked to evaluate the vehicles. The company employees, because of their extensive experience with the existing products, can provide useful information by participating in subjective evaluations. However, they may be biased, as they can be unduly harsh or lenient in their evaluations. Selection of employees who are not associated with the vehicle program can be a quick source of evaluators. Further, in many situations before the vehicle introduction, it is beneficial to ask only company employees to participate in the vehicle evaluation exercises, so that information about the upcoming products can be shielded from public exposure.

LABORATORY AND CONTROLLED FIELD TESTS

Whole-vehicle tests without the use of subjects (customers and experts) and using test equipment in static and dynamic conditions in laboratory or special test setups are conducted for the following purposes:

1. Dimensional measurements for verification of ride height and ground clearances, approach, departure and break-over angles, cargo volume, and so forth
2. Powertrain and fuel economy (using dynamometers)
3. Crash tests for front, side, and rear impact, roof crush, and fuel integrity
4. Field of view measurements (e.g., measurement of daylight openings, mirror field, detection fields of blind area sensors)

SOME EXAMPLES OF VALIDATION TESTS AND TEST DETAILS

This section provides some detail on the validation tests for evaluation of the vehicle and its systems for the following seven vehicle attributes: (1) vehicle performance, (2) comfort, (3) noise, vibrations, and harshness, (4) crash safety, (5) styling and appearance, (6) packaging and ergonomics, and (7) electrical and electronics. It should be noted that the actual validation test programs undertaken will vary greatly between different manufacturers and also between different vehicle programs of any given manufacturer. The details provided in this section are intended to provide a brief overview covering only a few important areas within the seven selected attributes.

VEHICLE PERFORMANCE

1. *Vehicle Dynamics and Handling*: The vehicle-level validation of vehicle dynamics and handling characteristics typically involves drive tests. Expert drivers are involved in a number of tests related to vehicle body motions under a variety of acceleration, deceleration, and steering maneuvers involving maximum lateral accelerations in handling courses involving curves with different radii, quick lane changes, and so forth. Objective measurements of stopping distances (e.g., braking from 60 to 0 mph) and time traces of velocity, accelerations (lateral and longitudinal), heading angle, and vehicle location are recorded. The customers are also asked to drive the vehicles on test courses involving a number of maneuvers. At the end of each test, the customers are asked to complete a questionnaire with ratings using 10-point scales about their perception of various vehicle handling considerations such as acceleration capability and vehicle lift, steering feel and sensitivity, brake pedal feel and braking effort, feel of suspension, stiffness and body motions (e.g., cornering stability, roll control, dive-in (i.e., braking pitch), and maneuverability (e.g., during parking, cornering, and brake-in-a-turn).

2. *Powertrain Performance*: Certification of the engine performance involves measurement of engine output, for example, horsepower and torque produced at different engine speeds (revolutions per minute) using a number of standard procedures (e.g., SAE J1349 and J2723 specify procedures to be used by the vehicle manufacturer to certify the net power and torque ratings of production engines [SAE, 2015]). These validation procedures are objective, and the customers are not involved in the evaluations. However, because performance feel is an important customer-desired attribute, additional

customer-based tests are performed to validate the engine and powertrain performance. A number of subjects are invited to test drive the vehicle along with competitors' vehicles on a predetermined test route that includes selected road and traffic conditions. The subject is asked to fill out a questionnaire regarding the performance of the vehicle with respect to its engine response and drivetrain capabilities. The rating procedure typically includes the use of a preselected set of scales to evaluate vehicle performance under different city and highway driving situations involving accelerating and decelerating maneuvers (e.g., passing, merging, decelerating at an intersection, hill climbing, and trailer towing). Expert drivers are also involved in whole-vehicle acceleration (i.e., powertrain pickup, wide open throttle 0–60 mph acceleration time), sound and shift feel characteristics, and so forth.

3. *Fuel Economy*: Fuel economy is an important attribute, because it affects recurring fuel expenses and government regulations. The validation plan involves in-house laboratory testing on dynamometers as well as road tests for a number of test vehicles using a number of different drivers to estimate the actual fuel economy of the vehicle. Certification of this attribute also involves independent government agencies (e.g., the Environmental Protection Agency [EPA]).

To test the fuel economy, the test vehicles are placed in a dynamometer and run by a trained operator. While on the dynamometer, the vehicles will be run using preselected speed–distance profiles that simulate city and highway driving. The profiles include set distances, speeds, and starts and stops (for city driving). Each test vehicle is subjected to the same profiles during testing. During the test, a hose is connected to the exhaust pipe to capture and measure the amount of substances (emissions) produced throughout the test. The test data can then be compared with similar data obtained for other benchmarked vehicles.

COMFORT

The following subattributes of the comfort attribute can be evaluated as follows:

1. *Interior Comfort*: A set of subjects who will be the potential users of the vehicle are invited to participate in the vehicle evaluation tests. The subjects are asked to drive the test vehicle on a preselected route and given precise instructions on tasks to perform and the test route to follow. The subjects are also asked to drive other benchmarked vehicles using the same procedure in a random order. All the vehicle identification markings, logos, and badges should be removed (or covered) during the tests to avoid any preconceived biases about different brands and models of vehicles. The subjects are given a questionnaire consisting of various questions related to the interior comfort, including the seating comfort, to be rated using specified scales for each question. The gathered data are analyzed, and the calculated values of performance measures are used to compare the vehicles. Some examples of the questions for interior comfort evaluations are

    a. How comfortable is the driver's seat? Provide an overall comfort rating using the 10-point scale, where 10 equals very comfortable and 1 equals very uncomfortable.

    b. How comfortable is the thigh support provided by the seat? Provide a comfort rating using the 10-point scale.

    c. While riding as a passenger, provide an overall comfort rating for the rear passenger seat using the 10-point scale.

2. *Interior Climate/Environment Comfort*: Each test subject is first asked to drive a test vehicle on a preselected route and then asked to rate the overall comfort provided by the interior climate. Some of the questions that can be included in the evaluation are

    a. Rate the cooling (or heating) capability of the climate control unit using the 10-point scale, where 10 equals very comfortable and 1 equals very uncomfortable.

    b. Rate the ability to control air flow rate on the 10-point scale.

    c. Rate the interior noise level at the current preselected speed (e.g., 70 mph) using the 10-point scale, where 10 equals very low noise level and 1 equals very high, discomforting noise level.

3. *Ride Comfort*: In this evaluation, a test subject riding the test vehicle in a specified seating position (as a driver and/or a passenger) is asked to provide ratings on ride comfort on preselected roads (with different road surface roughness) and speeds at preselected points during the trip. Some examples of the questions are

    a. Rate the overall comfort while riding in this vehicle using the 10-point scale, where 10 equals very comfortable ride and 1 equals very uncomfortable ride.

    b. Is the ride very smooth, about right, or too hard?

    c. While driving over this road segment, how noticeable are the road bumps—not at all noticeable, somewhat noticeable, or very noticeable?

NOISE, VIBRATION, AND HARSHNESS

There are three subattributes of NVH that are generally evaluated during the validation phase:

1. *BIW NVH*: The body-in-white (BIW) of the vehicle should be tested for the range of frequencies expected during driving. The BIW is tested on a testing machine where it is subjected to external excitations, and its NVH levels are measured from different seating locations.

2. *Powertrain NVH*: The noise and vibrations created by the powertrain and within the powertrain by external forces are measured by a testing machine. The simulator in the testing machine simulates the external forces experienced by the vehicle chassis from various sources (e.g., road surface roughness), and the test equipment measures the NVH levels in the engine and driveline components. The NVH levels should be lower than the target value set by the vehicle attribute requirements.

3. *Other sources of NVH*: Generally, there are a number of parts in a vehicle that are bolted, spot welded, or even joined using snap fits. These parts should not produce any squeaky noise (usually generated by rubbing of adjacent moving components) or any kind of rattling noise (generally created by loose or moving components) throughout the operation of the vehicle. These sources of NVH are tested using standard company procedures (under predefined input conditions related to road surfaces, wind gusts, powertrain, braking system operation, and so forth). The resulting NVH outputs in the vehicle must meet the subattribute requirements, which are typically based on customer perception and acceptance of the annoyance and/or discomfort caused by the NVH-related issues.

Customer drive tests involve ratings of the acceptability of vibrations and noise felt by the driver and passengers through vehicle suspensions, body structure, powertrain, seats, steering system and steering wheel, pedals, and braking system (e.g., brake roughness felt on the brake pedal) while driving on different roadways and at different speeds. The 10-point rating scale used for comfort (from lack of NVH) measurements can be defined as 10 equals no noticeable NVH and 1 equals very annoying NVH.

CRASH SAFETY

The safety requirements are specified in the FMVSS, and thus, compliance with the applicable FMVSS is mandatory (NHTSA, 2015). This attribute can be divided into the following subattributes:

1. *Frontal Impact*: The vehicle should pass the frontal crash tests specified in the FMVSS 208 (NHTSA, 2015). The full frontal fixed barrier crash test is an example of the tests mentioned in the standard. It is also called the *rigid barrier test*, which represents a vehicle-to-vehicle full frontal engagement crash with each vehicle moving at the same impact velocity. The test is intended to represent most real-world crashes (both vehicle-to-vehicle and vehicle-to-fixed object) with significant frontal engagement in a perpendicular impact direction. For FMVSS 208, the impact velocity is 0–48 km/h (0 to 30 mph), and the barrier rebound velocity, while varying somewhat from car to car, typically ranges up to 10% of the impact velocity for a change in velocity of up to 53 km/h. The head injury impact (HIC) in meeting the frontal crash test is 1000. The occupant chest decelerating requirement is 60 G, and chest deflection should be less than 76 mm (3 in). In addition, there should not be any protrusions or sharp broken parts that may be dangerous or even fatal to the occupants. FMVSS 212 requires that during front impact, at least 50% of the perimeter of the windshield must be retained. The steering column rearward displacement specified in FMVSS 204 should be less than 127 mm (5 in). Other tests include the oblique impact test (impact at a 30° angle with the object), the fuel system (fuel integrity/spillage) requirements in FMVSS 301, and the generic sled test in FMVSS 208.

2. *Side Impact*: This test evaluates the outcome of a crash of two vehicles in a "T shape" event-direction of crash. For the side-impact test, a 1000 kg mass collides with the side of the vehicle at a speed of 53 km/h. The passing criterion is that occupants should not incur any major injury risks to crucial body parts. The FMVSS 214 requires meeting many requirements on thoracic trauma index (TTI), pelvic acceleration, structural integrity, door opening, crush displacement and resistance force, and so forth (NHTSA, 2015).

3. *Rear Impact*: All the tests of frontal impact are performed on the rear body of the car. The requirements in FMVSS 223 and FMVSS 224 should be met (NHTSA, 2015). In addition, FMVSS 301 fuel integrity requirements must be met.

4. *Vehicle Roof Crush*: Here, the vehicle is tested for its safety in the event of vehicle rollover. The pillars of the vehicle are tested for their ability to hold up the roof (by maintaining the required headroom [survival space] within the required roof intrusion deformation limit) and to support themselves in a dynamic impact. FMVSS 216 requires that the vehicle's roof structure must withstand, in the specified test, from 1.5 times the vehicle's unloaded weight to 3.0 times the vehicle's unloaded weight. The requirements in FMVSS 201, 208, and 216 should be met (NHTSA, 2015).

STYLING AND APPEARANCE

The vehicle styling should be evaluated by considering its exterior and interior appearance subattributes as follows:

1. *Exterior Styling*: A set of subjects (representing the customer characteristics) are invited to rate the proposed vehicle alongside the competitors' vehicles. They are asked to rate the vehicle from various views (e.g., side, front, and rear). The evaluation setup and procedure are described in Chapter 11.

2. *Interior Styling*: For interior evaluations, the test subjects are led individually by an interviewer and asked to sit in different seat locations and to provide ratings on a number of interior characteristics, such as the overall styling of the vehicle interior, the shape, appearance, finish and touch feel of the interior panels, and the color and texture of the interior materials.

PACKAGING AND ERGONOMICS

The vehicle package and ergonomics attribute can be evaluated by considering the following areas:

1. *Occupant Space*: A set of representative subjects are invited to evaluate the vehicle interior package and give their comments on the spaciousness of the interior package. The feeling of spaciousness can be evaluated by asking questions related to space around the occupants, such as headroom, shoulder room, hip room, legroom, and space during entry/exit. The subjects are

asked a number of questions on the various rooms (spaces) using question-naires (see Table 22.4 and Chapters 11 and 22).

2. *Field of View*: The field of view issues can be better evaluated by ask-ing subjects to drive the vehicle under different road and traffic conditions. The locations and sizes of the obscurations caused by the pillars and fields of view available from the window openings and the inside and outside mirrors are evaluated. Some examples of the questions for the ergonomics evaluations using 10-point rating scales are

   a. Rate the size of the obscuration caused by the left A-pillar and the left outside mirror, where 10 equals very acceptable (small) obscuration and 1 equals very large and unacceptable obscuration.

   b. Rate the size of the field of view provided by the left outside mirror, where 10 equals very acceptable mirror field and 1 equals very unac-ceptable mirror field.

   c. Rate the size of the field of view provided by the inside mirror, where 10 equals very acceptable mirror field and 1 equals very unacceptable mirror field.

3. *Storage and Cargo Space*: Cargo space in the trunk can be measured using a standard test devised from a daily life scenario, such as the number of bags or boxes (of certain dimensions) it can accommodate. For proper engi-neering evaluation, the volume of the cargo space in liters or cubic meters can be measured using a coordinate measuring machine. The internal stor-age areas for items such as cups, coins, cell phones, and sunglasses can be evaluated by asking the subjects to store the items and rate their sizes and ease of storing. The storage spaces can also be evaluated by subjects using direction magnitude rating scales as follows:

   a. Is the storage space of the glove box generous, about right, or insufficient?

   b. Are the storage spaces in the instrument panel shelf generous, about right, or insufficient?

   c. Is the number of cup holders accessible from the front row seats more than needed, about right, or too few?

4. *Ease of Entry and Exit*: The subjects can be asked to get in and out of the vehicle and to provide ratings on a number of questions related to locations and size of items such as doors, seating reference point (SgRP) locations, rocker panel, seat, and steering wheel (see Chapter 20 for more details). In addition, a special group of subjects with varied anthropometric character-istics, such as tall males, short females, and obese and mature people, can be selected for the evaluations. The subjects can also be observed while entering and exiting to determine problems encountered during entry and exit (e.g., bumped head on the roof rail, hit foot on the rocker panel, dif-ficulty in moving the driver's right thigh under the steering wheel). Some of the questions asked during the evaluations are

   a. How easy is it for you to enter the vehicle? Please provide your rating using the 10-point rating scale, where 10 equals very easy and 1 equals very difficult to enter the vehicle.

    b. Rate the adequacy of the knee space during entering the vehicle using the 10-point scale, where 10 equals generous knee space and 1 equals too little knee space.

    c. How is the height of the top side of the rocker panel? Is it too high, about right, or too low?

    d. Is it easy to reach the inside door handle and close the door? Rate the door handle location using the 10-point scale, where 10 equals very acceptable location and 1 equals very unacceptable location.

5. *Locations of Major Controls and Items*: The controls and displays within a vehicle are designed considering a number of ergonomic design guidelines. The guidelines used by ergonomics engineers are described in detail by Bhise (2012). During validation, the subjects can be asked to use the controls and displays while driving. After each use of a control or a display, the subject can be asked to rate ease of use. In addition, the subjects can be asked to report any problems encountered during use, such as difficulty in reading labels or understanding how to operate a control. Some examples of the questions for the ergonomics evaluations are

    a. Are the controls easy to use? Provide your rating using the 10-point rating scale, where 10 equals very easy to use and 1 equals very difficult to use.

    b. Are the displays easy to read and use? Provide your rating using the 10-point scale, where 10 equals very easy to read and 1 equals very difficult to read.

    c. Can the steering wheel and seat be adjusted to your preferred driving position? Yes or no.

ELECTRICAL AND ELECTRONICS

The electrical system tests are performed to ensure that all electrical features are working normally under combinations of electrical loads by various vehicle systems in different road and environmental driving conditions. In addition, data loggers (for recording of data on circuit operations and fault/error detection) are installed on test vehicles to monitor the operation of all electrical and electronic equipment during all driving conditions. Using established measurement and calibration protocols, internal electronic control unit (ECU) signals can also be recorded. Logging periods may vary from a few hours to several days according to standards and procedures set by the vehicle manufacturers.

CONCLUDING REMARKS

Validation is the last important phase prior to the vehicle being sold to the customers. The vehicle validation tests must be carefully planned to ensure that they allow evaluation of the vehicle in areas that are important to the customers in satisfying their vehicle use needs. The vehicles are used under varied driving and environmental conditions by different users for a variety of trips. Therefore, the vehicle validation procedures must cover all important customer needs under all foreseeable vehicle usage situations and conditions to ensure customer satisfaction. The results of the vehicle validation evaluations provide information on how the product will be perceived by its customers. The information is used to further improve product acceptance by proposing refinements in (a) vehicle design issues (only minor product changes can be considered at this late stage), (b) vehicle assembly processes and procedures, (c) dealership training, and (d) marketing and sales plans.

REFERENCES

Bhise, V. D. 2012. *Ergonomics in the Automotive Design Process*. Boca Raton, FL: CRC Press. 978-1-4398-4210-2.

NHTSA. 2015. Federal Motor Vehicle Safety Standards. Website: www.nhtsa.gov/cars/rules/import/FMVSS/ (Accessed: July 5, 2015)

Society of Automotive Engineers, Inc. 2015. *The SAE Standards/Handbook*. Warrendale, PA: The SAE International. Website: https://global.ihs.com/standards.cfm?publisher=SAE&RID=Z56&MID=~SAE&gclid=CKSyxa7TqM0CFY-DaQodOykHUQ (Accessed: June 14, 2016)

15 Creating a Brochure and a Website for the Vehicle

INTRODUCTION

A vehicle brochure is typically about a 10–15-page printed booklet, which customers generally pick up at a vehicle dealership or at an auto show to take home and study the characteristics of the vehicle at a leisurely pace. Vehicle brochures are also available on most of the manufacturers' websites for browsing and downloading. The brochures serve the important function of providing necessary information about the products to help potential customers to make their decisions about their future vehicle purchases.

The vehicle manufacturer's website typically contains information on all vehicles and their models sold under the manufacturer's brand name. Thus, the website is like a virtual show room where different vehicle models can be compared, and details (technical, operational, colors, trim materials, and prices) about the availability and combinations of features can be reviewed. The website can also provide more information than the brochure by enabling search into deeper levels of menus. The website also has the advantage of including videos and animations that provide both visual and auditory (e.g., voice, engine sound, background music) information dynamically.

If created early during the concept development phase, the vehicle brochure can also serve as an important tool for the vehicle designers and product development engineers to understand the importance of the characteristics and features of different models (or options) of the proposed vehicle.

WHY CREATE A VEHICLE BROCHURE?

The purpose of creating a brochure for the vehicle is to provide information about the important characteristics, capabilities, and features of different models offered by the manufacturer to its prospective customers.

The brochure should include the following information:

1. Photographs and drawings to show exterior and interior views of the vehicle and its features, and illustrations of important capabilities of the vehicle and its models
2. Models available and standard features included in each model
3. Descriptions of vehicle features and their availability in different models
4. Vehicle exterior and interior dimensions
5. Capabilities and capacities (e.g., load-carrying capacity, acceleration and braking capabilities, engine type, size and outputs, fuel consumption)

6. Exterior and interior colors and interior materials offered
7. Key selling points (or vehicle characteristics very important to the customers)
8. Optional packages and features offered in each model of the vehicle
9. Technical superiority–related considerations (e.g., major engineering accomplishments; comparisons with leading competitors showing how the vehicle differs from other vehicles; characteristics of engines, transmissions, and suspensions; body construction and materials)

The vehicle data contained in the brochures are also included in the webpages of most vehicles in their manufacturers' websites. The websites also provide downloadable files of the brochures and videos illustrating various views and features of the vehicle.

Creating a brochure for a proposed vehicle during the early stages of the product development process is also a useful exercise for the vehicle development team. It forces the team members to think about many characteristics and features that should be incorporated into the vehicle. It also forces the marketing, design, and engineering team members to brainstorm about its customer preferences and satisfaction issues.

VEHICLE WEBSITE VERSUS BROCHURE

With the widespread availability of the Internet, most consumers find that they can obtain the information they need more quickly by accessing the vehicle manufacturer's website. The website has many advantages over the traditional brochure:

1. The website generally includes information on all vehicles and their models sold by the manufacturer, as compared with the brochure, which typically provides information on the selected vehicle and its models.
2. The website can also provide a side-by-side comparison of many selected vehicles.
3. The website can show 360° easy-to-adjust views of the vehicle exterior and interior in customer-selected colors and trim combinations from customer-selected viewing angles.
4. The website can provide more information through videos via dynamic presentation of visual and auditory modes.
5. With interactive menus, the website can provide the required information on many vehicle features quickly and in greater detail.
6. The information provided in the websites can be easily and quickly updated at much lower cost as compared with reprinting new brochures with updated information.
7. The information displayed on the website can be easily shared with others by sending them the webpage address.
8. The information transmission costs are very low as compared with the costs associated with printing and distribution of the brochures.
9. A vehicle brochure can be obtained by downloading from the website.

10. The customer can build a vehicle on the website by selecting a vehicle model, packages, and features, and get information on the estimated price (monthly installments for purchase or lease with different down payments and installment payment options).
11. The websites of most manufacturers also allow customers to browse through the vehicle inventories of dealers at locations in the vicinity of the zip code provided by the customer.

The advantages of the brochure over the website are

1. A printed brochure can be viewed without the use of the Internet and an accessing device (e.g., a computer, a smartphone, or a tablet).
2. A brochure is a physical object and can therefore be taken and made available anywhere. It is visible under most lighting conditions, indoors or outdoors.
3. Brochures can be printed in different styles, colors, and formats on varied sizes and types of paper to provide the perception of luxury, as compared with the website, which transmits virtual images of vehicle characteristics that are limited by the capabilities of the display screen (e.g., screen resolution, color, and brightness).

CONTENTS OF THE BROCHURE

VEHICLE MODELS, PACKAGES, AND THEIR FEATURES

Types of Model and Optional Packages of Features

The brochure provides information on all available vehicle models, features, and options. A vehicle is generally marketed by creating different models with different levels of standard and optional features. The models can differ by variations in body-style (e.g., two-door sedan, four-door sedan, five-door hatchback) and/or variation in combinations of available powertrains (e.g., a base engine with a manual transmission or automatic transmission, or one or more higher-power and more sophisticated engines, each with a manual transmission or an automatic transmission). More expensive (or luxury) models also have more standard features with increasing levels of technology and more sophisticated options. The vehicles are thus designed with different models, packages of options that are offered with certain models, and optional features.

Vehicle Models

Manufacturers create different models of a vehicle (e.g., S, SE, SEL, Titanium, Limited, and Platinum in Ford cars; LX, Sport, EX, EX-L, and Touring in Honda cars). Each of the models is produced with a specified set of standard and optional packages of features. The vehicle brochures and websites generally provide charts (as tables or in other equivalent formats) showing combinations of standard features, optional packages, and optional features included in each vehicle model.

Standard Features

Standard features are the features that are included in a specified vehicle model. Thus, the vehicle model comes with the standard features as the minimum set of features. The standard features may be grouped into a number of categories, such as mechanical, exterior, interior, and safety. Any other features that can be added by the original equipment manufacturer (OEM) are ordered as optional features. Many optional features are grouped in packages (called *optional packages*). The optional packages can have different content depending on the manufacturer and its models.

Optional Features

These features may be included in certain packages (as standard packaged features) and can also be added at an extra price in addition to the price of the vehicle model with the standard features. The optional features may vary depending on vehicle brand and model. Some examples of optional features are optional engines, type of transmission (manual or automatic), sunroof/panoramic roof, and navigation system.

Many manufacturers bundle several optional features in packages (or groups) to reduce the number of combinations of optional features. Some examples of optional packages are light group, technology group, convenience group, comfort group, and driver assistance group. Thus, the vehicle brochure must contain information on combinations of models, packages, and optional features that can be ordered, including the ordering code numbers.

Vehicle Packages

The vehicle features are packaged in different optional packages for convenience in selecting different options. The concept of offering different packages also reduces the complexity of assembling vehicles with different combinations of features ordered by the customers. Some examples of packages are

1. Value group: Includes air-conditioning, power mirrors, power door locks and windows, and remote keyless entry
2. Premium group: Includes leather trimmed seats, heated front seats, heated steering wheel, dual zone automatic temperature control, remote start, and universal garage door opener
3. Popular equipment group: Includes leather-wrapped steering wheel, speed control, steering wheel–mounted audio controls, front passenger in-seat storage, remote start, overhead console, illuminated visor/vanity mirrors, front seatback map pockets, 12 V auxiliary power outlet, illuminated front cup holders, electronic vehicle information center, trip computer, and tire-pressure monitoring display
4. Technology group: Includes keyless entry, push-button start, remote proximity entry, blind spot monitoring system, rear cross-path detection, rain-sensitive windshield wipers, parking assist system, and intrusion alarm
5. Premium audio group: Includes high-definition multi-speaker audio system, touch screen controls, display for rear view camera in the center stack, USB port, and secure digital (SD) card slot

Exterior and Interior Colors and Materials

Each model can be ordered with an exterior color from a group of colors assigned to each vehicle model. (Note that some vehicles may be sold with two colors; i.e., there is a two-tone option for certain models.) The interior colors and materials (e.g., for seats, instrument panel, and trim parts) can also be selected to go with certain exterior colors. Thus, information on which colors and materials can be ordered in which combination of models and packages should also be provided in the vehicle brochure.

Thus, each vehicle can be ordered with a unique combination of model, package, optional features, exterior color, interior color, and interior materials. The number of combinations can be very large, and to reduce the complexity of creating different components in different optional features and packages, manufacturers purposely limit (i.e., restrict) the configurations in which the vehicle can be ordered.

PICTURE GALLERIES

Many pictures of exterior and interior views and features of the vehicle from selected viewing locations are presented in the vehicle brochures and on websites. In addition, the selection of available exterior and interior colors and interior materials of seats, instrument panel, and trim parts is also included. The website of the vehicle allows dynamic presentations with and without audio clips via inclusion of videos and 360° viewers of key vehicle areas and features.

VEHICLE PRICE

The brochures of the vehicles typically do not include price information. The websites of most vehicle manufacturers do provide prices of each vehicle model with standard features. (Note: changes in vehicle prices can be easily adjusted on the websites.) Nowadays, many vehicle websites allow the reader to access and browse vehicle inventory data of dealers located in different parts of the country by entering the zip code and then selecting model year, vehicle name, and model. The dealer inventory data provide detailed information of standard and optional features of vehicles in the dealer inventory including pictures of their window stickers which also contain prices.

EXAMPLES OF BROCHURE CONTENTS

To provide a better insight into the usefulness of creating a vehicle brochure during the early part of the vehicle development process, the author asked graduate students taking the AE 500 course ("Automobile: An Integrated System") to create a brochure for a "target" vehicle that they would like to develop for introduction into the market as a 2020 model year vehicle (see Appendix V). Several outputs from a project created to design a 2020 model year mid-size sports utility vehicle (SUV) are illustrated in this section.

VEHICLE DIMENSIONS: EXTERIOR AND INTERIOR

A vehicle brochure should provide a few important exterior and interior dimensions of the target vehicle. The vehicle dimensions provide a better understanding of the size of the vehicle in terms of its key exterior and interior dimensions. The dimensional data can also be used by the reader for comparison with dimensions of other vehicles.

Table 15.1 provides an example of important exterior and interior dimensions of a mid-size SUV. (Note: These dimensions are described and illustrated in Chapter 20.)

POWERTRAIN AND FUEL ECONOMY

The brochure should include information on available combinations of engines and transmissions. It was proposed to equip the target vehicle with the following three powertrains:

1. 2.0 L four-cylinder engine with six-speed automatic transmission (standard powertrain)
2. 2.7L V6 engine with eight-speed automatic transmission (optional powertrain)
3. 3.5 L V8 engine with five-speed manual transmission (optional powertrain)

Fuel economy numbers in miles per gallon and emissions in grams of CO_2 (equivalent) per mile should be provided for each vehicle model and engine–transmission

---

TABLE 15.1

Exterior and Interior Dimensions of a Mid-Size SUV

Vehicle Dimensions	Target SUV Dimensions
Exterior Dimensions	
Overall length (in)	179.2
Overall width (except mirrors) (in)	73.4
Overall width (including side view mirrors) (in)	84.1
Overall height (in)	65.2
Wheelbase (in)	105.9
Front tread width (in)	62.4
Rear tread width (in)	62.5
Interior Dimensions	
Headroom—front (in)	39.6
Headroom—rear (in)	38.7
Legroom—front (in)	42.8
Legroom—rear (in)	36.8
Hip room—front (in)	54.4
Hip room—rear (in)	52.8
Shoulder room—front (in)	56.0
Shoulder room–rear (in)	55.3
Curb weight (lb)	3791

combination for city driving, highway driving, and combined city and highway driving cycles according to the U.S. Environmental Protection Agency (EPA) requirements.

KEY VEHICLE ATTRIBUTES

Key attributes are the attributes of the vehicle that its customers consider to be the most important to provide the characteristics that are essential for using and purchasing the vehicle. These key attributes provide the vehicle with the characteristics that cause it to fit their market segment in terms of its body-style, size, class, performance, and level of luxury.

For example, a sports car must have two doors, very low overall height, high engine power, and precise and stiffer suspension. On the other hand, an SUV must have a command seating position (provides ample visibility over the hood and side windows), high seating reference point (to allow easy entry/exit), two or three rows of seating, and a large storage space in the rear with a higher floor height (at slightly below the standing knuckle height of a shorter female customer) with a hatchback door.

The vehicle brochure should be designed to emphasis such key attributes of the vehicle.

SAFETY FEATURES

Safety is an important vehicle attribute. Customers expect the vehicle to be safe for use in all possible foreseeable situations and environments. The vehicle brochure should emphasize features that prevent the driver from getting involved in an accident (called the *crash-avoidance* characteristics of the vehicle) and reduce the severity of injury in the case of unavoidable accident situations (called the *crashworthiness* characteristics of the vehicle).

Safety characteristics can be also categorized into passive and active. Passive safety features are vehicle features (or mechanisms) that reduce the driver's chances of getting involved in a crash and reducing severity in the case of an accident without the driver taking any action related to the unsafe situation. Thus, the driver is considered to be "passive" in such situations. With "active" safety features, the driver is alerted to the impending safety situation, is typically required to use the feature, and must be engaged in making a decision and performing an action to avoid getting into the unsafe situation. Examples of passive safety systems are air bags and automatic braking systems. Blind area sensing systems, rear view cameras, and lane-departure warning systems are examples of active safety systems.

SPECIAL FEATURE CATEGORIES

Product development teams are typically asked to prepare lists of engineering details (such as values of relevant vehicle dimensions and capabilities of vehicle features) to help the company's marketing department prepare a vehicle brochure for prospective customers.

For example, to understand the importance of vehicle features to customers, during the project work for the author's automotive systems engineering class, the students were asked to prepare three separate features lists for the vehicles they were asked to design. The three lists were defined and categorized as follows: (a) items that would create a "Wow" impression among prospective customers (i.e., the customers have not seen such features in vehicles of the market segment of your target vehicle), (b) items that would be considered to be absolutely important for the customers (i.e., "Must Have" for their buying decision), and (c) items that would be on the customers' "Nice to Have" list. It should be noted that these three categories of features also relate to three types of features, namely, customer "delighters" (or Wows), "dis-satisfiers" (if not provided), and "satisfiers" (to provide more of), included in the Kano model of quality (Yang and El-Haik, 2003).

Table 15.2 presents examples of such lists of features included in the student projects.

TABLE 15.2

Three Types of Features Expected in 2020 Model Year Vehicles

Market Segment	"Wow" Features	Must Have Features	Nice to Have Features
Mid-size SUV	High fuel economy (29/38 mpg city/highway)	Reversing camera view in the center stack display	Eight-speed transmission
	Very large driving range on full gas tank (over 600 miles)	Fold-flat rear seats	Lane-departure warning and assist system
	Very spacious interior (large headroom, shoulder room, and legroom)	Power liftgate	Adaptive cruise control
	Very large cargo compartment (over 36 cu. ft.)	Stability and traction control	Panoramic moon-roof
	Advanced front lighting system (self-leveling, bi-xenon)	USB and AUX audio inputs	Sports handling
Entry-luxury mid-size car	Adaptive suspension and super handling performance	Memory seating options for user comfort	Large storage spaces
	Navigation system with weather and traffic inputs	Active safety features, for example, lane-departure assist	Panoramic moon-roof
	Automatic driver body comfort reconfigurable seats	Traction control system for snow driving	360 degree view camera
	Customizable body styling accessories	Programmable emergency calling for help	Messaging seats

	Smart headlighting with GPS map data inputs	Superior craftsmanship, soft leather and wood trim	Internet access and automatic travel advisories
Full-size pickup truck	10-speed transmission	150,000 miles durability	Comfortable seats and ride
	4.1 L V6 Diesel DI, twin turbo, and cylinder deactivated engine	Five-star safety rating	Large storage spaces
	Autonomous driving capability	Spacious interior—large headroom, shoulder room, and legroom	Large mirrors
	Lifestyle integration features	Superior off-road performance	Long (12,000 mile) oil change intervals
	Best in class fuel economy (21/29 mpg)	Smartphone/device/ infotainment integration	Easy to enter and egress

Note: GPS, global positioning system.

"Wow" Features

These are features of the vehicle that the customer in the class and market segment of the vehicle has not expected, but is pleasantly surprised and delighted to see in the vehicle. Such features are new and generally only available in high-end vehicles.

"Must Have" Features

These are features that customers clearly consider to be very desirable and useful. If these features are not available in the vehicle, the customer most likely will not purchase the vehicle, and will look at buying other competitors' vehicles that have these features.

"Nice to Have" Features

These features are considered by the customer to be desired but not absolutely essential in making his/her buying decision. The presence of more features in this category will make the vehicle more attractive to the customer.

CONCLUDING REMARKS

The vehicle brochure and website are important tools for connecting the new automotive product with prospective customers. They generate an impression of the vehicle in terms of degree of advances in vehicle design, styling, level of luxury, availability of features, and vehicle capabilities. The creation of a brochure or a website during the early phases of a vehicle development program is also an excellent exercise for the design team, helping them to understand many models, features, and options needed in marketing the vehicle.

REFERENCE

Yang, K. and B. El-Haik. 2003. *Design for Six Sigma*. New York: McGraw-Hill. ISBN 0071412085.

Section II

Tools Used in the Automotive Design Process

16 Tool Box for Automotive Product Development

INTRODUCTION

This chapter is intended to provide an overview of a number of tools used in various phases of the vehicle development process. The tools provide information in the form of data or their presentation to visualize and understand certain situations. The inputs to the tools are generally obtained from previously collected data, assumptions, or data obtained from prior analysis conducted internally within the tool. The data are mostly analyzed to determine relationships between certain independent variables or to predict the effects of certain variables on a dependent (or response) variable used to make decisions (see Chapter 17).

The tools (techniques or methods) can be organized in many different ways. The tools can be categorized as (a) general-purpose tools and (b) special-purpose tools. Generic (or general-purpose) tools provide some basic capabilities that are used to conduct analyses of common or similar processes, whereas there are a number of specialized tools used for analyzing problems in specialized disciplines.

Software applications for collecting, organizing, recording/storing, and displaying, and data manipulation tools such as spreadsheets, database management, data plotting tools, project management, and computer-aided design (CAD) tools for visualization of products and processes are examples of general-purpose tools. These tools are used by engineers, designers, and specialists in many disciplines.

Many specialized engineering disciplines or application areas require special-purpose tools, such as computer-aided engineering (CAE) tools, which can be further classified into areas such as computational fluid dynamics (CFD) tools for aerodynamics and fluid flow analysis, thermodynamics tools for heat transfer analysis, power load evaluation for electrical architectural design, and finite element analysis for structural design.

Measurement and test equipment primarily used in engineering verification of entities (i.e., vehicle systems and their lower-level subsystems down to component level) also involves many specialized tools. For example, many measurement and recording tools require specialized hardware, that is, coordinate (or dimensional) measurement machines (CMM) and sensors to convert changes in physical characteristics into analog or digital signals (e.g., photocells, motion sensors, accelerometers, temperature sensors, and pressure sensors). Other examples of specialized tools or test equipment used in specialized laboratory and/or field tests are engine dynamometers, vibration testing machines, gas emissions analyzers, crash test dummies, crash testing sleds, and wind tunnels.

Tools used in manufacturing and assembly operations also involve much specialized equipment, and they are not covered here, as the focus of this chapter is on the tools used during the early engineering phases related to the left side of the systems engineering "V" model (see Figure 2.2).

The systems engineering management plan (SEMP) will generally include a schedule of tasks to be performed and tools to be used to meet the objectives of the vehicle program and requirements for the vehicle(s) developed in the program. The SEMP is covered in Chapter 12.

TOOLS USED DURING VEHICLE DEVELOPMENT PHASES

This section presents an overview of important tools used in the vehicle development process. The tools are presented in the order typically used in the sequence of applications in the vehicle development process.

SPREADSHEETS

Spreadsheets, such as those created by Microsoft Excel, are probably the most commonly used tools to display and summarize tabular or matrix formatted data. All product planning, scheduling, engineering, and financial activities use spreadsheets. The spreadsheets allow data to be organized in columns and rows and enable the computation of a number of mathematical and statistical functions. The data plotting functions further enable the data to be displayed using various types of charts: scatter diagrams, line charts, bar charts, pie charts, 3-D charts, spider charts, and so forth. Many of the tools presented in this chapter can be easily implemented by the use of a spreadsheet. Some examples of application areas of the spreadsheet are (a) benchmarking table (see Table 4.1), (b) requirements table (see Table 9.1), (c) interface matrix (see Table 8.1), (d) failure modes and effects analysis (see Table 18.6), (e) decision analysis table (see Table 17.4), (f) Pugh diagram (see Table 17.5), and (g) financial analysis (see Table 19.2).

DESIGN STANDARDS AND GUIDELINES

The teams involved in designing a vehicle and its systems must gather and familiarize themselves with all available design standards and the design requirements and design guidelines on vehicle attributes and associated vehicle systems. Most systems design standards and guidelines are generally documented and maintained by the attribute engineering and/or functional engineering offices of the vehicle manufacturers. These standards usually refer to other applicable standards, such as those developed by government agencies, professional societies (e.g., Society of Automotive Engineers [SAE] standards and recommended practices [SAE, 2009]), automotive suppliers, and other industries. The standards include rationale, background information, terminology, design and performance requirements, test procedures, and guidelines (for design and/or installation) to achieve the required level of performance. They also include tips (or issues to consider) on avoiding many past mistakes from lessons learned. The Federal Motor Vehicle Safety Standards (FMVSS) are available from the National Highway Traffic Safety Administration (NHTSA 2015). The fuel economy and emissions requirements are available from the Environmental Protection Agency (EPA) and NHTSA (2012), and the SAE standards are available in SAE (2009) (also see Chapter 3).

PRODUCT PLANNING TOOLS

The important tools used during the early stages of product planning include (1) benchmarking and breakthrough, (2) Pugh diagram, (3) quality function deployment, (4) failure modes and effects analysis, and (5) other product development tools such as a business plan, program status charts, and CAD.

Benchmarking

Benchmarking is probably the most commonly used and popular tool to understand design and construction issues within various automotive systems. During the early phases of design of any system and its subsystems, benchmarking is generally used to study and compare similar systems used in many existing vehicles, including those produced by leading competitors. Benchmarking provides a quicker understanding into the strengths and weaknesses of different designs and helps identify good design features that should be included and refined, and poor features and mistakes to avoid (e.g., especially when used in conjunction with the Pugh diagram, covered in the next section). Chapter 4 presents the benchmarking technique in more detail.

Pugh Diagram

The Pugh diagram is a tabular formatted tool consisting of a matrix of product attributes (or characteristics) and alternate product concepts along with a benchmark (reference) product called the *datum*. The diagram helps to undertake a structured concept selection process and is generally created by a multidisciplinary team to converge on a superior product concept. The process involves creation of the matrix by inputs from all the team members. The rows of the matrix consist of product attributes based on the customer needs, and the columns represent different alternate product concepts.

The evaluations of each product concept on each attribute are made with respect to the *datum*. The process uses classification metrics of "same as the datum" (S), "better than the datum" (+), or "worse than the datum" (−). The scores for each product concept are obtained by simply adding the number of plus and minus signs in each column. The product concept with the highest net score (the "sum of plus signs" minus the "sum of minus signs") is considered to be the preferred product concept. Several iterations are employed to improve product superiority by combining the best features of highly ranked concepts on each attribute till a superior concept emerges and becomes the new benchmark. Chapter 4 presents more information and examples of Pugh diagrams.

Quality Function Deployment (QFD)

Quality Function Deployment (QFD) is a technique used to understand customer needs (voice of the customer) and to transform customer needs into engineering characteristics of products or processes in terms of functional or design requirements. It relates "what" (what are the customer needs) to "how" (how should engineers design to meet the customer needs) and "how much" (magnitude of design variables, i.e., their target values), and also provides competitive benchmarking information—all in one diagram. QFD was originally developed by Dr Yoji Akao of Japan in 1966 (Akao, 1991). QFD is used in a wide variety of applications and is considered a key tool in the design for six sigma (DFSS) process. It is also known as the "House of Quality" because the correlations matrix drawn on the top of the QFD matrix diagram resembles the roof of a house. Chapter 18 presents more detail of QFD and an example.

Failure Modes and Effects Analysis

The failure modes and effects analysis (FMEA) method was initially developed in the 1960s as a systems safety analysis tool. It was used in the early days of defense and aerospace systems design to ensure that the product (e.g., an aircraft, a spaceship, or a missile) was designed to minimize the probabilities of all failures. The failures are discovered by brainstorming and evaluating all possible causes leading to failures that could possibly occur. The resulting list of failures is prioritized, and corrective actions are developed. For more than 20 years, the method has been routinely used by product design and process design engineers to reduce the risk of failures in the designs of products and processes used in the production, operation, and maintenance of various systems used in many industries (e.g., automotive, aviation, utilities, and construction). FMEA conducted by product design engineers is typically referred to as DFMEA (where D stands for design), and FMEA conducted by process designers is referred to as PFMEA (where P stands for process).

FMEA is a proactive and qualitative tool used by quality, safety, and product/process engineers to improve reliability (i.e., to eliminate failures, thus improving quality and customer satisfaction). The development of a FMEA involves the following basic tasks: (a) identify possible failure modes and failure mechanisms, (b) determine the effects or consequences that the failures may have on the product and/or process performance, (c) determine methods of detecting the identified failure modes, (d) determine possible means of prevention, and (e) develop an action plan to reduce the risks due to the identified failures. It is very effective when performed early in the product or process development by experienced multifunctional team members as a team exercise. Chapter 18 presents more details on FMEA and an example.

CAD and Packaging Tools

Computer graphics tools are used in the design studios for 3-D modeling (e.g., surfacing [surface representation]) and visualization. (See Chapters 10 and 13 for more information on CAD tools and applications.) Studio outputs are also used to create physical models or bucks using surface scanning and digitizing machines, plotting devices, and milling machines (i.e., for milling clay or wood).

The packaging engineering and analysis work also includes the use of CAD tools as well as other tools such as virtual reality (VR) simulations using digital vehicle models (i.e., virtual vehicle builds) and digital human models (e.g., the Jack model). A human simulation system originally developed at the University of Pennsylvania in the 1980s and 1990s (Badler et al., 1993). VR tools are also used for applications such as evaluations of vehicle assembly problems (e.g., interferences, insufficient clearances) and other ergonomic considerations such as reach distances, allowable weights that can be lifted and carried, and steps required in assembly tasks. In addition, driving simulators and programmable vehicle bucks are used in ergonomic evaluations (Bhise, 2012).

ENGINEERING ANALYSIS TOOLS

A vast number of CAE tools are used in many design, engineering, and manufacturing areas in the automotive industry. Their use has enabled automakers to reduce product development costs and time while improving the safety, weight, comfort, and durability of the vehicles. Examples of such applications are (a) finite element analysis (FEA) for structural analysis, (b) occupant impact analyses, (c) vehicle dynamics and suspension system analyses, (d) aerodynamic and thermodynamic analyses, (e) electrical load and electronic data transfer analyses, (f) field of view and visibility analyses, and (g) optical analyses (e.g., headlamp beam pattern design and night visibility analyses). In addition, systems engineers use database and management tools for requirements management (e.g., traceability, functional allocation, systems interfacing, and cascading), verification, and validation for attribute management.

The predictive capability of CAE tools has progressed to the point where much of the design verification is now done by using computer simulations rather than physical prototype testing. However, physical testing is still used, especially in the final verification and validation of vehicles and their subsystems, because the CAE tools cannot predict effects of all variables in complex assemblies (e.g., due to manufacturing variations such as unpredictable warping, stretching, and thinning of materials). Chapter 13 presents more information on CAE tools.

QUALITY TOOLS

Seven new tools and seven traditional tools are the basic quality tools in the total quality management (TQM) field. The seven traditional tools are (1) Pareto chart, (2) cause-and-effect diagram, (3) check sheet, (4) histogram, (5) scatter diagram, (6) stratification chart, and (7) control chart. The seven new tools are (1) relations diagram, (2) affinity diagram, (3) systematic diagram, (4) matrix diagram, (5) matrix data analysis, (6) process decision program chart (PDPC), and (7) arrow diagram. The above tools, along with experimental design are used in six sigma projects to solve quality problems in the auto industry (see Bhise [2014] for more information and examples of these tools).

HUMAN FACTORS AND ERGONOMICS TOOLS

Human factors engineers use a number of tools during the vehicle development process. The tools are used for the following purposes: (a) to obtain information about characteristics, capabilities, and limitations of populations of users, (b) to apply the collected information to design the products, and (c) to evaluate the products during different phases of the product programs. The goal of the human factors engineer is to ensure that the product is designed such that most individuals in its intended user population are able to use the product easily, comfortably, and safely.

The human factors methods can be categorized as follows:

1. Databases on human characteristics and capabilities
2. Anthropometric and biomechanical human models
3. Checklists and score cards
4. Task analysis
5. Human performance evaluation models
6. Laboratory, simulator, and field studies
7. Human performance measurement methods

More information and examples of these tools are available in Bhise (2012).

SAFETY ENGINEERING TOOLS

Safety engineers use a variety of tools to solve various safety problems, ranging from identifying hazards inherent in a product to analyzing accidents occurring during product use and monitoring safety performance during the product life cycle. Hazard identification and risk reduction involve tools such as hazard analysis, FMEA, and fault tree analysis. These tools help in reducing potential occurrences of accidents during product use. More information and examples of these tools are available in Bhise (2014).

MEASUREMENT TOOLS

CMMs are used to check and verify the dimensions and locations of various components in physical properties (e.g., bucks, prototypes, and other competitive vehicles included for benchmarking) used during the product development process. Scanners are used to digitize surfaces initially developed on clay models. The digitized data can be used to create dies for producing parts using sheet metal stamping, hydroforming, die casting, forging, and casting processes.

PROGRAM/PROJECT MANAGEMENT TOOLS

A number of tools are used for project management. Chapter 12 describes the following tools used to perform project management functions: (1) Gantt chart, (2) critical path method (CPM), (3) program (or project) evaluation and review technique (PERT), (4) work breakdown structure (WBS), (5) project management software (e.g., Oracle, Microsoft Project, and Project Pro for Office 365, developed and sold by the Microsoft Corporation [Microsoft, 2016]), and (6) other tools. Many other tools are available for specialized analyses, such as investment analysis, cost–benefits analyses, expert surveys, simulation models and predictions, risk profile analyses, surcharge calculations, milestone trend analysis, cost trend analysis, target versus actual comparisons of dates, time used, and costs incurred and head count.

FINANCIAL ANALYSIS TOOLS

Many different software applications are available to perform product life cycle costing and to create various reports (e.g., by systems, program phases, and months; comparisons with budgeted costs). Many of the applications are integrated with other functions such as management information systems, product planning, and supply-chain management. The software systems are used for production scheduling, component ordering, inventory control, product control, shop floor management, cost accounting, and so forth. Some examples of such software systems are manufacturing resource planning (MRP) and enterprise resource planning (ERP). The software systems are available from a number of developers (e.g., SAP, Oracle, Microsoft, EPICOR, and Sage). Chapter 19 provides additional information on financial analysis and provides examples.

MARKET RESEARCH TOOLS

The marketing research department will use a number of tools, such as personal interviews, focus group sessions, and mail and telephone surveys to obtain data on customer needs, product evaluations, complaints, and satisfaction. These tools are covered in Chapter 11 and by Zikmund and Babin (2009).

CONCLUDING REMARKS

Many specialized tools are being used to conduct the necessary analyses during the vehicle development process. The tools must be used at the right time, and the outputs of the analyses should be reviewed and discussed at various design and program review meetings to ensure that proper trade-offs between different attributes are made to meet the vehicle requirements. The schedule and the details of the analyses and evaluation process are specified in the SEMP (see Chapter 12).

REFERENCES

Akao, Y. 1991. Development History of Quality Function Deployment. In: Mizuno, S., Akao, Y., and Ishihara, K. *The Customer Driven Approach to Quality Planning and Deployment*. Minato, Tokyo 107, Japan: Asian Productivity Organization.

Badler, N. I., C. B. Phillips, and B. L. Webber. 1993. *Simulating Humans: Computer Graphics, Animation, and Control*. UK: Oxford University Press.

Bhise, V. D. 2012. *Ergonomics in the Automotive Design Process*. Boca Raton, FL: CRC.

Bhise, V. D. 2014. *Designing Complex Products with Systems Engineering Processes and Techniques*. Boca Raton, FL: CRC.

EPA and NHTSA. (2012). 2017 and Later Model Year Light-Duty Vehicle Greenhouse Gas Emissions and Corporate Average Fuel Economy Standards. *Federal Register*, 77.199: 62623–63200. Environmental Protection Agency, 40 CFR Parts 85, 86, and 600.

Department of Transportation National Highway Traffic Safety Administration, 49 CFR Parts 523, 531, 533, 536, and 537. [EPA–HQ–OAR–2010–0799; FRL–9706–5; NHTSA–2010–0131]. RIN 2060–AQ54; RIN 2127–AK79.

Microsoft Corporation. 2016. Project Pro and Office 365. Website: www.microsoft.com/project/en-us/demos.aspx (Accessed: June 15, 2016).

NHTSA. (2015). Federal Motor Vehicle Safety Standards and Regulations, U.S. Department of Transportation. Website: www.nhtsa.gov/cars/rules/import/FMVSS/(Accessed: June 14, 2015).

SAE. 2009. *SAE Handbook*. Warrendale, PA: Society of Automotive Engineers.

Zikmund, W. G. and B. J. Babin. 2009. *Exploring Market Research*. 9th edn. Publisher: Cengage Learning.

17 Decision-Making Tools

INTRODUCTION

Decisions are made throughout the life cycle of a vehicle program. Decisions are also made during each phase and at each milestone of a vehicle development program when the management decides whether to proceed to the next phase, make changes to the vehicle being designed, or even scrap the program.

Early decisions are related to the type of vehicle to be designed, requirements to be selected for the vehicle, and its characteristics (e.g., 0–60 mph acceleration time). Later, the decisions are related to the number of systems in the vehicle, their functions, and how the systems should be configured and packaged within the vehicle space. Early decisions have a major impact on the overall costs and timings of the program, because the later decisions depend on the design-specific parameters and their values selected in the earlier phases of the program. For example, powertrain type, size, its location in the vehicle space (e.g., front-wheel drive or rear-wheel drive), and the technologies to be implemented in the new powertrain will affect decisions related to the design of its other systems (e.g., fuel system, cooling system, space available to package suspensions).

After the vehicle is introduced into the market, customer feedback is received. The reasons for customer dissatisfaction need to be understood, and the manufacturer needs to decide whether to make any changes. In some situations, the decisions may involve recalling the vehicle and further deciding on how and when to fix any defects in the vehicle. After the vehicle has been marketed for an extended period of time, additional decisions need to be made on what vehicle characteristics should be revised, how to revise them, and when to revise them.

Decisions are made whenever alternatives exist and the best, that is, the most desired alternative needs to be selected. The selected alternative should result in reduced risks and increased benefits. Many different criteria can be used in selecting one or more alternatives. Systems engineering involves decision making, such as what needs to be done, when, how, and how much, and taking into account the trade-offs between possible design considerations.

In a decision-making situation, the decision maker (e.g., engineer, designer, or program manager) is faced with the task of deciding on an acceptable alternative among several possible available alternatives. The decision maker also needs to consider possible future outcomes (i.e., what will happen in the future) and the costs and benefits (called the *payoffs*) associated with each combination of an alternative and an outcome. Further, each possible outcome may or may not occur in the future.

All these decisions involve risks. For example, adding more features (or capabilities) than the customers want and over-designing will waste resources. Conversely, failure to incorporate any customer-desired major changes and under-designing the product will result in loss of sales or even degrade the reputation of the manufacturer and its brand image in the marketplace.

This chapter covers various decision-making approaches and models and also provides an understanding of the issues related to risks and methods to analyze the risks.

AN AUTOMAKER'S DECISION-MAKING PROBLEM: AN EXAMPLE

Let us assume that a large automaker currently produces about a million vehicles annually in various sizes and body-styles. One of the major decisions that the automaker wants to make is to decide on the type of vehicle program to undertake to maintain its profitability. Table 17.1 presents 12 possible alternative programs to replace its existing products that the automaker can consider. For example, the first alternative, shown in the first row of Table 17.1, is to consider the vehicle program

TABLE 17.1

Alternative Vehicle Programs Considered during Program Selection Decision Making

Vehicle Program	Vehicle Type	Selling Price	Profit per Vehicle	Annual Sales Volume	Total 5 Year Volume	Total Profit	Total Revenue	Profit as % of Revenue
P1	Small "B" size car	$17,000	$300	50,000	250,000	$75,000,000	$4,250,000,000	1.8
P2	Small "C" size car	$19,000	$450	120,000	600,000	$270,000,000	$11,400,000,000	2.4
P3	Mid-size "C/D" size car	$22,000	$1,500	120,000	600,000	$900,000,000	$13,200,000,000	6.8
P4	Large "D" size car	$26,000	$2,000	60,000	300,000	$600,000,000	$7,800,000,000	7.7
P5	Small "C" size SUV	$21,000	$700	70,000	350,000	$245,000,000	$7,350,000,000	3.3
P6	Mid-size SUV	$32,000	$3,000	100,000	500,000	$1,500,000,000	$16,000,000,000	9.4
P7	Large SUV	$50,000	$8,000	25,000	125,000	$1,000,000,000	$6,250,000,000	16.0
P8	Small pickup	$20,000	$1,000	60,000	300,000	$300,000,000	$6,000,000,000	5.0
P9	Large pickup	$28,000	$5,000	150,000	750,000	$3,750,000,000	$21,000,000,000	17.9
P10	Mid-size luxury car	$30,000	$9,000	35,000	175,000	$1,575,000,000	$5,250,000,000	30.0
P11	Large luxury car	$45,000	$14,000	20,000	100,000	$1,400,000,000	$4,500,000,000	31.1
P12	High-performer sports car	$80,000	$30,000	2,000	10,000	$300,000,000	$800,000,000	37.5
			Sum -->	812,000	4,060,000	$11,915,000,000	$103,800,000,000	11.5

P1, which involves making a small "B" size car (such as a Ford Fiesta) with a selling price of about $17,000, expecting to generate about $4.25 billion revenue by selling 250,000 vehicles over the next 5 years and make $75.0 million profit. Other alternatives to be considered are programs P2 to P12 with different body-styles (see Table 17.1).

The problem of selecting a vehicle program needs consideration of many issues. If the corporate goal is to maximize profits, then the total profit column shows that program P9, to produce 750,000 large pickup trucks over 5 years, has the potential to produce the maximum profit of $3.75 billion. The $3.75 billion profit amounts to 17.9% of the total revenue generated by the program. On the other hand, if the goal is to maximize the profits as a percentage of total revenue generated by a vehicle program, then program P12, to produce a high-performer sports car, is the best alternative, as its profits amount to 37.5% of the revenue generated by the program. The profit per vehicle column is also worth looking at, as it shows that the high-performer sports car can generate a net profit of $30,000 per vehicle as compared with only $300 for a "B" size car in program P1. Further, one needs to have reliable estimates of the sales volume needed (e.g., total 5 year sales volume) to make the profits shown in the table. Here we can assume that the selling price minus profit per vehicle represents the cost per vehicle, which should include per vehicle costs due to (a) product development, (b) purchasing parts from the suppliers, (c) manufacturing and assembly operations of the auto manufacturer, and (d) vehicle marketing. These costs estimation issues are covered in Chapter 19.

The decision to select a program will also depend on the sales rate of current products, new products introduced by other vehicle manufacturers in each market segment, future trends in design, technologies, and government regulations, and so forth. Changing economic and political conditions in the markets also affect future sales volumes and add uncertainty and risks to the revenue generation.

This example provides some insight into the complexity associated with selecting and undertaking a vehicle program, the ability to sell the new vehicles at the projected sales volumes, and the risks associated with creating a successful product that customers will want.

DECISION MAKING IN PRODUCT DESIGN

KEY DECISIONS IN PRODUCT LIFE CYCLE

Some key decisions made in a product program typically involve

1. *Program Kick-Off*: The top management of the organization decides that a new product (or revisions to an existing product) should be developed and a program should be kicked off to develop the product.
2. *Program Confirmation*: The top management confirms the decision to select a product concept for detailed engineering based on (a) additional information obtained from the design team's presentation of the created product concepts, (b) market research results, (c) a review of the latest trends in new technologies and design, and (d) the competitors' capabilities. Additional

decisions are also made to allocate budget and dates for the product introduction into the selected markets.

3. *Product Concept Freeze*: Management decides that the selected product concept is sufficiently developed (i.e., all design and engineering managers feel confident that the product can be produced [i.e., it is feasible] within the planned budget and schedule). Thus, the concept will be frozen (i.e., no major changes will be made), and succeeding program activities will be continued.

4. *Engineering Sign-Off*: All key managers of engineering activities sign a document stating that the product "as designed" will meet all applicable requirements with a high probability (e.g., 90%).

5. *Production Release*: All product testing (verification and validation tests) is completed, and the product is determined to be ready for the market. The product is released for production; that is, the factories begin production of units for sale.

6. *Periodic Reviews*: Periodic (monthly, quarterly, or annual) reviews of the product sales, customer satisfaction, and comparisons with data from the competitors' products are conducted to decide whether any changes in the product volume or product characteristics are needed.

7. *Product Discontinuation and Replacement*: Based on the market data and the customer feedback, the management decides to terminate the production of the product on a certain date and requests the marketing department to plan for future product(s) or model(s) for its replacement.

TRADE-OFFS DURING DESIGN STAGES

Teams involved in designing any product need to make a number of decisions involving trade-offs between a number of conflicting design considerations (e.g., product characteristics and attributes). Some examples of trade-off considerations in designing passenger cars are

1. *Space for Vehicle Systems versus Space for Occupants*: The space within the vehicle is occupied by various vehicle systems and their components, and space is used to accommodate occupants in the passenger compartment and other items in the trunk (or cargo areas). To provide more space for the occupants (interior passenger space), the space occupied by vehicle systems (e.g., vehicle body structure sections, engine, chassis and suspension components, and fuel tank) needs to be reduced. Thus, a vehicle designer can make trade-offs by designing more compact vehicle systems to allow more space for the occupants. This trade-off is commonly referred to in the auto industry as "Machine Minimum and Man Maximum" (i.e., minimizing the space for mechanical components and maximizing the space for the occupants).

2. *Vehicle Performance versus Fuel Economy*: A vehicle with a high performance (i.e., acceleration capability, commonly measured by the minimum time required for accelerating from 0 to 60 mph [called the 0-to-60 time in

seconds]), requires higher engine power, which in turn reduces fuel economy (measured in miles per gallon of gasoline consumed).

3. *Vehicle Performance versus Vehicle Weight*: This trade-off is commonly referred to by considering the horsepower-to-weight ratio. Any increase in vehicle weight will reduce the acceleration capability of a vehicle with the same engine power.

4. *Ride Comfort versus Handling*: Better-riding cars require softer suspensions, which generally reduce the handling (maneuvering) capability of the vehicle.

5. *Lightweight Materials versus Cost*: Lightweight materials (e.g., aluminum, magnesium, high-strength steels, and carbon-fiber materials) can reduce vehicle weight. However, these lightweight materials are more costly than the commonly used sheet steel (mild steel) material.

6. *High Raked Windshield versus Costs*: Windshields with a higher rake angle (more sloping windshields; see Figure 17.1) can reduce aerodynamic drag, increase fuel economy, and provide a sleeker, more aerodynamic appearance than conventionally styled vehicles with a more upright windshield. The high raked windshields are longer (see Figure 17.1, where $L_1 > L_0$) than more conventional low raked windshields. The longer length (L_1) also requires a larger windshield and thicker glass, longer wipers, a more powerful wiper motor, a higher-capacity windshield defroster, and higher-capacity air-conditioning (due to higher heat/sun load on the larger windshield). The thicker glass also reduces the light transmission of the windshield, which in turn, reduces driver visibility. The thicker glass also increases vehicle weight, which in turn, can reduce fuel economy. Higher raked windshields can thus increase vehicle costs.

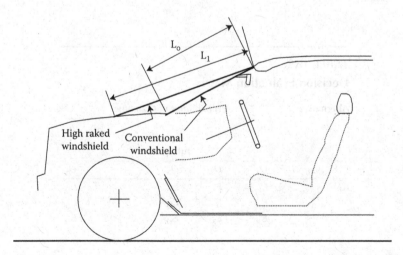

FIGURE 17.1 Comparison of conventional versus high raked windshield. (*Note*: The conventional windshield has L_0 length. The higher raked windshield has L_1 length.)

WHAT IS INVOLVED IN DECISION MAKING?

ALTERNATIVES, OUTCOMES, PAYOFFS, AND RISKS

Systems engineering involves decision making, such as what needs to be done, when, how, and how much, and taking into account the trade-offs between possible design considerations. In a decision-making situation, the decision maker (e.g., engineer, designer, or program manager) is faced with the task of deciding on an acceptable alternative among several possible alternatives. The decision maker also needs to consider possible future outcomes (i.e., what possible events will happen in the future) and the costs or benefits (called the *payoffs*) associated with each combination of an alternative and an outcome. Further, each possible outcome may or may not occur in the future. There are many different decision principles to determine a desired or an acceptable alternative (Blanchard and Fabrycky, 2011). The decision problems and the principles that can be applied will now be described.

Let us assume the following:

A_i = ith alternative where $i = 1, 2, ..., m$
O_j = jth outcome where $j = 1, 2, ..., n$
P_j = the probability that jth outcome will occur where $j = 1, 2, ..., n$
E_{ij} = evaluation measure (positive for benefit [profit] and negative for cost [loss]) associated with the ith alternative and the jth outcome

The decision evaluation matrix associated with this problem is presented in Table 17.2. Many principles can be used to select a desired alternative. The principles are described in the following subsections.

TABLE 17.2
Decision Evaluation Matrix

Alternative	Probabilities of Outcomes					
	P_1	P_2	P_3	.	.	P_n
	Outcomes					
	O_1	O_2	O_3	.	.	O_n
A_1	E_{11}	E_{12}	E_{13}	.	.	E_{1n}
A_2	E_{21}	E_{22}	E_{23}	.	.	E_{2n}
A_3	E_{31}	E_{32}	E_{33}	.	.	E_{3n}
.		.	.	.	.	.
.		.	.	.	.	.
A_m	E_{m1}	E_{m2}	E_{m3}	.	.	E_{mn}

MAXIMUM EXPECTED VALUE PRINCIPLE

One commonly used principle to select an alternative is based on the maximum expected value. The expected value of the ith alternative, $A_i = \{E_i\}$, can be computed as $\sum_j [P_j \times E_{ij}]$.

If the expected value is negative, that is, a loss, it can be regarded as a risk. Consideration of risks is important during decision making. In general, a prudent decision maker will strive to reduce risks in selecting alternatives and during the program management. The risk-related issues are covered in a later section of this chapter (see Figure 17.5).

Thus, under this principle, the decision maker will select the alternative with the maximum expected value, which is defined as max $\{E_i\}$ for $i = 1, 2, ..., m$. The maximum expected value is equivalent to minimum risk.

The selection of the alternative and the application of the above principle are illustrated in the following example.

Let us assume that an automotive manufacturer wants to select a powertrain for its new small vehicle. The manufacturer is considering the following five alternatives:

A_1 = Design a new small car using a state-of-the-art gasoline powertrain.

A_2 = Do not design a new small car—continue with the present model with minimum modifications.

A_3 = Design a new small car with an electric powertrain.

A_4 = Design a new small car with a turbo-diesel powertrain.

A_5 = Design a small car with all three (gasoline, diesel, and electric) powertrain options.

Six possible outcomes assumed by the manufacturer are

O_1 = Economy does not change, oil prices remain low, and the battery technology does not improve.

O_2 = Economy improves by 5%, oil prices remain low, and the battery technology does not improve.

O_3 = Economy degrades by 5%, oil prices increase by 30%, and the battery technology does not improve.

O_4 = Economy does not change, oil prices remain low, and the battery technology improves by 50%.

O_5 = Economy improves by 5%, oil prices remain low, and the battery technology improves by 50%.

O_6 = Economy degrades by 5%, oil prices increase by 30%, and the battery technology improves by 50%.

The evaluation measures (estimated total net profit over the 5 years of sale, in dollars) associated with the combinations of the five alternatives and the six outcomes are provided in Table 17.3. The table also provides probabilities for each of the outcomes assumed by the manufacturer.

TABLE 17.3
Data for Powertrain Selection Decision Problem

Alternatives			Probability of Outcome			
	0.2	0.15	0.1	0.2	0.3	0.05
			Outcomes			
	O_1	O_2	O_3	O_4	O_5	O_6
A_1	$4,000,000,000	$5,000,000,000	$2,000,000,000	$3,000,000,000	$4,500,000,000	$1,750,000,000
A_2	$2,000,000,000	$2,500,000,000	$1,500,000,000	$1,750,000,000	$2,000,000,000	$1,200,000,000
A_3	−$50,000,000	$300,000,000	$25,000,000	$100,000,000	$400,000,000	$600,000,000
A_4	$200,000,000	$250,000,000	$150,000,000	$200,000,000	−$100,000,000	−$250,000,000
A_5	$3,500,000,000	$4,500,000,000	$2,250,000,000	$3,750,000,000	$5,500,000,000	$2,000,000,000

Note:
O_1 = Economy does not change, oil prices remain low, and the battery technology does not improve.
O_2 = Economy improves by 5%, oil prices remain low, and the battery technology does not improve.
O_3 = Economy degrades by 5%, oil prices increase by 30%, and the battery technology does not improve.
O_4 = Economy does not change, oil prices remain low, and the battery technology improves by 50%.
O_5 = Economy improves by 5%, oil prices remain low, and the battery technology improves by 50%.
O_6 = Economy degrades by 5%, oil prices increase by 30%, and the battery technology improves by 50%.
A_1 = Design a new small car using state-of-the-art gasoline powertrain.
A_2 = Do not design a new small car—continue with the present model with minimum modifications.
A_3 = Design a new small car with an electric powertrain.
A_4 = Design a new small car with a turbo-diesel powertrain.
A_5 = Design a small car with all three (gasoline, diesel, and electric) powertrain options.

The following computation illustrates the computation of the expected value of A_1:

$$\text{Expected value of } A_1 = \{E_1\} = (0.2 \times 4,000,000,000) + (0.15 \times 5,000,000,000)$$

$$+ (0.1 \times 2,000,000,000) + (0.2 \times 3,000,000,000)$$

$$+ (0.3 \times 4,500,000,000) + (0.05 \times 1,750,000,000)$$

$$= \$3,787,500,000$$

The expected values of A_2, A_3, A_4, and A_5 are $1,935,000,000, $207,500,000, $90,000,000, and $4,100,000,000, respectively. Thus, the alternative A_5 has the maximum expected value of $4,100,000,000 among the five alternatives, and it will be selected under the maximum expected value principle (see the column labeled "Expected Value Principle" in Table 17.4).

OTHER PRINCIPLES IN SELECTING ALTERNATIVES

Six additional principles that can be used to select an alternative are described in this section.

TABLE 17.4
Alternatives Selected by Five Principles

Alternatives	Probability of Outcome						Expected Value Principle	Laplace Principle (Average Value)	Maximin Principle (Minimum Values)	Maxmax Principle	Hurwicz Principle (with $\alpha = 0.5$)
	0.2	0.15	0.1	0.2	0.3	0.05					
	Outcomes										
	O_1	O_2	O_3	O_4	O_5	O_6					
A_1	$4,000,000,000	$5,000,000,000	$2,000,000,000	$3,000,000,000	$4,500,000,000	$1,750,000,000	$3,787,500,000	$3,375,000,000	$1,750,000,000	$5,000,000,000	$3,375,000,000
A_2	$2,000,000,000	$2,500,000,000	$1,500,000,000	$1,750,000,000	$2,000,000,000	$1,200,000,000	$1,935,000,000	$1,825,000,000	$1,200,000,000	$2,500,000,000	$1,850,000,000
A_3	–$50,000,000	$300,000,000	$25,000,000	$100,000,000	$400,000,000	$600,000,000	$207,500,000	$229,166,667	–$50,000,000	$600,000,000	$275,000,000
A_4	$200,000,000	$250,000,000	$150,000,000	$200,000,000	–$100,000,000	–$250,000,000	$90,000,000	$75,000,000	–$250,000,000	$250,000,000	$0
A_5	$3,500,000,000	$4,500,000,000	$2,250,000,000	$3,750,000,000	$5,500,000,000	$2,000,000,000	$4,100,000,000	$3,583,333,333	$2,000,000,000	$5,500,000,000	$3,750,000,000

Note: The selected alternatives are shown in the underlined cells of this table.

1. *Aspiration Level*: The principle of aspiration level is based on the assumption that the decision maker needs to meet a certain aspiration (or desired) level, such as a minimum acceptable profit level or a maximum amount of tolerable loss. If we assume that the decision maker in this example (Table 17.3) wants to make at least $4,000,000,000 profit, then he or she would consider alternatives A_1 and A_5 (because these two alternatives include the outcomes with payoff of $4,000,000,000 or more). On the other hand, if he or she does not want to incur any loss, he or she would not consider alternatives A_3 and A_4 (as these two alternatives can incur a loss in at least one outcome).

2. *Most Probable Future*: The decision maker may decide based on the most likely outcome (which has the highest probability of occurrence). In our example (Table 17.3), the outcome O_5 has the highest probability (0.3) of occurrence. Under this situation (outcome O_5), the selection of alternative A_5 will provide the maximum profit of $5,500,000,000.

3. *Laplace Principle*: The Laplace principle assumes that the decision maker does not have any information on the probability of occurrences of any of the outcomes, and thus, he or she assumes that all the outcomes are equally likely. In our example, under this principle, all the occurrence probabilities will be equal to 1/6. Thus, the decision maker can simply take the average value of all E_{ij} for each alternative (i.e., over each i) and select the alternative with the maximum profit. In our example above, under this principle, the decision maker would select alternative A_5 with the maximum average profit of $3,583,333,333 (see the column labeled "Laplace Principle" in Table 17.4).

4. *Maximin Principle*: This principle is based on the "Extremely Pessimistic View" of the decision maker (i.e., that nature will do its worst under every alternative). Therefore, the decision maker will select the alternative that maximizes the value of the proceed (profit) among the minimum values of all alternatives (i.e., the decision maker will reduce his or her loss by selecting the alternative with the lowest loss [or selecting the alternative with the highest profit among the minimum values]). The profit (R_i) in the ith alternative can be defined as

$$R_i = \max_i \{ \min_j E_{ij} \}$$

Table 17.4 shows that under this principle, the decision maker will select alternative A_5, which has the highest value ($2,000,000,000) among the lowest possible values of the evaluation measure among all the alternatives (see the column labeled "Maximin Principle" in Table 17.4).

5. *Maxmax Principle*: This principle is based on the "Extremely Optimistic View" (think about the best possible) of the decision maker. The decision maker will select the alternative that maximizes the maximum values in each alternative, that is, taking the maximum of the maximum values in each alternative. The profit (R_i) in the ith alternative can be defined as

$$R_i = \max_i \{ \max_j E_{ij} \}$$

Table 17.4 shows that under this principle, the decision maker will select alternative A_5, which has the highest value (\$5,500,000,000) among the highest possible values of the evaluation measure among all the alternatives (see the column labeled "Maxmax Principle" in Table 17.4).

6. *Hurwicz Principle*: This principle is based on a compromise between optimism (Maxmax principle) and pessimism (Maxmin principle). The profit (R_i) in the ith alternative is computed based on the selection of value of index of optimism (α) as follows:

$$R_i = \alpha \, [\max_i \, (\max_j \, E_{ij})] + (1 - \alpha) \, [\max_i \, (\min_j \, E_{ij})]$$

where α = index of optimism, which can vary as follows: $0 \le \alpha \le 1$
Note: $\alpha = 1$ indicates that the decision maker is extremely optimistic; $\alpha = 0$ indicates that the decision maker is extremely pessimistic.

The value of R_i should be computed for each alternative using this formula, and the alternative with the maximum value of R_i should be selected.

The last column of Table 17.4 illustrates that for $\alpha = 0.5$, alternative A_5 will be selected, because it has the highest value (\$3,750,000,000) in the last column when the values are computed using this expression for R_i.

Thus, observing the alternatives selected in each of the principles described, alternative A_5 is the best alternative, as it was selected under all the principles.

In real situations, it is possible that different alternatives may be selected using different principles. In that case, the decision maker will need to use all the results to guide his or her decision process and will make the final decision based on a few key principles that he or she believes (maybe from gut feel) to be the most appropriate for the situation.

DATA GATHERING FOR DECISION MAKING

Decision makers need information to help decide on all the values of the basic parameters (e.g., the variables covered in the previous section, such as number of alternatives, possible outcomes, probabilities of the outcomes, and costs or benefits associated with each alternative in each outcome) of the problem. Without the availability of reliable information, the decisions made by the decision maker might not be very useful or could even be very misleading. Further, care must be taken in selecting the decision maker to ensure that he or she is not biased and does not have any misconceptions or preconceived notions related to the product concepts, technologies considered in the concepts, customer expectations, and so forth.

Many techniques are available to gather information and display the information in a format that will help the decision maker to understand the problem. An overview of the tools is provided by Bhise (2014).

IMPORTANT OF TIMELY DECISIONS

IMPORTANCE OF TIMELY DECISIONS

In real decision-making situations, the estimates of numbers such as those shown in Table 17.3 will vary with time. Thus, it is possible that different alternatives will turn out to be the best at different points in time. However, since vehicle development is a long and complex process, once the decision about the basic product configuration is made, for example, to make the base vehicle a front-wheel-drive vehicle, all other subsequent decisions will be based on this early decision. If this early decision is changed (e.g., to make it a rear-wheel-drive vehicle) at a later point in the program, a lot of design work may need modifications. Modifications to the vehicle program are time consuming and costly. The decision maker thus needs to be very careful and to consider many possible scenarios (alternatives) and outcomes in making early decisions to avoid costly changes in the later phases of the vehicle program.

ROBUSTNESS EVALUATION THROUGH SENSITIVITY ANALYSIS

Another important consideration in decision making is to ensure that the selected decision is robust; that is, it will be relatively insensitive to changes in the input assumptions (e.g., the assumed values in Table 17.2). A good decision maker should conduct sensitivity analyses by changing the values of different inputs (e.g., by ± 10% or 20%) under different possible combinations and redoing the computations to determine whether the selected decision is robust (i.e., it will not change). A Monte Carlo simulation approach (whereby simulated iterations are defined by random values of parameters from their input distributions) can be used here to determine the percentage of simulations in which the selected decision does not change.

MULTI-ATTRIBUTE DECISION MODELS

Many decision situations include consideration of many (or multiple) attributes and their levels. Thus, in selecting an alternative, combinations of all attributes and their levels must be considered. Three techniques—the Pugh diagram, weighted attributes ratings, and analytical hierarchy method—allow consideration of multiple attributes. These three techniques are described in this section.

PUGH DIAGRAM

The Pugh diagram is a simple but effective tool for understanding how many attributes can be used to study the comparison between different products (or product concepts) to select the best product. The tool can also help in improving the selected product in additional iterations of comparisons. (Note that the Pugh diagram was introduced earlier in Table 4.4 and Chapter 16.)

The Pugh diagram is a tabular formatted tool consisting of a matrix of product attributes (or characteristics) and alternate product concepts along with a benchmark (reference) product called the *datum*. The diagram helps to undertake a structured concept selection process and is generally applied by a multidisciplinary team to converge on a superior product concept. The process involves creation of the matrix using inputs from all the team members. The rows of the matrix consist of product attributes based on customer needs, and the columns represent different alternate product concepts.

The evaluations of each product concept on each attribute are made with respect to the datum. The process uses classification metrics of "same as the datum" (S), "better than the datum" (+), or "worse than the datum" (−). The scores for each product concept are obtained by simply adding the number of plus and minus signs in each column. The product concept with the highest total score ("sum of plus" signs minus "sum of minus" signs) is considered to be the preferred product concept. Several iterations are employed to improve the preferred product concept by combining the best features of highly ranked concepts for each attribute till an acceptable concept emerges and becomes the new benchmark.

Table 17.5 presents a Pugh diagram created to develop a new concept (target vehicle) for a 2020 model of the Jeep Cherokee (mid-size sports utility vehicle [SUV]) by comparison with three 2015 model year SUVs. The table shows how each vehicle compares with the 2015 Jeep Cherokee vehicle (used as the datum) for each vehicle attribute. (Note that +, S, and − symbols, respectively, indicate that the vehicle in the column is better, the same as, or worse than the datum). The sum of number of attributes receiving a + sign minus the sum of attributes receiving a − sign provides

TABLE 17.5
Pugh Diagram for Evaluating a New Mid-Size SUV Concept

Vehicle Attribute	2020 Jeep Cherokee (Target)	2015 Jeep Cherokee (Trailhawk) (Datum)	2015 Ford Escape (4WD Titanium)	2015 Honda CRV (Touring AWD)
Durability	+		S	+
Off-road capability	+		−	−
Fuel economy	+		+	+
Noise, vibrations and harshness	S		+	+
Handling and dynamics	S		S	+
Towing capacity	S		S	−
Ride comfort (on road)	−		S	S
Ease of maintenance and modification	+		S	S
Cost	S		−	−
Weight	+		+	+
Safety	S		S	+
Aesthetics	+		S	S
Aero/thermal	S		+	+
Sum of +	6		3	7
Sum of −	1		2	3
Sum of S	6		8	3
Total score	5		1	4

a total score, which is a measure of improvement in each vehicle over the datum vehicle. Thus, the 2020 Jeep Cherokee received a total score of 5, which is higher than the corresponding scores of the other two benchmarked vehicles.

It should be noted that this Pugh diagram was created with the assumption that all vehicle attributes are equally important to the customer. Hence, scores of sum of plus signs and sum of minus signs were simply obtained by adding the number of attributes in which the vehicle corresponding to a column was better or worse than the datum. However, in reality, some attributes may be more important to customers than other attributes. Thus, the analysis can be modified to take into account different importance weights for each attribute. The weighted Pugh analysis will be covered in the next subsection.

WEIGHTED PUGH ANALYSIS

There are many different methods for including importance weights for each attribute. Table 17.6 shows a modified Pugh diagram for this problem. Here, an additional column for importance rating is included in the Pugh diagram. The importance weighting is based on a 10-point scale in which a rating of 10 is assigned as most

TABLE 17.6
Weighted Pugh Diagram

Vehicle Attribute	Importance Rating	2020 Jeep Cherokee (Target)	2015 Jeep Cherokee (Trailhawk) (Datum)	2015 Ford Escape (4WD Titanium)	2015 Honda CRV (Touring AWD)
Durability	10	1		0	1
Off-road capability	7	1		−1	−1
Fuel economy	8	1		1	1
Noise, vibrations and harshness	8	0		1	1
Handling and dynamics	8	0		0	1
Towing capacity	3	0		0	−1
Ride comfort (on road)	6	−1		0	0
Ease of maintenance and modification	6	1		0	0
Cost	8	0		−1	−1
Weight	5	1		1	1
Safety	9	0		0	1
Aesthetics	7	1		0	0
Aero/thermal	4	0		1	1
Weighted sum		37		10	34

important and a rating of 1 is assigned as least important. The +, S, and − signs in each column (except the datum) are replaced by 1, 0, and −1 weights, respectively. The final score, called the *weighted sum*, is computed for each column by adding the sums of importance rating multiplied by the 1, 0, or −1 score in the column for each vehicle.

A comparison of weighted sum values for the three vehicles shows that the 2020 Jeep Cherokee received the highest score of 37, and the 2015 Ford Escape received the lowest score of 10. The design team will study the numbers and find further opportunities to improve the weighted score for the new vehicle.

WEIGHTED TOTAL SCORE FOR CONCEPT SELECTION

During the product development process, the decision makers (e.g., usually top management) are faced with the decision to select a concept and proceed with its detailed design and engineering work. The selection is complicated, because the product concepts need to be evaluated by considering many attributes of the product. The attributes are generally developed from the customer needs, obtained from extensive interactions with the customers (e.g., by conducting market surveys or from customer feedback). The customers can also be asked to provide importance ratings (or weights) for each of the attributes. The weights can also be developed by using the analytical hierarchy process, covered in the next section. The customers (or the design team members) can also be asked to rate each product concept on each attribute. All this information can then be used to determine the total weighted score of each product concept. The product concepts can be compared based on the total weighted score, and the concept with the highest total score can be selected.

The computation of the total weighted score (T_j) of the jth product concept is described by the following mathematical expression:

$$T_j = \sum_{i=1}^{n} w_i R_{ij}$$

where:

T_j = total weighted score of jth product concept considering all the n attributes
Note: $i = 1, 2, ..., n$
w_i = weight of ith product attribute
R_{ij} = ratings of concept C_j on ith attribute

Table 17.7 provides an example of this weighting scheme. Each product concept is defined as C_j, where $j = 1$ to 4, and each product attribute is defined as A_i, where $i = 1$ to 5. The ratings (R_{ij}) are provided using a 10-point scale, where $1 =$ poor and $10 =$ excellent. The attribute weights (w_i) were obtained by using a 5-point scale, where $1 =$ not important and $5 =$ very important. The total weight score $(T_1 = 119)$ of concept C_1 was the highest, and concept C_3 $(T_3 = 92)$ scored the lowest. Thus, the product concept C_1 can be selected for further development, or the ratings data can be used to come up with further modifications of the product concepts. New ratings can be obtained after the modifications to iterate the above procedure till an acceptable product concept is achieved.

TABLE 17.7

Illustration of Total Weighted Score of Product Concepts Based on Attribute Weights and Ratings of Product Concepts by Attributes

Attribute	Attribute weight (w_i)	Product Concepts			
		C_1	C_2	C_3	C_4
A_1	5	10	8	5	7
A_2	3	5	8	9	4
A_3	5	7	4	5	7
A_4	1	5	8	3	6
A_5	2	7	9	6	8
Total weighted score		T_1	T_2	T_3	T_4
		119	110	92	104

This method is used in quality function deployment (QFD) (to compute absolute importance scores of functional specifications), and it can be considered as a modified scoring method for the Pugh diagram. The QFD tool is described in Chapter 18.

The following section describes the procedure for application of a technique called the *analytical hierarchy method*. The technique is used to extract the judgments of experts in decision-making situations.

ANALYTICAL HIERARCHY METHOD

The analytical hierarchy method (also called the *analytical hierarchy process* [AHP]) is a simple technique to determine the relative preference (or relative importance) of different alternatives. It is based on subjective judgments made by one or more decision makers. Each decision maker is assumed to be an expert in the problem area and is free from any biases. The method is described by Satty (1980) and Bhise (2012).

The decision maker's task is further simplified by paired comparisons of the alternatives. For example, if there are n possible alternatives, then there will be $n(n-1)/2$ possible pairs of alternatives. The decision maker is given (or shown) each pair separately and is asked to select the better (or preferred, or more important) of the two alternatives and assign a relative importance rating (weight) to the selected alternative based on a preselected criterion (or a product attribute). The preference ratings are then used to compute the relative weights of each of the alternatives. The alternative with the highest weight is selected as the most preferable alternative.

In this method, the products (or alternatives) are compared in pairs. The better product in each pair is also rated in terms of the strength of the attribute (used for evaluation) it possesses in relation to the strength of the same attribute in the other product in the pair. The strength of the attribute is expressed using a ratio scale. The scale (or the weight) value of 1 is used to denote equal strength of the attribute in both the products in the pair, and the scale value of 9 is used to indicate extreme or absolute strength of the attribute in the better product. The product with the weaker strength is assigned the inverse of the scale value of the better product. The following example will illustrate this rating procedure.

Let us assume that there are two products, M and L, in a pair, and the attribute to compare the products is "exterior styling." The scale values assigned to the products using the ratio scale would be as follows:

1. If product M is considered to be "absolutely better styled" as compared with product L, then the weight of M preferred over L will be 9, and the weight of L preferred over M will be 1/9.
2. If product M is "very strongly better styled" as compared with product L, then the weight of M preferred over L will be 7, and the weight of L preferred over M will be 1/7.
3. If product M is "strongly better styled" as compared with product L, then the weight of M preferred over L will be 5, and the weight of L preferred over M will be 1/5.
4. If product M is "moderately better styled" as compared with product L, then the weight of M preferred over L will be 3, and the weight of L preferred over M will be 1/3.
5. If product M is "equally styled" as compared with product L, then the weight of M preferred over L will be 1, and the weight of L preferred over M will also be 1.

Satty (1980) described the nine-point scale with the following adjectives to indicate the level of preference (or importance) for comparing the two items in each pair:

1 = Equal preference
2 = Weak preference
3 = Moderate preference
4 = Moderate plus preference
5 = Strong preference
6 = Strong plus preference
7 = Very strong or demonstrated preference
8 = Very, very strong preference
9 = Extreme or absolute preference

From the viewpoint of making the scales more understandable and easier to apply, usually only the odd-numbered scale values (shown in bold case in the list) are described and presented to the subjects. To allow the subjects to decide on the weight, the author found that the scale presented in Figure 17.2 works very well. Here, the subject will be asked to put an "X" mark on the scale on the left side if product M is preferable over product L. The higher numbers on the scale indicate higher preference. If both products are equally preferred, then the subject will be asked to place the "X" mark at the mid-point of the scale with value equal to 1. If product L is preferred over product M, then the subject will use the right side of the scale.

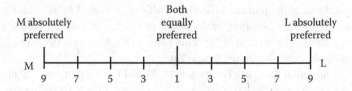

FIGURE 17.2   Scale used to indicate strength of the preference when comparing two products (M and L).

TABLE 17.8

Matrix of Paired Comparison Responses for One Evaluator

	M	P	W	J	L	T
M	1	1/3	3/1	1/9	1/3	1/1
P	3/1	1	3/1	1/5	1/3	2/1
W	1/3	1/3	1	1/7	1/3	2/1
J	9/1	5/1	7/1	1	3/1	7/1
L	3/1	3/1	3/1	1/3	1	3/1
T	1/1	1/2	1/2	1/7	1/3	1

Note: The value in a cell (of Table 17.8) indicates the preference ratio for comparing the product in a row with the product in a column corresponding to the cell.

Let us assume that we have to compare six products, M, P, W, J, L, and T, using the analytical hierarchy method. A subject will be asked to compare the products in pairs. The 15 possible pairs of the six products will be presented to the subject, one pair at a time, in a random order. (Note: for $n = 6$, $n(n-1)/2 = 15$.) The subject will be given a preselected attribute (e.g., exterior styling) and asked to provide the strength of preference ratings for the preferred product in each of the 15 pairs using scales such as the one presented in Figure 17.2. The data obtained from the 15 pairs will then be converted into a matrix of paired comparison responses, as shown in Table 17.8. Each cell of the matrix indicates the ratio of preference weight of the product in the row over the product in the column. Thus, the ratio 1/3 in the first row and second column indicates that the product in the column (P) was "moderately preferred" (i.e., considered to have moderately better exterior styling = rating weight of 3) over the product in the row (M).

To compute the relative weights of preference of the products, the fractional values in Table 17.8 are first converted into decimal numbers, as shown in the left-side matrix in Table 17.9. All the six values in each row are then multiplied together and entered in the column labeled "Row Product" in Table 17.9. The geometric mean of each row product is computed. It should be noted that the geometric mean of the product of n numbers is the $(1/n)$th root of the product (e.g., the 1/6th root of 0.03703 is

TABLE 17.9
Computation of Normalized Weights of the Products

	M	P	W	J	L	T	Row product	Geometric mean	Normalized weight
M	1.0000	0.3333	3.0000	0.1111	0.3333	1.0000	0.03703	0.57734	0.06698
P	3.0000	1.0000	3.0000	0.2000	0.3333	2.0000	1.19988	1.03084	0.11960
W	0.3333	0.3333	1.0000	0.1429	0.3333	2.0000	0.01058	0.46853	0.05436
J	9.0000	5.0000	7.0000	1.0000	3.0000	7.0000	6615.00000	4.33266	0.50267
L	3.0000	3.0000	3.0000	0.3333	1.0000	3.0000	26.99730	1.73202	0.20095
T	1.0000	0.5000	0.5000	0.1429	0.3333	1.0000	0.01190	0.47784	0.05544
						Sum ----->		8.619231538	1.00000

0.57734. Note: $0.57734^6 = 0.03703$). All the six geometric means in the column labeled "Geometric Mean" are then summed. The sum, as shown in Table 17.9, is 8.6192. Each of the geometric means is then divided by their sum (8.6192) to obtain the normalized weight of the products (see last column of Table 17.9). It should be noted that due to the normalization, the sum of the normalized weights over all the products is 1.0.

The normalized weights (also called the *normalized preference values*) are plotted in Figure 17.3. The figure, thus, shows that the most preferred product (based on the attribute "exterior styling") was J (with its normalized weight of 0.50267), and the least preferred product was W (with its normalized weight of 0.05436).

This example was based on data obtained from one subject. If more subjects are available, then the normalized weights for each subject can be obtained by using the same procedure, and the average weight of each product can be obtained by averaging over the normalized weights of all the subjects for each product.

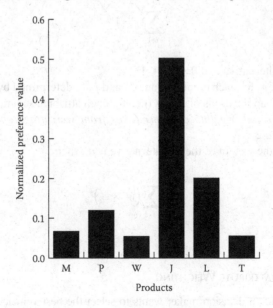

FIGURE 17.3   Normalized preference values of the six products.

AHP APPLICATION FOR MULTI-ATTRIBUTE DECISION MAKING

AHP can also be applied when multiple attributes are used to select an alternative. The problem of selecting an alternative is solved in three steps:

1. Obtain weights of each of the attributes using AHP
2. Obtain weights of each of the alternatives for each of the attributes using AHP
3. Obtain weights for alternatives from the weights obtained in the above two steps.

The procedure is described in this section. The alternatives and attributes are first defined as follows:

Alternatives $\{A_i\} = A_1, A_2, \ldots \ldots, A_n$. Thus, i ranges from 1 to n.
Attributes $\{T_j\} = T_1, T_2, \ldots, T_m$. Thus, j ranges from 1 to m.
Step 1: Weighting of attributes
In this step, weights of each attribute (T_j) are determined using AHP. Assume that $w_1, w_2, \ldots, w_m$ are weights of attributes, where $0 \leq w_j \leq 1$ and

$$\sum_{j=1}^{m} w_j = 1$$

Step 2: Weighting of alternatives based on each attribute separately
Assume that v_{ij} is the weight of the ith alternative on the jth attribute, such that

$$\sum_{i=1}^{n} v_{ij} = 1$$

for each ith alternative and $0 \leq v_{ij} \leq 1$.
The values of v_{ij} for each combination of i and j are determined by conducting m separate applications of AHPs (i.e., for each attribute separately).
Step 3: Determine weights of alternatives from weightings obtained from Steps 1 and 2
Assume that the weight of the ith alternative is a_i, then

$$a_i = \sum_{j=1}^{m} \left(w_j \times v_{ij} \right)$$

Example: Multiattribute Weighting

Let us assume that a decision maker wants to select the best vehicle among a set of four vehicles produced by different manufacturer in a market segment. The vehicles are identified as H, T, E, and R. Five attributes to be considered in the selection are (a) quality (overall quality by considering vehicle exterior and interior surface characteristics and features), (b) price (manufacturer's suggested retail price of the vehicle), (c) styling (exterior styling and appearance of the vehicle), (d) comfort (seating, ride, and interior climate-related comfort), and (e) service (service experience provided by the dealer).

The AHP method was used in the evaluation process using the following scale:

9 = Extremely or absolutely important (or preferred)
7 = Very strongly important (or preferred)
5 = Strongly important (or preferred)
3 = Moderately important (or preferred)
1 = Equally important (or preferred)

Step 1: These five attributes were compared to obtain their importance weights by asking a decision maker to make paired comparisons between the five attributes. The resulting matrix is shown in Table 17.10. Table 17.11 presents the calculations and weightings of the five attributes.

Step 2: The four vehicles are first rated for each attribute separately. Table 17.12 presents the ratings matrix (on the left side) and the calculations for obtaining normalized weights (on the right side) for each vehicle for the five attributes consecutively.

Step 3: The final normalized weights for each product are computed by summing the multiplications of weight of each vehicle for each attribute (computed in Step 2)

TABLE 17.10
Matrix of Paired Comparison Ratings of the Five Attributes

	Quality	Price	Styling	Comfort	Service
Quality	1/1	1/1	5/1	5/1	1/1
Price	1/1	1/1	2/1	2/1	5/1
Styling	1/5	1/2	1/1	3/1	1/1
Comfort	1/5	1/2	1/3	1/1	2/1
Service	1/1	1/5	1/1	1/2	1/1

TABLE 17.11
Weightings of the Five Attributes

	Quality	Price	Styling	Comfort	Service	Row product	Geometric mean	Normalized weight
Quality	1.000	1.000	5.000	5.000	1.000	25.0000	1.9037	0.3326
Price	1.000	1.000	2.000	2.000	5.000	20.0000	1.8206	0.3181
Styling	0.200	0.500	1.000	3.000	1.000	0.3000	0.7860	0.1373
Comfort	0.200	0.500	0.333	1.000	2.000	0.0666	0.5817	0.1016
Service	1.000	0.200	1.000	0.500	1.000	0.1000	0.6310	0.1103

Sum-->   5.7229   1.000

TABLE 17.12
Step 2 Calculations to Obtain Weights for Vehicles for Each Attribute

1 Quality

Importance of criterion in row over criterion in column

	H	T	E	R	Row product	Geometric mean	Normalized weight
H	1.000	1.000	5.000	3.000	15.0000	1.9680	0.3937
T	1.000	1.000	5.000	3.000	15.0000	1.9680	0.3937
E	0.200	0.200	1.000	0.500	0.0200	0.3761	0.0752
R	0.333	0.333	2.000	1.000	0.2222	0.6866	0.1374
				Sum -->		4.9986	1.0000

2 Price

Importance of criterion in row over criterion in column

	H	T	E	R	Row product	Geometric mean	Normalized weight
H	1.000	1.000	5.000	3.000	15.0000	1.9680	0.3937
T	1.000	1.000	5.000	3.000	15.0000	1.9680	0.3937
E	0.200	0.200	1.000	0.500	0.0200	0.3761	0.0752
R	0.333	0.333	2.000	1.000	0.2222	0.6866	0.1374
				Sum -->		4.9986	1.0000

3 Styling

Importance of criterion in row over criterion in column

	H	T	E	R	Row product	Geometric mean	Normalized weight
H	1.000	2.000	1.000	1.000	2.0000	1.1892	0.2995
T	0.500	1.000	2.000	3.000	3.0000	1.3161	0.3314
E	1.000	0.200	1.000	0.500	0.1000	0.5623	0.1416
R	1.000	0.333	2.000	1.000	0.6667	0.9036	0.2275
				Sum -->		3.9712	1.0000

4 Comfort

Importance of criterion in row over criterion in column

	H	T	E	R	Row product	Geometric mean	Normalized weight
H	1.000	1.000	0.500	1.000	0.5000	0.8409	0.1988
T	1.000	1.000	0.500	2.000	1.0000	1.0000	0.2364
E	2.000	2.000	1.000	2.000	8.0000	1.6818	0.3976
R	1.000	0.500	0.500	1.000	0.2500	0.7071	0.1672
				Sum -->		4.2298	1.0000

5 Service

Importance of criterion in row over criterion in column

	H	T	E	R	Row product	Geometric mean	Normalized weight
H	1.000	1.000	5.000	8.000	40.0000	2.5149	0.4660
T	1.000	1.000	5.000	3.000	15.0000	1.9680	0.3647
E	0.200	0.200	1.000	0.500	0.0200	0.3761	0.0697
R	0.125	0.333	2.000	1.000	0.0833	0.5373	0.0996
				Sum -->		5.3962	1.0000

TABLE 17.13
Calculations of Final Weights

		$j=1$	$j=2$	$j=3$	$j=4$	$j=5$	
Attributes -->		Quality	Price	Styling	Comfort	Service	Final
Normalized Weights of Attributes-->		0.3326	0.3181	0.1373	0.1016	0.1103	Weights
Normalized weights of vehicles by	H	0.3937	0.3937	0.2995	0.1988	0.4660	0.3689
attributes	T	0.3937	0.3937	0.3314	0.2364	0.3647	0.3660
	E	0.0752	0.0752	0.1416	0.3976	0.0697	0.1165
	R	0.1374	0.1374	0.2275	0.1672	0.0996	0.1486
Sum-->		1.0000	1.0000	1.0000	1.0000	1.0000	1.0000

by the normalized weight of the attribute (computed in Step 1). Table 17.13 presents the final weights of products obtained from the calculations in Step 3. For example, the final weight of vehicle H was computed as follows:

$$(0.3937 \times 0.3326) + (0.3937 \times 0.3181) + (0.2995 \times 0.1373) + (0.1988 \times 0.1016)$$

$$+ (0.4660 \times 0.1103) = 0.3689$$

Figure 17.4 presents a bar chart showing the final weights of the four products. It shows that vehicles H and T are rated higher than vehicles E and R. Thus, considering the five attributes, the final weights show that vehicle H is the best and vehicle T is very close to it. Vehicle E was the worst vehicle.

INFORMATIONAL NEEDS IN DECISION MAKING

The key to making good decisions is to ensure that the decision makers have sufficient information and good understanding of issues related to the alternatives, outcomes, trade-offs, and payoffs associated with the decision situation. Therefore, it is important to select decision makers carefully and to make sure that the individuals are familiar with the products and their uses.

In most product evaluation situations, customers who have used similar products (i.e., similar to those that they currently own) are asked to provide their ratings on each product (or alternative) used in the evaluation. On the other hand, the use of experts who are very familiar with the product and have extensive knowledge about the product can be very discriminating (much more so than even the most familiar customers) and can provide unbiased evaluations.

Experts generally have additional information obtained through other methods, such as (a) benchmarking other products, (b) literature surveys, (c) exercising available analytical models (e.g., models to predict the performance of products under different situations) and using the information obtained from the model outputs/

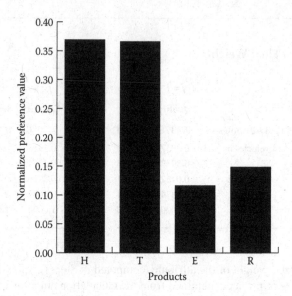

FIGURE 17.4   Final weights of the four products.

results, and (c) conducting experiments (see Chapters 16 and 21 on more methods and issues in the product evaluation area).

Exercising available models (e.g., product performance evaluation models and computer-aided engineering [CAE] methods) under various "what if" scenarios (i.e., conducting sensitivity analyses) can also provide more insights into the variability (or robustness) in the performance of the products and thus can prepare the decision maker to make more informed decisions. Design reviews with different groups and experts representing different disciplines can also generate information on strengths and weaknesses of the product (or product concept) being reviewed.

RISKS IN PRODUCT DEVELOPMENT AND PRODUCT USES

Product programs involve many risks. All important decisions in business and life involve some level of risk. A risk is considered to be present when an undesired event (which generally incurs substantial loss) is probable (i.e., likely to occur with some level of probability). Risks are possible anytime during or after the product development process. If the decision maker takes too little risk by over-designing (e.g., using too high a safety factor), the product will be more costly, and the extra cost most likely will be wasted. On the other hand, if the decision maker takes too much risk by under-designing (e.g., the product will have insufficient strength or use cheap low-quality materials), then the program will be too costly due to high costs resulting from product failures and/or rejections by the customers. Product failures can cause accidents, which can incur additional costs due to the occurrence of (a) injuries, (b) property damage, (c) loss of income, (d) interruptions or delays in work situations, (e) product litigation, and so forth.

DEFINITION OF RISK AND TYPES OF RISKS IN PRODUCT DEVELOPMENT

A risk is generally associated with the occurrence of an undesired event, such as a financial loss and/or an injury resulting from a product-related failure. The risk can be measured in terms of the magnitude of the consequence due to the occurrence of an undesired event. The consequence due to a risk can be measured by costs associated with customer dissatisfaction, loss due to product defects or resulting accidents, loss due to interruption of work, loss of revenue, loss of reputation, and so forth.

The risk is generally assessed by consideration of the following variables:

1. Probability of occurrence of the undesired event
2. Consequence (or severity) of the undesired event (e.g., amount of loss or severity of injuries)
3. Probability of detection of the undesired event before or when it occurs
4. Preparedness of the risk-fighting unit (e.g., fire department, emergency response units, and police) that can attempt to contain the severity of the loss or injury

The risks during the product development process can be categorized as follows:

1. *Technical Risk*: This type of risk occurs due to one or more technical problems with the design of the product. For example, a design flaw may be discovered during testing of an early production component. Such a problem may prevent the product from achieving the required technical capability or performance. To eliminate the technical problem or the flaw, additional analyses, engineering changes (with or without changes in technology to be implemented), and additional testing may be needed. These additional tasks usually result in an increase in the costs and delays in the schedule. Adoption of new technologies before adequate developmental work often leads to serious delays (e.g., problems in developing carbon-fiber components, manufacturing problems with lightweight materials to improve fuel consumption in automotive products).
2. *Cost Risk*: This risk is associated with cost overruns due to technical problems, changes in management decisions, changes in supplier capabilities, delays due to unforeseen situations, and so forth. The risk also may be due to under-budgeting caused by assuming optimistic estimates or under-estimation of required tasks, time, and costs (e.g., not providing sufficient allowance for rework).
3. *Schedule Risk*: This risk is related to not being able to meet the schedule due to delays for a number of possible reasons (e.g., parts not delivered on time by suppliers, late changes made in the design due to failures uncovered in testing, or planned schedule too optimistic).
4. *Programmatic Risk*: This risk is associated with the product development program (e.g., being over budget, delayed, modified, or even cancelled due to a number of reasons). Since most complex products have many components

that are made and supplied by various suppliers, selection of suppliers with unproven or low technical capabilities often leads to program delays, lower quality, and cost overruns.

These four categories of risks are generally interrelated; that is, a risk in any one of the categories also affects the associated risks in other categories. The risks also cause backward cascading effects from problems in the work completed in the early phases but discovered in the later phases. These problems affect progress in the succeeding phases due to factors such as redesign, rework, retests, delays, and cost overruns.

TYPES OF RISKS DURING PRODUCT USE

The risks after the product has been introduced into the market and used by end-users can be categorized as

1. *Loss of User Confidence in the Product*: The end-users may be afraid to use a product because of a defect in the product. The defect may be caused by a design or manufacturing defect or some "hidden danger" that can cause an undesired event (e.g., sudden loss of control, a fire or explosion, an accident, or exposure to toxic substances).
2. *Loss in Future Sales*: The likelihood of an undesired event can cause loss in the reputation of the producer and thus can affect future sales.
3. *Excessive Repair or Recall Costs*: The producer will need to fix the product problems by repairing under warranty or by initiating a product recall.
4. *Product Litigation Costs*: The costs of defending the product in product liability cases and costs related to settlements before the court trials or payments of penalties, fines, and so forth.

RISK ANALYSIS

A risk analysis can be defined as a decision-making exercise conducted to determine the next course of action after a potential undesired event has been identified and the magnitude of the consequence of the undesired event has been estimated. The risk analysis is usually preceded by risk identification and risk assessment phases. The phase of identification of the undesired event can be termed the *risk identification* phase, and the phase of estimation of the magnitude of the consequence due to the undesired event can be termed the *risk assessment* phase.

Some commonly used methods for risk identification, risk assessment, and risk analysis are listed here.

1. *Risk Identification Methods*: Brainstorming, interviewing experts, hazard analysis, failure modes and effects analysis (FMEA; see Chapter 18), checklists, and historic data (e.g., past records of product defects, warranty problems, customer complaints) (see Bhise, 2014).

2. *Risk Assessment Methods*: Estimation of probability (or frequency) of occurrence of undesired events, magnitude of the consequence (or severity) of the undesired event, and probability of detection of the undesired event by brainstorming, interviewing experts, safety analysis (e.g., fault tree analysis [see Bhise, 2014] and FMEA), and historic data (e.g., costs of past product failures).
3. *Risk Analysis Methods*: Risk matrix, risk priority number (RPN), nomographs, existing design and performance standards, and specialized risk models (Floyd et al., 2006).

RISK MATRIX

The risk matrix involves simply creating a matrix with combinations of relevant variables associated with the degree of risk due to the undesired outcomes. A risk matrix is a simple graphical tool. It provides a process for combining (a) the probability of occurrence of an undesired event (usually an estimate) and (b) the consequence if the undesired event occurred (usually cost estimates in dollars).

The risk (in dollars) can be computed as

$$\text{Risk}(\$) = [\text{Probability of occurrence}] \times [\text{Consequence of the undesired event}(\$)]$$

A simplified form of this relationship between the probability of occurrence and the magnitude of the consequence can be presented in a matrix format. Figure 17.5 presents an example of a risk matrix. The cells of the matrix represent different risk levels, increasing from low risk in the lower left corner of the matrix to high risk in the top right corner of the matrix. The risk matrix thus allows for a quick assessment of the risk level after the occurrence probability and the magnitude of the consequence due to an undesired event have been estimated.

RISK PRIORITY NUMBER

RPN is another method used to assess the level of risk. It is based on multiplication of three ratings: (a) severity, (b) occurrence, and (c) detection. This method is used in FMEA, which is presented in Chapter 18. Examples of rating scales used for severity, occurrence, and detection are presented in Tables 18.3 through 18.5, respectively. Different definitions of the rating scales are generally used by different companies, industries, and government agencies.

Nomographs are also used as an alternate method for estimating the RPN. An example of a nomograph is provided by Bhise (2014).

Other methods, such as modeling and simulations, are used to facilitate decision making. Exercising models under different assumptions (conducting sensitivity analysis) can provide a good understanding of the underlying variables and their effects on risks and subsequent decisions. Analyses under a range of possibilities with different levels of optimistic and pessimistic assumptions are also useful to

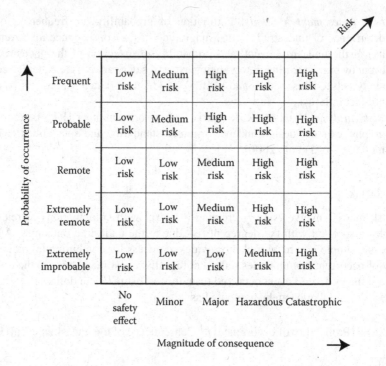

FIGURE 17.5 An example of a risk matrix. (*Source*: Redrawn from Federal Highway Administration, Systems Engineering Guidebook for ITS, 2015. www.fhwa.dot.gov/cadiv/segb/views/document/sections/section3/3_9_4.cfm.)

estimate the limits of risks. Historic data and judgments of experts can also play a major role in the decision making.

PROBLEMS IN RISK MEASUREMENTS

Assessing the risks to users/customers involves identifying the hazards and assessing the potential consequences and the occurrence probability of such consequences. Identification of hazards is particularly difficult when both the potential customers and product uses (i.e., how different users will use a given product under a variety of situations) are difficult to predict. Products involving new technologies are also difficult to evaluate, because failure data are generally very scarce. It is especially hard to predict the risks during the early stages of product development when the product concept is also not fully developed.

Problems in risk measurements occur due to many reasons. Most problems occur due to (a) lack of data on different types of hazards and risks, (b) subjectivity involved in identification and quantification of the data, and (c) differences in assumptions made during the design phases about how the customers or users will use the product versus the actual uses of the product. The risk assessment models used in this area are therefore not precise. But they can be used as guides, along with the recommendations of multiple experts and discussions between the decision makers and the experts.

Subjective assessments of the three component areas (occurrence, severity, and detection) are also difficult and subject to a number of questions, such as: Who would collect the data and conduct evaluations? Should the evaluations be conducted by experts, product safety advisory boards, or teams or individuals involved in the design process? Further, the level of understanding and awareness of risks varies considerably between different evaluators. Costs are another problem in collecting failure-related data, as product tests are generally costly, and funds for undertaking costly data collection studies are usually limited.

There are trade-offs in the application of risk assessment methods between the consistency in the data, the level of detail related to the outcomes, and the time and resources (particularly human and financial) required for the analyses. Apparently simple methodologies may contain implicit weightings that may not be appropriate for every product being assessed. Judgments may be intuitive, based on implicit assumptions, especially in relation to the boundaries between categories (or ratings). Taken together, these factors can result in a high degree of subjectivity in risk assessment, although the subjectivity can be reduced by the extent of guidance provided to the assessors in applying the various scales and ratings. In general, the potential for inconsistency in the results will be directly related to the level of subjectivity involved in the risk measurement process.

IMPORTANCE OF EARLY DECISIONS DURING PRODUCT DEVELOPMENT

"Designing right the first time" is very important, as reworking any product design in the later phases is always very time consuming and costly. Early in the product development, key decisions are generally made on what technologies to use and how the product should be configured. Any changes to these early assumptions in the later stages of product development can increase costs substantially, because such changes may require throwing away much of the early design work (and even some hardware development work) and redoing all the analyses with a different set of assumptions and requirements.

The involvement of specialists from all key technical areas is a very important aspect of the systems engineering process, as it ensures that all possible technologies and design configurations are considered as possible alternatives before converging on one or a few alternatives. The subsequent decisions are dependent on the selected technologies and design configurations.

CONCLUDING COMMENTS

This chapter covered several basic models and issues in decision making. Decision making in the real world involves consideration of many issues (both internal and external to the organization), many variables and their effects, and likelihoods of outcomes and associated costs that cannot be well quantified for reasons such as missing facts, uncertainties over the readiness of new technologies, unknown future developments, and the global economy. Many complex models (involving varied levels of complexity and using many independent variables) can be created to analyze the effects of many risk-related variables. The models can be exercised under different sets of assumptions (conducting sensitivity analysis) to get a good understanding of the effects of decisions on product performance and associated risks. However, a good decision maker will also inject some subjectivity based on his/her intuition or judgment to make the final decisions. The decisions are never final but can be revisited once new and more reliable information is available. However, later changes in the decisions made earlier in the program generally result in cost increases due to resulting rework.

REFERENCES

Bhise, V. D. 2012. *Ergonomics in the Automotive Design Process*. Boca Raton, FL: CRC Press.

Bhise, V. D. 2014. *Designing Complex Products with Systems Engineering Processes and Techniques*. Boca Raton, FL: CRC Press.

Blanchard, B. S. and W. J. Fabrycky. 2011. *Systems Engineering and Analysis*. 5th edn. Upper Saddle River, NJ: Prentice Hall PTR.

Federal Highway Administration. 2015. *Systems Engineering Guidebook for ITS*. V3.0.Website: www.fhwa.dot.gov/cadiv/segb/views/document/sections/section3/3_9_4.cfm (Accessed September 7, 2015).

Floyd, P., T. A. Nwaogu, R. Salado, and C. George. 2006. Establishing a Comparative Inventory of Approaches and Methods Used by Enforcement Authorities for the Assessment of the Safety of Consumer Products Covered by Directive 2001/95/EC on General Product Safety and Identification of Best Practices. Final Report dated February 2006 prepared for DG SANCO, European Commission by Risk & Policy Analysts Limited, Farthing Green House, 1 Beccles Road, Loddon, Norfolk, NR14 6LT, UK.

Satty, T. L.1980. *The Analytic Hierarchy Process*. New York, NY: McGraw Hill.

18 Product Planning Tools

INTRODUCTION

The function of product planning department is to plan the development of the "right" product. The characteristics of the "right" product must be identified and described in terms of vehicle parameters and features to be included in the new vehicle. The information should be communicated and presented to the senior company management to obtain their approval to undertake more detailed work in planning the vehicle program and developing the vehicle concept.

The steps involved in vehicle product planning include (a) attending auto shows, nonautomotive product shows (e.g., other consumer product shows), technology shows, exhibitions, and conferences to search for new product ideas and applications of new technologies and concepts, (b) keeping abreast of changes in the government regulations that would apply to future vehicle programs, (c) benchmarking recent vehicles and vehicle concepts introduced by leading competitors all around the globe, (d) studying available information on future product plans of your company as well as those of your leading competitors, (e) studying market trends, and (f) developing product plans for future vehicle products. It should be noted that all specialized engineering offices (e.g., body engineering, powertrain engineering, and electrical and electronics engineering) also continuously search for trends in design and technologies, new features, new materials, and new manufacturing processes. The information obtained is discussed in advanced product planning and technologies meetings with the heads of the engineering, design, and marketing offices.

The primary tools involved in the early stages of product planning include (1) benchmarking and breakthrough, (2) Pugh diagram, (3) timing charts and gateways, (4) quality function deployment (QFD), (5) failure modes and effects analysis, and (6) other product development tools such as business plan and computer-aided design (CAD).

Benchmarking is used to compare currently available competitors' products with the manufacturer's current product concepts to understand the "gaps" between the product concepts and the benchmarked products. The knowledge gained can be used to incorporate some of the best ideas learned from the competitors. Breakthrough methodology enables large improvements in product designs to be achieved by thinking beyond the present product design and production capabilities. The Pugh diagram is used to select and improve a product concept by using a number of product attributes to compare various alternate product concepts with a reference called the *datum*. Timing charts and gateways allow us to plan and communicate schedules for various tasks to be accomplished by all people involved in the vehicle program.

QFD is used to translate the customer needs into engineering functional specifications and to determine the engineering product specifications that are critical to customer satisfaction. The failure modes and effects analysis method is used to improve the reliability and safety of the product by tabulating its failure modes and their effects on other entities within the product and prioritizing the uncovered product problems by the application of a risk priority number assessment method. Business plan and CAD are used in communication of the program and product information within the design, engineering, and program management activities.

The following sections of this chapter describe and provide examples of these tools.

BENCHMARKING AND BREAKTHROUGH

Benchmarking and breakthrough methods are generally used during the very early stages of a product development program. From the information gathered during the benchmarking exercises, the product designers can realize the "gaps" between the characteristics and capabilities of the products of their competitors and their own new product concepts, whereas the breakthrough approach forces the design teams to look beyond the existing products and technologies and thus think about developing a totally new product or features and achieve huge improvements over the existing product designs.

BENCHMARKING

Benchmarking is a process of measuring products, services, or practices against the toughest competitors or those companies recognized as the industry leaders. Thus, it is a search for the industry's best products or practices that can lead to superior performance.

A multifunctional team within the product development community is usually selected to perform the product benchmarking activities. The benchmarking exercise typically starts with identifying the toughest competitors (e.g., very successful and recognized brands as the industry leaders) and their products (models) that serve similar customer needs to the manufacturer's proposed product. The selected competitor products are used to compare against the target product. The target product is the product considered by the manufacturer to be its future product (or an existing model of the future product). The team gathers all important competitive products and information about the products and compare the competitors' products with their target product through a set of evaluations (e.g., measurements of product characteristics; tearing down the products into their lower-level entities for close examination; and evaluations by experts for the study of performance, capabilities, unique features, materials and manufacturing processes used by the competitors, estimates of costs to produce the benchmarked products, and so forth.

The information gathered from the comparative evaluations is usually very detailed. However, the depth of evaluations included in benchmarking can vary between problem applications and companies. For example, the benchmarking of an automotive disc brake may involve comparisons based on part dimensions, weights, materials used, surface characteristics, strength characteristics, heat dissipation characteristics, processes needed for its production, estimated production costs, features that would be "liked very much" by the customers, features that may be "hated" by the customers, features that would create a "Wow" reaction among the team members and potential customers, special performance tests such as part temperatures during severe braking torque applications, brake "squealing" sound, and so on. In addition, digital pictures and videos can be taken to help visualize differences in the benchmarked products and their components.

The gathered information is generally summarized in a tabular format, with product characteristics listed as rows and different benchmarked products represented in columns. Chapter 4 provides more information on benchmarking. Tables 4.1 to 4.4 present examples of benchmarking of vehicle characteristics. Figure 4.1 presents an example of photo-benchmarking of vehicles. Similar benchmarking exercises are also conducted at system, subsystem, and component level by various specialized engineering activities.

BREAKTHROUGH

The breakthrough approach involves throwing away all the existing product designs (and processes) and brainstorming to develop a totally new design, to obtain huge potential gains in terms of performance improvements, costs, and added customer satisfaction. Breakthrough designs typically require radically new thought dimensions and lead to the adoption of new technologies. Thus, the implementation of a breakthrough design creates new problems in systems integration and management. Some examples of breakthrough product designs and comparison of benchmarking versus breakthrough are provided in Chapter 4.

PUGH DIAGRAM

The Pugh diagram is a tabular formatted tool consisting of a matrix of product attributes (or characteristics) and alternate product concepts along with a benchmark (reference) product called the *datum*. The diagram helps to conduct a structured concept selection process and is generally created by a multidisciplinary team to converge on a superior product concept. The process involves creation of the matrix by inputs from all the team members. The rows of the matrix consist of product attributes based on customer needs, and the columns represent different alternate product concepts.

The evaluations of each product concept for each attribute are made with respect to the *datum*. The process uses classification metrics of "same as the datum" (S), "better than the datum" (+), or "worse than the datum" (−). The scores for each product concept are obtained by simply adding the number of plus and minus signs in each column. The product concept with the highest total score ("sum of pluses" minus "sum of minuses") is considered to be the preferred product concept. Several iterations are employed to improve product superiority by combining the best features of highly ranked concepts till a superior concept emerges and becomes the new benchmark.

AN EXAMPLE OF PUGH DIAGRAM APPLICATION

An automotive powertrain engineer wanted to determine whether the performance of a transient turbo-charged gasoline engine could be improved over the gasoline turbo direct injection (GTDI) methodology by employing the following three concepts: (a) Concept #1: an electric turbo-boost (e-Turbo), (b) Concept #2: a hybrid turbo using an electric motor assist in parallel with the turbo operated by the exhaust gases, or (c) Concept #3: use of electrical compressor only (Black, 2011). The engineer created a Pugh diagram to compare these three technologies with the GDTI as the *datum*. Table 18.1 presents the Pugh diagram. The product attributes used to compare the three technologies are presented in the second column from the left. The four right-hand columns represent product concepts #1, #2, #3, and the datum as the last column.

All the three product concepts improve the "performance" attribute (Attribute #4) as compared with the datum by eliminating the turbo-lag (a transient condition during quick accelerations). This is shown by the + signs in all the three product concepts (columns) of the row corresponding to Attribute #4. However, they introduce additional negatives into the system due to additional cost (Attribute #2), Weight (Attribute #10), noise (Attribute #11), electrical and electronics (Attribute #9), product process compatibility (Attribute #13), and life-cycle durability (Attribute #1). The

TABLE 18.1
Pugh Diagram for Product Concept Selection

Attribute No.	Customer-based Product Attribute	Product Concept #1	Product Concept #2	Product Concept #3	Datum
1	Life-cycle durability	–	–	–	
2	Cost	–	–	–	
3	Package and ergonomics	–	+	+	
4	Performance	+	+	+	
5	Fuel economy	S	+	S	
6	Safety/security	–	–	–	
7	Vehicle dynamics	+	+	+	
8	Emissions	+	+	+	
9	Electrical and electronics	–	–	–	
10	Weight	–	–	–	
11	Noise, vibrations, and harshness	–	–	–	
12	Styling and appearance	S	S	S	
13	Product process complexity	–	–	–	
	Sum of pluses	3	5	4	
	Sum of minuses	8	7	7	
	Total score	–5	–2	–3	

bottom row of total scores ("sum of pluses" minus "sum of minuses") shows that none of the three product concepts was better than the datum, since the total scores of all were negative. Concept #2 (Hybrid Turbo) is the least negative based on the net score. Life-cycle durability, cost, safety/security, electrical and electronics, weight, and noise, vibrations, and harshness (NVH) are all additional problem issues with Concept #2 compared with the datum (traditional turbo [GTDI]). In this analysis, all the product attributes were considered to have equal weight; that is, the number of plus and minus signs was simply added to obtain the total score.

The section entitled "Weighted Total Score for Concept Selection" in Chapter 17 illustrated a method of using different weights for the attributes. This method was also used here, as follows. Table 18.2 presents the problem using importance weighting for each of the product attributes. The importance of each product attribute was rated using a 10-point scale, where 10 = most important and 1 = least important (see

TABLE 18.2
Pugh Analysis with Ratings

Customer-based Product Attribute	Importance Rating	Importance Weight	Preference Ratings Using −5 to +5 Scale Compared with the Datum			
			Product Concept #1	Product Concept #2	Product Concept #3	Datum
Customer life-cycle durability	5	0.06	−3	−3	−3	
Cost	10	0.11	−3	−5	−5	
Package and ergonomics	3	0.03	−3	3	3	
Performance	10	0.11	5	5	5	
Fuel economy	10	0.11	0	5	0	
Safety/security	10	0.11	−3	−3	−3	
Vehicle dynamics	8	0.09	5	5	5	
Emissions	3	0.03	5	5	5	
Electrical and electronics	5	0.06	−3	−5	−5	
Weight	5	0.06	−3	−5	−5	
Noise, vibrations, and harshness	9	0.10	−3	−5	−5	
Styling and appearance	1	0.01	0	0	0	
Product process complexity	10	0.11	−1	−3	−4	
Sum or weighted sum	89	1.00	−0.52	−0.63	−1.30	

column with heading "Importance Rating" in Table 18.2). The importance scores were converted to "Importance Weight" (by dividing the importance rating of each attribute by the sum of the all the importance ratings). The importance weights are shown in the column to the right side of the importance rating column. Each product concept was evaluated with respect to the datum (the current product GTDI system) for each attribute by using another 10-point scale ranging from -5 to $+5$. Here, a $+5$ score indicates that a given product concept is very much better than the datum, and a -5 score indicates that the product concept is very much inferior to the datum. The sum of the weighted scores of each product concept was obtained by summing the multiplied values of importance weight and product rating over the entire set of product attributes. The weighted sums of the three product concepts are -0.52, -0.63, and -1.30 (see last line of Table 13.3). Concept #1 had the largest value of the weight sum (-0.53). Thus, Concept #1 (e-turbo) emerged as the winner among the three concepts. However, it is still worse than the datum. If fuel economy becomes more important in the future, then this concept has the potential to be implemented. The adoption of a 42 V electrical system in the future could aid in the implementation of the concept.

The benefit of the hybrid turbo is that it enables a completely independent intake and exhaust, permitting many modes of operation, including additional fuel savings. This also helps eliminate some of the air intake system routing and packaging issues. The penalties in the trade-offs are due to (a) higher electrical load to drive the electric motor–driven compressor, (b) poor reliability and durability of the electrically driven compressor, (c) added complexity due to extra parts, and (d) additional costs of extra hardware.

Additional examples of Pugh diagrams are provided in Chapters 17, 23, 24, and 25.

TIMING CHARTS AND GATEWAYS

Timing charts and gateways (milestones) are probably the most important product planning tools after the decision on the type of vehicle to be developed is made. The timing chart provides dates and durations of various phases of the vehicle program. The timings are developed after inputs (on time required to perform various tasks) from all the major product design and engineering activities have been incorporated into the concurrent engineering plan to ensure that many of the design and engineering activities can be performed in parallel to reduce the overall program duration. Figure 1.2 shows a Gantt chart of a vehicle program, and Figure 2.4 provides a time chart showing various gateways (called *milestones* or *decision points*. See Table 2.1 for definitions of gateways). The timings of all major program phases and gateways are defined and reviewed with all major engineering activities to ensure that all the required tasks that need to be performed in each program phase can be accommodated. The timing charts are then used to plan manpower needs and budgets for all activities of the vehicle program. Chapter 12 provides a more detailed description of various program management activities and tools used to manage the vehicle program. The financial analyses used by program planning and management are described in Chapter 19.

QUALITY FUNCTION DEPLOYMENT

Quality function deployment (QFD) is a technique used to understand customer needs (voice of the customer) and transform the customer needs into engineering characteristics of a product (or a process) in terms of functional or design requirements. It relates "what" (what are the customer needs) to "how" (how should engineers design to meet the customer needs) and "how much" (magnitude of design variables, i.e., their target values), and also provides competitive benchmarking information—all in one diagram. QFD was originally developed by Dr Yoji Akao of Japan in 1966 (Akao, 1991). QFD is applied in a wide variety of applications and is considered a key tool in design for six sigma (DFSS) projects. It is also known as the "House of Quality," because the correlations matrix drawn on the top of the QFD matrix diagram resembles the roof of a house.

Figure 18.1 illustrates the basic structure (or regions) of the QFD. The contents of each region of the QFD are described in the following list. (Note: An example of a completed QFD chart is presented in the next section [Figure 18.3].)

1. *Customer Needs (What)*: The needs of the customers, as expressed by "what the customer wants in the product," are listed sequentially in the rows of this leftmost region. Each customer need should be described using the customer's words (as the customer would describe it in his/her own words, e.g., give me a vehicle that would last for a long time, vehicle that looks good, vehicle that works flawlessly, energy efficient, fun to drive). The list of customer needs should be sorted into different categories, and duplicated needs (even with different wordings) should be removed. The categorized needs should be organized in a hierarchical order, such as primary needs, secondary needs (within each primary need), and tertiary needs (within each secondary need). The ith customer need (tertiary) is defined as C_i. (Note: the mathematical definitions of all the QFD variables are described later in this section.)

2. *Importance Ratings*: This column provides importance ratings for each of the customer needs. The importance rating column is provided to the right of the customer needs column. The importance ratings (or weights) can be obtained using a number of different weighting techniques (e.g., rating scales, analytical hierarchical process). However, a 10-point rating scale is commonly used, where 10 = extremely important and 1 = not at all important. The importance rating for the ith customer need is defined as W_i.

3. *Functional Specifications (How)*: The functional specifications are created by the engineers involved in the product development to define how the product (or the entity for which the QFD is prepared) "should function" or "should be designed to meet its customer requirements." The functional specifications describe "How" the engineers will address the customer needs. Thus, these specifications should be described using technical terms and variables that are used and selected by the engineers, such as functions to be performed, types of mechanisms, materials, dimensions, strengths

(capabilities), manufacturing processes, and test requirements. Here, the engineer should list each functional (or engineering) specification in a separate column. The functional specification columns are provided to the right of the importance ratings column. The jth functional specification is defined as F_j.

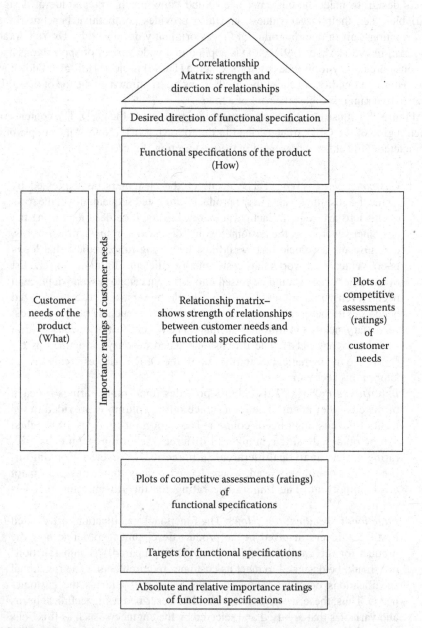

FIGURE 18.1   Structure of a QFD diagram showing its regions and contents.

The functional specifications should cover engineering considerations, methods, or variables that need to be considered during the product development. Some examples of the variables used for the functional specifications are engineering requirements that must be met (e.g., maximum force in a specified direction and number of cycles of force applications), type of construction (e.g., welded vs. assembled using fasteners, material to be considered to make the entity [e.g., steel, high-strength steel, aluminum, carbon fiber]), type of production process to be used (e.g., extrusion vs. cast), locations (e.g., installation locations expected by the customers for operation and service), physical space (e.g., envelope size, volume) needed for the specified entity, product characteristics (e.g., maximum achievable acceleration), capacity or capabilities of the entity, durability (e.g., works without a failure for 100,000 cycles under specified conditions), and so forth.

4. *Relationship Matrix*: The relationship matrix is formed by customer needs as its rows and functional specifications as its columns. Each cell of the matrix represents the strength of relationship between the customer need and the functional specification defining the cell. Weights of 9, 3, or 1 are commonly used to define a strong, medium, or weak relationship, respectively. The following coded symbols are used to illustrate the strengths of relationships: two concentric circles (for 9 = strong), one open circle (for 3 = medium), and a triangle (for 1 = weak). A cell is left blank when no relationship exists between the customer need and the functional specification defining the cell. The relationship in a cell of the relationship matrix is defined as R_{ij} (i.e., relationship between ith customer need and jth functional specification).

5. *Desired Direction of Functional Specification*: This row (placed above the functional specifications row) shows an up-arrow, a down-arrow, or a 0 (zero) to indicate the desired direction of the value of each engineering specification (defined in the column). The up-arrow indicates that a higher value is desirable. The down-arrow indicates that a lower value is desirable. Zero indicates that the functional specification is not dependent on either an increase or a decrease in its value. Thus, a quick visual scan of this row gives graphic information on whether the desired values of the functional specifications included in different columns need to be larger or smaller or are not dependent on their values.

6. *Correlation Matrix (Roof)*: The correlation matrix is formed by the relationships between combinations of any two engineering specifications defined by the cells of the matrix. The direction and strength of the relationship are indicated in the cell by a positive or a negative number (defined as I_{jk}) in the cell. Coded symbols are also used to indicate the direction and strength of the relationships. Only half of the matrix above the diagonal is shown as the roof of the QFD chart (see the roof in Figure 18.1).

7. *Absolute Importance Ratings of Functional Specifications*: The absolute importance rating of each functional specification (defined as A_j) is computed by summing weighted relationships between customer needs and

the functional specification. The weighting of relationships is based on the importance rating (W_i) of each customer need and the strength of the relationships (R_{ij}s). The expression for computation of values of A_j is given later in this section. The absolute importance ratings are presented in a row just above the last row of the QFD chart.

8. *Relative Importance Ratings of Functional Specifications*: The relative importance rating (defined as V_j) of each functional specification is expressed as a percentage of the ratio of its contribution (A_j) to the sum of all A_js. The expression for computation of values of V_j is given later in this section. The relative importance ratings are presented in the last row of the QFD chart.

9. *Competitive Assessment of Customer Needs*: Each product used in benchmarking (along with the manufacturer's target product) is rated to determine how well each customer need (C_i) is satisfied by the product. The rating is usually given by the customers (or by the product development team members once they are very familiar with the customer needs and the products) by using a five-point scale, where 1 = poor (the product poorly meets the customer need) and 5 = excellent (The product meets the customer need at a high excellence level). The ratings of each product are plotted on the right side of the relationship matrix.

10. *Competitive Assessment of Functional Specifications*: Each product used in benchmarking (along with the manufacturer's target product) is rated to determine how well each functional specification is satisfied by the product. The rating is usually provided by the technical experts or the product development team members (once they are very familiar with each functional specification and the products) by using a five-point scale, where 1 = poor (the product poorly meets the functional specification) and 5 = excellent (The product meets the functional specification at a high excellence level). The ratings of each product are plotted on the bottom side just below the relationship matrix.

11. *Targets for Functional Specification*: The target values for each of the functional specifications (provided in the columns of the QFD) are provided below the competitive assessment plot of the functional specifications. The target values are determined by the team after extensive discussions on all the collected data. The target should be specified precisely. Examples of targets are (a) specification of material to be used, for example, aluminum, (b) class level to be achieved (e.g., best-in-class, among the leaders, slightly above the average, average, or below average product among all the current products in its class), (c) minimum value to be achieved; for example, minimum engine torque output should be 300 ft-lb at 3000 RPM, or (d) minimum rating level to be achieved; for example, ratings value greater than or equal to 4 on a five-point scale, where 5 = excellent and 1 = poor.

The following mathematical definitions will clarify these variables and their relationships:

$C_i = i$th customer need, where $i = 1, 2, ..., m$

$F_j = j$th functional specification, where $j = 1, 2, ..., n$

W_i = Importance rating of ith customer need, where $i = 1, 2, ..., m$
(The value of importance rating can range from 1 to 10, where 10 = extremely important and 1 = not at all important)

R_{ij} = Relationship between ith customer need and jth functional specification
(The values assigned to R_{ij} will be 9, 3, or 1 to define a strong, medium, and weak relationship, respectively. The ijth cell in the relationship matrix will be left unfilled (i.e., blank) if there is no relationship between C_i and F_j)

I_{jk} = Relationship between jth functional specification and kth functional specification, where j and $k = 1, 2, ..., n$ and $k \neq j$ (thus, it is the interrelationship between two functional specifications shown in the roof of the QFD diagram)

The values of I_{jk} can be as follows:

+9 = Strong positive relationship between jth functional specification and kth functional specification

+3 = Positive relationship between jth functional specification and kth functional specification

0 = No relationship between jth functional specification and kth functional specification

−3 = Negative relationship between jth functional specification and kth functional specification

−9 = Strong negative relationship between jth functional specification and kth functional specification

A_j = Absolute importance rating of jth functional specification

$$= \sum_i \left[W_i \times R_{ij} \right]$$

V_j = Relative importance rating of jth functional specification (%)

$$= 100 \times A_j / \left[\sum_j A_j \right]$$

An Example of the QFD Chart

A driver's side front door trim panel had to be designed for a new mid-sized four-door sedan. The door trim panel covers the inner side (occupant side) of the steel doors and includes items such as inside door opening handle, door pull handle, armrest, door mounted switches (e.g., mirror, window, and door lock switches), courtesy lights, speakers, and map pocket storage (which typically stores bottles and an umbrella) (see Figure 18.2). The appearance of the door trim panel is important, as it

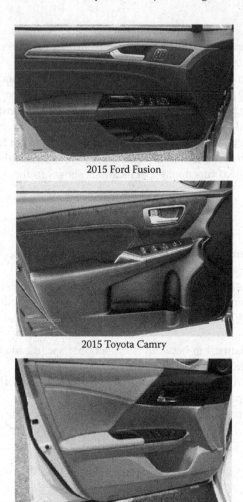

2015 Ford Fusion

2015 Toyota Camry

2015 Honda Accord

FIGURE 18.2   Door trim panels of three mid-sized vehicles.

should match in color and materials with the instrument panel and other interior trim parts and seats. The manufacturer formed a team of interior designers, package engineers, body engineers, electrical engineers, ergonomics engineers, market researchers, and engineers from the suppliers of the door trim panel, switches, speakers, and window raising and lowering mechanism to create a QFD chart for the front door trim panel.

The team interviewed a number of customers (i.e., the owners of their current vehicle model and two of their best competitors) and asked them about their needs and expectations about the door trim panel of their future vehicle. The customers first started telling the team what they wanted: a good door trim panel. The team members kept asking a number of probing questions, such as What do you mean by

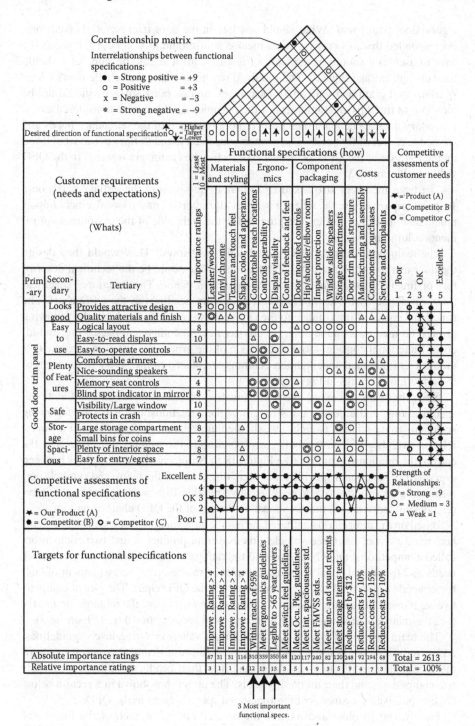

FIGURE 18.3   QFD chart for an automotive door trim panel.

a good door trim panel? What would you like in the door trim panel? The customers responded that a good trim panel means: (a) it should look good, (b) it should be easy to use, (c) it should have plenty of features, (d) it should be safe, (e) it should have enough storage capacity, and finally, (f) it should make the vehicle interior look spacious (not crammed with small clearances, and its outer surfaces should not be too close to the hips and shoulders of the occupants). These were considered as the secondary needs. The team then probed more into each of the secondary needs and created lists of tertiary needs. The primary, secondary, and tertiary needs were listed on the left side of the QFD as "whats" (i.e., what the customers wanted) in the QFD shown in Figure 18.3.

The team members also asked customers to rate each of the tertiary needs on a 10-point importance scale, where 1 = not at all important and 10 = extremely important. The importance ratings are provided to the right side of the customer requirements column in Figure 18.3.

The team brainstormed and created a list of "Hows:" How would they design the door trim panel? How should it function? What would be the technical descriptors or functional specifications of the door trim panel? The functional specifications developed by the team are listed as column headings in the QFD chart. The functional issues of importance to the engineers were categorized into the following groups: (a) materials and styling, (b) ergonomics, (c) component packaging, and (d) costs. Functional considerations under each of the groups are listed in separate columns. The functional specification columns are placed immediately to the right of the importance ratings in Figure 18.3.

Next, the team discussed every combination of customer needs and functional specifications, and assessed the strength of their relationships using the following scale: (a) strong relation (weight = 9), (b) medium relation (weight = 3), and (c) weak relation (weight = 1). The symbols corresponding to the weights were placed in the cells of the relationship matrix (see Figure 18.3). Similarly, the relationships between every pair of functional specifications were discussed by the team, and symbols corresponding to very positive to very negative relationships were placed in the interrelationship (correlation) matrix shown on the top of the QFD chart as its roof.

Based on the information gathered during the customer interviews, the team members rated each of the three vehicles (their current product A and two competitors called Competitor B and C), and plotted the ratings on each of the tertiary customer needs and functional specifications. The plot of the competitive assessment of customer needs is provided as the rightmost part of the QFD chart. The plot of competitive assessment of functional specifications is provided below the relationship matrix (i.e., the matrix of customer needs and functional specifications) (see Figure 18.3).

The team developed targets (i.e., what target values or rating levels, guidelines, and/or requirements to use for designing their future instrument panel) for each of the functional specifications by discussing how their product compared with their two competitors and their marketing goals. The targets are shown in a section below the competitive assessments of the functional specifications in the QFD.

Finally, the absolute and relative importance ratings of each of the functional specifications were computed and entered in the last two rows of the QFD chart. The three functional specifications that received the highest importance ratings are (1) display visibility (i.e., displays visible to the driver without any obscurations and legible labels and graphics), (2) control operability (i.e., how the controls are configured and their operating motions), and (3) comfortable reach locations (i.e., placement of the controls/items within the driver's hand reach zone). These three functional specifications must be given very high priority in designing the door trim panel.

CASCADING QFDs

The QFD technique can be cascaded in multiple steps to link the customer needs for the product (i.e., from its product- or vehicle-level customer needs) to its component-level production specifications. The cascading is illustrated in Figure 18.4. The figure shows a series of five QFD charts linked such that outputs of a preceding chart become inputs to the next or succeeding QFD chart. Here, the first QFD chart on the left translates the customer needs of the product (labeled A) into the functional specification of the product (labeled B). The second QFD chart takes the functional specifications as inputs (described in rows and labeled B) and translates them into systems specifications (shown in the columns, labeled C). (Note: the systems can be defined here as the vehicle systems that form the vehicle.) The third QFD translates the systems specifications (labeled C) into specifications of its components of the systems (labeled D). (Note: this corresponds to cascading system-level requirements to component-level requirements. See Chapter 9). The fourth QFD translates each component's specifications (labeled D) into its manufacturing process specifications (i.e., how the component should be produced using manufacturing processes and machines (labeled E). And the last (fifth) QFD translates the manufacturing process specifications (labeled E) to the component production specifications (i.e., characteristics of the component after it is produced), labeled F. Thus, the component production specifications can be traced back to the original customer needs. Such a series of QFD cascades ensures that components, when produced, will indeed function to meet the customer needs for the product.

ADVANTAGES AND DISADVANTAGES OF QFD

Developing a single QFD chart can be very time consuming, as it takes many hours of teamwork, involving meetings, discussions, customer visits, benchmarking of the competitive products, development of targets, and so on. The advantage is that it exposes the entire product design team to all aspects of the decisions to be taken during the product development process. The process of developing a QFD thus educates the team, documents the collected information, and prioritizes information needed during its development. Thus, when the team actually starts developing the product, the subsequent decisions generally take much less time, as the team members have already discussed all the issues and are very aware of most of the interfacing and trade-off considerations. A product developed using QFD, therefore, will have a better chance of being the right product and satisfying its customers.

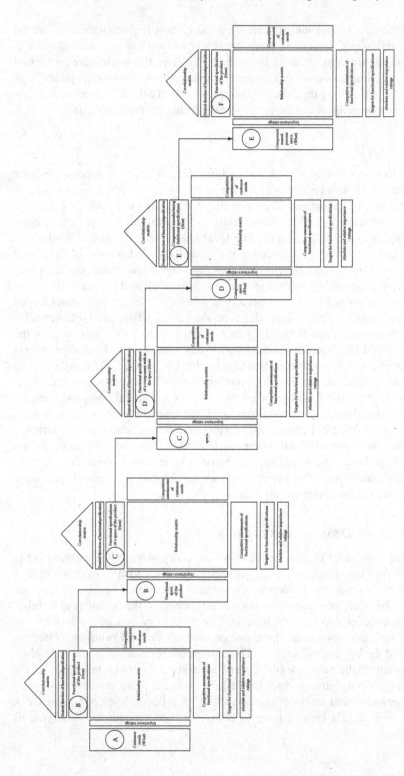

FIGURE 18.4 Example of cascading of customer needs of the product to component manufacturing process specifications using a series of five QFD analyses.

FAILURE MODES AND EFFECTS ANALYSIS

The failure modes and effects analysis (FMEA) method was used in the 1960s as a systems safety analysis tool. It was used in the early days of defense and aerospace systems design to ensure that the product (e.g., an aircraft, a spaceship, or a missile) was designed to minimize the probabilities of all major failures by brainstorming and evaluating all the possible failures that could occur and acting on the resulting prioritized list of corrective actions. For more than 20 years, the method has been routinely used by product design and process design engineers to reduce the risk of failures in the design of products and processes used in many industries (e.g., automotive, aviation, utilities, and construction). FMEA conducted by product design engineers is typically referred to as DFMEA (where D stands for design), and FMEA conducted by process designers is referred to as PFMEA (where P stands for process). In many automotive companies, product (or process) design release engineers are required to perform the task of creating the FMEA chart and to demonstrate that all possible failures with a risk priority number (RPN) above a certain value are prevented.

FMEA is a proactive and qualitative tool used by quality, safety, and product/ process engineers to improve reliability (i.e., to eliminate failures, thus improving quality and customer satisfaction). The development of a FMEA involves the following basic tasks:

1. Identify possible failure modes and failure mechanisms
2. Determine the effects or consequences that the failures may have on the product and/or process performance
3. Determine methods of detecting the identified failure modes
4. Determine possible means for prevention of the failures
5. Develop an action plan to reduce the risks due to the identified failures

FMEA is very effective when performed early in the product or process development and conducted by experienced multifunctional team members as a team exercise.

The method involves creating a table with each row representing a possible failure mode of a given product (or a process) and providing information about the failure mode using the following columns of the FMEA table:

1. Description of a system, subsystem, or component
2. Description of a potential failure mode of the system, subsystem, or component
3. Description of potential effect(s) of the failure on the product/system, its subsystems, components, or other systems
4. Potential causes of the failure
5. Severity rating of the effect due to the failure
6. Occurrence rating of the failure
7. Detection rating of the failure or its causes
8. RPN (the multiplication of the three ratings in Items 5, 6, and 7)
9. Recommended actions to eliminate or reduce the failures with higher RPNs
10. Responsibility of the persons or activities assigned to undertake the recommended actions and target completion date

11. Description of the actions taken
12. Resulting ratings (severity, occurrence, and detection) and RPNs (after the action is taken) of the identified failures in Item 2

Examples of rating scales used for severity, occurrence, and detection are presented in Tables 18.3 through 18.5, respectively. The definitions of the scales generally vary between different organizations depending on the type of industry, the product or process, the nature of the failures, associated risks to humans, and costs due to the failures.

AN EXAMPLE OF AN FMEA

An automatic transmission in an automobile will not operate properly if the transmission fluid leaks out. An engineer designing a transmission fluid hose conducted an FMEA to evaluate possible failures caused by the hose. The FMEA is presented in Table 18.6. The hose involved in this example consists of a nylon tube with

---

TABLE 18.3

Example of a Rating Scale for Severity

Rating	Effect	Criteria: Severity of Effect
10	Hazardous/ fatalities—without warning	Very high severity rating when potential failure mode affects safe product operation and/or involves noncompliance with government regulations without warning. Multiple fatalities possible.
9	Hazardous/ fatalities—with warning	Very high severity rating when potential failure mode affects safe product operation and/or involves noncompliance with government regulations with warning. Fatalities possible.
8	Very high/injurious	Product inoperable with loss of primary function. Severe injuries possible.
7	High/injurious	Product operable with reduced level of performance. Customer dissatisfied. Minor to moderate injuries possible.
6	Moderate	Product operable but usage with reduced level of comfort or convenience. Customer experiences discomfort or minor injuries.
5	Low/discomfort	Product operable but usage without comfort or convenience. Customer experiences discomfort.
4	Very low	Minor product defect (e.g., noise, vibrations, poor surface finish) only noticed by most customers.
3	Minor	Minor product defect only noticed by average customer.
2	Very minor	Minor product defect only noticed by discriminating customer.
1	None	No effect.

TABLE 18.4
Example of a Rating Scale for Occurrence

Rating	Probability of Failure	Possible Failure Rates
10	Very high: Failure is almost inevitable	≥ 1 in 2
9		1 in 4
8	High: Repeated failures	1 in 10
7		1 in 25
6	Moderate: Occasional failures	1 in 100
5		1 in 1000
4	Low: Relatively low failures	1 in 2000
3		1 in 10,000
2	Remote: Failure is unlikely	1 in 100,000
1		≤ 1 in 1000,000

TABLE 18.5
Example of a Rating Scale for Detection

Rating	Detection Level (Detection Rate)	Criteria: Likelihood of Detection by Design Control
10	Absolutely uncertain (1 in 1,000,000)	Design control cannot detect a potential cause or mechanism for the failure mode; or there is no design control.
9	Very remote (1 in 100,000)	Very remote chance that the design control will detect a potential cause/mechanism for the failure mode.
8	Remote (1 in 10,000)	Remote chance that the design control will detect a potential cause/mechanism for the failure mode.
7	Very low (1 in 1000)	Very low chance that the design control will detect a potential cause/mechanism for the failure mode.
6	Low (1 in 100)	Low chance that the design control will detect a potential cause/mechanism for the failure mode.
5	Moderate (1 in 50)	Moderate chance that the design control will detect a potential cause/mechanism for the failure mode.
4	Moderately high (1 in 25)	Moderately high chance that the design control will detect a potential cause/mechanism for the failure mode.
3	High (1 in 10)	High chance that the design control will detect a potential cause/mechanism for the failure mode.
2	Very high (1 in 5)	Very high chance that the design control will detect a potential cause/mechanism for the failure mode.
1	Almost certain (1 in 2)	Design control will almost certainly detect a potential cause/mechanism for the failure mode.

TABLE 18.6
Example of an FMEA

FAILURE MODES AND EFFECTS ANALYSIS

Part Number: 1008-92   Original Date: 1/03/2015

Description: Date Revised: 3/03/2016
Powertrain Fluid Hose
Assembly

Model Year: 2017   Prepared by: T. James

No.	Item/Function	Potential Failure Mode	Potential Effect(s) of Failure	Severity Rating (S)	Potential Causes of Failure	Occurrence Rating (O)	Current Design Controls	Detection Rating (D)	Risk Priority Number (RPN)	Recommended Action	Responsibility	Action Results				
												Action	S	O	D	RPN
1	Hose—allows transmission fluid to travel and exert pressure to point of use	Leak in the hose	A slow leak would not be noticed immediately. The driver would notice gear shifting delays over time and finally the vehicle will stop shifting gears.	8	Hole in the hose. Defective hose material. Hose material degrades.	3	100% leak test with air.	1	24	No action required.						
2		Hose burst at low pressure	A quick and fast leak with immediate loss of shifting function.	8	Defective hose material. Hose damaged by debris on road.	3	Material burst test.	7	168	Investigate if variations in the burst test results.	T. James 07/20/16	Stronger exterior hose casing. Reduced hose extrusion temperature for consistency.	6	1	7	42
3	Connectors—attach hose to pressurized oil reservoir	Leaks around connector	A slow leak would not be noticed immediately. The driver would notice gear shifting delays over time and finally the vehicle will stop shifting gears.	8	Loose connector or omitted washer or uneven connector mating surface.	3	100% leak test with air.	1	24	No action required.						

	Function	Failure mode	Effect	Sev	Cause	Occ	Current controls	Det	RPN	Recommended action	Responsibility	Action taken	Sev	Occ	Det	RPN
4	Conduit—protects hose during operation from moving parts and stones sprayed by tires	Connector disconnected from hose or reservoir	A quick and fast leak with immediate loss of shifting function.	8	Connector not fully seated in the assembly.	2	Pull out testing at final inspection. 100% visual checks during service.	4	64	Investigate if a sensor could detect if the connector was fully seated at the hose end.	T. Jack 07/20/16	Improved connector threads. Sensor not possible currently.	3	2	4	24
5		Insufficient conduit wear strength	Possible damage to tube in service.	6	Conduit material defective.	1	Design verified during sign-off.	7	42	No action required.						
6		Ferrule—holds the connector on to the hose	Connector disconnected from hose or reservoir	A quick and fast leak with immediate loss of shifting function.	8	Ferrule or hose not properly positioned in the crimping die or press failure.	2	Pull out testing at final inspection. 100% visual checks during service.	4	64	Investigate if parts presence sensor can be located on the die.	T. James 07/20/16	Sensor installed on the die and press stops if sensor is not activated.	8	1	2	16
7		Barcode label—lot and part traceability	Missing label	Loss of traceability and identification.	4	Label fell off, never affixed, not affixed properly.	4	Visual inspection of label.	2	32	No action required.						
8		Packing—protect the part while shipping	Damaged part	Returned by the customer.	4	Insufficient packing strength.	1	Packaging standard used for all parts shipping.	4	16	No action required.						

connectors inserted into each end. Ferrules are crimped onto each end to help hold the connectors on. A conduit made of plastic covers the hose to protect it from heat and moving parts near the engine. The hose carries transmission fluid from the reservoir to the clutch actuation system. The transmission fluid in the hose is pressurized during operation to about 6.5 bars (94.3 psi). The hose must also be able to withstand temperatures above 60 °C (140 °F). During the development phase, the design itself is proven in a series of tests referred to as the design verification plan and report (DVP&R). The DVP&R is conducted once for each design, so it does not take into account all the sources of variability that the product and materials are exposed to during the life of the part. There were three failures with RPN over 50 in the FMEA. The actions taken by the engineer reduced the RPNs of all the three failure modes (see Table 18.6).

FAILURE MODES AND EFFECTS AND CRITICALITY ANALYSIS

Failure modes and effects and criticality analysis (FMECA) is very similar in format and content to FMEA. It contains an additional column of criticality. The criticality column provides a rating illustrating the level of criticality of failure (in each row) in accomplishing the major goal (or mission) of the product. The technique is also called failure modes and criticality analysis (Hammer, 1980).

Criticality can be rated using different scales for different products. The criticality ratings typically cover a range from low criticality, involving stoppage of equipment (requiring minor maintenance), to high criticality levels, involving failures resulting in potential loss of life.

OTHER PRODUCT DEVELOPMENT TOOLS

During the product development process, tools from many areas, such as systems engineering, specialty engineering areas, and program and project management, are used to manage both the technical and the business activities of the program. The systems engineering and program management tools are covered in Part I of this book. Parts II of the book cover additional tools covered in decision making, product development, vehicle packaging, and financial analyses. Part III provides additional applications of many of these tools. Among other remaining tools, the following part of this section covers the business plan, program status chart, design standards, CAD tools, and other evaluation tools. These tools are covered in this chapter because they provide important information in management and technical decision-making activities during the product planning and development process.

BUSINESS PLAN

A business plan is a proposal for creating or developing a new product. It is usually prepared internally within a company to obtain concurrence from the top management to approve the product program. The business plan is thus a document prepared to describe the details of a proposed product, the product program timing plan, and corporate resource needs to develop the product. It is typically prepared jointly by the company's product planning, engineering, marketing, and finance activities.

The business plan should include

1. Description of the proposed product:
   - Product configuration (e.g., for an automotive product, the body-style of the vehicle, such as a sedan, a coupe, a crossover, a sports utility vehicle [SUV], a pickup, or a multi-passenger vehicle [MPV] and its variations)
   - Size class (e.g., subcompact, compact, intermediate, large)
   - Markets where the product will be sold and used (countries)
   - Market segment (e.g., luxury, entry-luxury, or economy)
   - Manufacturer's suggested retail price (MSRP) and price range with different models and optional equipment
   - Production capacity and estimated sales volumes over the product life cycle
   - Makes, models, and prices of leading competitors in the proposed market segment
2. Attribute rankings (i.e., how the product would be positioned in its market segment, such as best-in-class, above the class average, average in the class, or below the class average, by considering each product attribute)
3. Pugh diagram showing how the proposed product will compare with its current model (used as the datum) and other leading competitors' products by considering all important product attributes and changes in the proposed product
4. Dimensions and options:
   - Overall exterior dimensions (e.g., product package envelope with length, width, height, wheelbase, and cargo/storage volume)
   - Interior dimensions (e.g., people package with legroom, headroom, shoulder room, number of seating locations, luggage/cargo volume, and so forth)
   - Major changes in the product's systems as compared with the previous (outgoing) model, for example, drive options (front-wheel drive [FWD], rear-wheel drive [RWD], all-wheel drive [AWD]), powertrains (types, sizes, and capacities of engines/motors and transmissions), descriptions of unique features (e.g., type of suspension) and optional equipment
5. One-paragraph description of the proposed product with several adjectives to describe its image, stance, and styling characteristics (e.g., futuristic, traditional, retro, fast, dynamic, aerodynamic, tough, or chunky—like a Tonka truck)
6. Program schedule:
   - Program kick-off date, timings of major milestones
   - Job#1 date (i.e., date when first production unit will be out of the assembly plant) and model year
7. Projected sales volumes:
   - Quarterly or yearly sales estimates of each model in each market segment

8. Financial analysis:
- Curves of estimated cumulative costs, revenues, and cash flow during product life cycle for different scenarios (e.g., best case, average, and worst case)
- Anticipated quarterly funding needed during product development and during revenue buildup
- Anticipated date of breakeven point
- Return on investment

9. Product life cycle:
- Estimated lifespan
- Possible product refreshments, future models, and variations
- Recycling of the plant and products

10. Proposed plant location and plant investments
11. Potential sources of risks in undertaking the program
12. Justification (reasons) for approval of the proposed plan versus other alternatives

Additional information on the business plan is presented in Chapter 5.

PROGRAM STATUS CHART

Product planners and program managers keep track of progress on product development programs by using a number of different techniques, such as program timing charts and Gantt charts, presented in Chapter 2, and cash flows, covered in Chapter 19. One chart that is very popular is the program status chart, which is typically used to track the status of problems encountered during a program (or a project).

Status charts are also called *Red-Yellow-Green charts*, as they indicate the problem status by use of colors: (a) red indicates that the problem is not yet solved and is a "job stopper" (i.e., it will stop progress on the entire program till it is solved); (b) yellow indicates that the problem can introduce significant delays into the program unless it is solved quickly; and (c) green indicates that the problem is no longer a timing threat to the program.

Table 18.7 presents an illustration of a program status chart. The status column in the chart uses the letters R, Y, and G to indicate the red, yellow, and green colors when colored charts have to be prepared on a normal "black-ink only" printer. Such charts are typically used in senior management–level meetings to draw attention and get fast resolution on "job stopper" problems.

STANDARDS

Product design standards serve as very useful tools in reducing the time required to make design decisions. Properly developed design standards that incorporate the rationale and assumptions used in their development can provide basic knowledge on whether the standard can be applied during the design process. When the standard meets the needs of the customers for the product, then use of the design requirements

TABLE 18.7
Program Status Chart

Program Status Chart

Program: XM25			Program Manager: RJW			Date: 08/30/15	
Problem No.	Problem Description	Status*	Target Date	Expected Completion Date	System/ Subsystem/ Component	Product Attribute	Responsibility: Organization and Manager
1	Center stack screen freezes unexpectedly during navigation operation	R	3/2/2016	4/2/2016	Navigation— software	Driver communication	Driver interface—JLM
2	Gloss level on the top resurface of the instrument panel should be reduced to meet veiling glare standard	Y	9/15/2015	10/30/2015	Instrument panel—crash pad	Safety	Body and trim—JBM
3	Squealing noise during braking	Y	11/10/2015	12/30/2015	Braking system— brake pad material	Safety	Chassis—brakes— WLV
4	Premature wear in front bearings of turbo-boosters	R	11/15/2015	Unresolved	Powertrain— turbo boosters	Performance	Powertrain—EER
5	Noticeable jerk during transmission shifting	Y	11/25/2015	12/30/2015	Powertrain—X5 transmission	Performance	Transmission—JJT
6	Wiper flutter at speeds over 70 mph	G	7/15/2015	8/26/2015	Body—visibility	Safety	Body electrical—RGK

* R = Red: critical issue (job stopper— must be resolved before next program review meeting) Y = Yellow: important customer need—must be resolved by next milestone
G = Green: issue resolved—no action needed.

and design procedures provided in the standards can reduce the time required to get the necessary information and decide on how to design.

Some standards may only specify how the product should perform, and thus, they provide design flexibility (i.e., the designer can design using any appropriate solution as long as it meets the required performance). Such performance standards can promote innovative product designs, as they are not restricted by compliance to any given design configuration and its specifications. The advantages and disadvantages of design versus performance standards, along with other types of standards and issues related to the standards, are covered in Chapter 7.

CAD Tools

A number of CAD tools are used to create three-dimensional solid models using software such as AutoCAD, CATIA, Pro/Engineer, SolidWorks, and Rhino. These tools do not only perform the traditional engineering drawing and drafting work; they allow visualization of the product model from different eye points to evaluate issues such as (a) exterior and interior appearance (e.g., shapes, continuity/discontinuity between adjacent surfaces, tangents, reflections), (b) spaces (clearances) between different entities within the product, (c) postures of human occupants/operators (with digital human models) in the products (e.g., cars, airplanes, and boats) or workplaces, (d) feeling of interior spaciousness and storage spaces, layouts of hardware placements (mechanical packaging), (e) comparisons of alternate designs (by superimposition or side-by-side viewing of different product concepts and competitive products), (f) assembly analyses to evaluate assembly feasibility (e.g., by detecting interferences between parts being assembled), and (g) alternate assembly methods and fit (e.g., gaps) between parts. The newer CAD models can also simulate movements of parts within the product and movements of the product in its work environment to aid in visualizing how the product will look and fit within other existing systems.

CAD models are also very useful for communication between different design studios, product engineering offices, and supplier facilities. CAD files for the products can also be used as inputs to a number of other sophisticated computer-aided engineering (CAE) analyses to evaluate structural/mechanical (e.g., strength, dynamic forces, deflections, vibrations during simulated operating environments), aerodynamic (using computer-aided fluid dynamics), and thermal (temperature, heat buildup, and heat transfer) aspects of the products. CAD files can also be used to facilitate manufacturing operations. For example, CAD files serve as inputs to computer-aided process planning (CAPP) as well as for creating machining programs for computerized numerical control (CNC) machines.

CAD has become an especially important technology within the scope of computer-aided technologies, with benefits such as lower product development costs and a greatly shortened design cycle. CAD enables designers to create layouts and develop their work on a display screen, print it out, and save it for future editing— thus saving time in creating variations in designs and their drawings.

Additional information on CAD tools is also presented in Chapter 13.

PROTOTYPING AND SIMULATION

Virtual and physical prototyped parts can be created for visual evaluations and physical mock-ups for use in design reviews. A number of computer simulation systems are also available for human factors testing of user interfaces. Three-dimensional parametric solid modeling requires the design engineer to input values of key parameters— what can be referred to as the "design intent." The objects and features created can be shown to customers for their feedback and adjusted by creating many design iterations till an acceptable design is achieved. Further, any future modifications can be easily made by inputting parameter changes in a computer-controlled prototype. Many automotive manufacturers use computer-controlled adjustable vehicle models (or programmable vehicle bucks) during early concept phases to compare and evaluate a number of automotive designs by quick changes in many key vehicle package parameters (Richards and Bhise, 2004; Ford Motor Company, 2008; Prefix, 2012).

A number of specialized computer software systems are increasingly used to simulate product testing and evaluation. For example, CAD is used to create accurate photo simulations that are often required in the preparation of environmental impact reports, in which computer-aided designs of intended buildings, vehicles, and other products are superimposed into photographs or videos of existing environments to represent what that locale will be like when the proposed facilities are allowed to be built. Visible fields and potential obstructions along various sight lines and shadow studies are also frequently made through the use of CAD. Vehicle designers often use such simulations to compare exterior designs of various vehicle concepts with their competitor products in various simulated road environments.

PHYSICAL MOCK-UPS

Physical mock-ups of product concepts for products such as cars, trucks, airplanes, and boats are useful during design reviews to get a better feel of the size, space, and configuration of the product in its early phases. The mock-ups can also be shown to potential customers and users for their feedback during informal quick evaluations as well as structured market research clinics (see Chapters 11 and 13 for more information).

TECHNOLOGY ASSESSMENT TOOLS

Using new technologies to improve product designs has been a continuous process of achieving improvements in performance, efficiencies, safety, and costs. However, most new technologies cannot be immediately applied. It can take many years, or even decades, to solve the problems in bringing new technology applications to a state of readiness and implementation. Technical experts in various specialized areas generally follow advances in new technologies. Progress in the most promising technologies is closely followed, and research departments are asked to perform evaluations and undertake development projects to improve the technologies so that they can be quickly implemented in future products.

Many methods to assess technologies have been developed. Forgie and Evans (2011) have provided an excellent review of available techniques for technology assessments.

CONCLUDING REMARKS

The product planning process involves the integration of many ideas, product features, and technologies. It is important to use tools that can help in searching, developing, and evaluating ideas that can be implemented to develop balanced product concepts. Systems engineering, along with other specialized engineering and management disciplines, allows simultaneous consideration of many inputs from multidisciplinary teams. Developing the right product at the right time is important. The tools presented in this chapter and their applications can aid in selecting the "right" product concept and then refining it during the early stages of product development.

REFERENCES

Akao, Y. 1991. Development History of Quality Function Deployment. In: Mizuno, S., Akao, Y., and Ishihara, K. *The Customer Driven Approach to Quality Planning and Deployment.* Minato, Tokyo 107, Japan: Asian Productivity Organization. ISBN 92-833-1121-3.

Ford Motor Company. 2008. Ford Uses Leading-Edge Virtual Lab to Deliver Increased Comfort, Visibility and Quality. Website: http://ophelia.sdsu.edu:8080/ford/10-21-2008/about-ford/news-announcements/press-releases/press-releases-detail/pr-ford-uses-leadingedge-virtual-lab-29037.html (Accessed: May 27, 2016).

Forge, C. C. and G. W. Evans. 2011. Assessing Technology Maturity as an Indicator of Systems Development Risk. In: Kamarani, A. K. and Azimi, M. *Systems Engineering Tools and Methods.* Boca Raton, FL: CRC.

Hammer, W. 1980. *Product Safety Management and Engineering.* Englewood Cliffs, NJ: Prentice-Hall.

Prefix Corporation. 2012. Programmable Vehicle Model (PV). Website: www.prefix.com/PVM/ (Accessed: June 20, 2012).

Richards, A. and Bhise, V. 2004. Evaluation of the PVM Methodology to Evaluate Vehicle Interior Packages, SAE Technical Paper 2004-01-0370. SAE International, Warrendale, PA.

19 Financial Analysis in Automotive Programs

INTRODUCTION

One of the key objectives of any automotive company is to make money by selling its products over a long time. The net income (or profits) earned by a company can be determined using a simple formula of revenue minus costs (or expenses). The revenue that a company generates is typically from selling its products (i.e., product volume times selling price) plus any investment income from its accumulated cash. The costs of developing, producing, distributing, and maintaining the products are very important. And the goal of the automotive company is thus to minimize the total costs during the entire life cycle of its products—from the conception of the products to the retirement and disposal of production equipment and facilities. At the early stages, accurate estimates of the costs are required to develop a plan and budget for every product program and to get it approved. The actual costs should be continuously compared with the budgeted costs to ensure that the program is meeting its budgetary requirements. Differences between the budgeted costs and the actual costs may signal over– or under–expenditures or errors in estimating the budgeted costs.

The costs are incurred over time. The costs during the product development phases are primarily "nonrecurring;" that is, they do not recur but are one–time costs associated with the product concept development, product design, detailed engineering, testing, and building tools and facilities. Once production begins, the costs associated with purchasing raw materials, parts purchased from suppliers, plant running costs, direct labor costs, insurance costs, and so forth are "recurring" and are generally proportional to the volume of products manufactured. As products are sold and the revenues are generated, the need for additional funds to sustain the program (i.e., production) decreases.

The objective of this chapter is to understand different types of costs associated with the various tasks involved during the product life cycle. This chapter also presents how the financial analyses are performed by determining costs, revenues, and profits as functions of time and how the present values of different cash flows are considered to take account of interest (or discount factor) and inflation in the evaluation of different alternatives.

TYPES OF COSTS AND REVENUES IN VEHICLE PROGRAMS

The costs are estimated by breaking down a large product program into a series of manageable tasks. Based on the work content in each task, the availability of cost information from previously conducted similar tasks, and adjustments for the prevailing and future economic and technological conditions, experienced cost estimators usually develop time and cost estimates to complete the tasks. The costs of all tasks are then added, along with some allowances for errors, interest, inflation, and other unknown or unforeseen problems. The projected cost estimates are also refined several times during the program execution as some of the less predictable tasks and unknown issues (e.g., technology development and competitors' new products) are resolved or better understood.

NONRECURRING AND RECURRING COSTS

The costs are incurred throughout the life cycle of a product program. The total life–cycle costs of a product can be divided into (1) nonrecurring costs and (2) recurring costs. These costs are described in this section.

Nonrecurring Costs: These costs represent expenses and investments that are made during product development and creation of the production systems, and also to retire and dispose of the systems after the product is terminated. These costs are incurred before the beginning of production and at the end of production, that is, the retirement (disposal) stages in the life cycle of a product. The early costs incurred to reach operational status of the program include product design, development, and refinement costs. These costs include personnel costs (salaries and benefits) of the design team as well as the costs for the development of models and prototypes, market research, verification tests, tools and fixtures design and build, plant and facilities building, and equipment/tooling installation and prove–out. These nonrecurring costs do not vary as a function of the quantity of products produced. Thus, they are also referred to as the *fixed costs*.

Recurring costs: These costs continue to occur throughout the production, sales, and service/maintenance of the products. These costs include personnel costs of production and distribution (direct and indirect labor), parts and materials purchases, plant and equipment maintenance, utilities, insurance, marketing and sales costs, and warranty costs. The recurring costs vary as a function of the quantity of products produced. Thus, they are also referred to as the *variable costs*.

COSTS AND REVENUES IN PRODUCT LIFE CYCLE

As the products are sold, the generated revenues (positive values) are tracked and added to the total costs (negative values). The revenues are also affected by a number of factors, such as selling price (manufacturer's suggested retail price [MSRP] minus discounts and/or rebates), changes in product volumes due to increase in popularity or obsolescence of products over time, emergence of new trends in design and technologies, introduction and availability of new products from competitors, and changes in economic conditions (e.g., state of the economy, interest, inflation, and currency exchange rates).

Figure 19.1 shows two charts. The upper chart shows various costs (which have negative values, as they represent money spent or lost) as they are incurred as functions of time during various life–cycle stages of a typical product program for a manufactured product. The top chart in Figure 19.1 also shows the revenue. The revenue has positive value, as it represents income. The revenue is only generated after the products are sold. (Note: revenue = units sold × unit price.) The lower chart

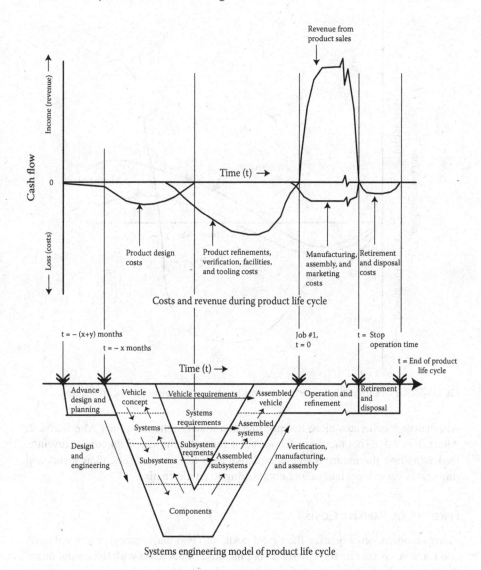

FIGURE 19.1 Costs and revenues in the product life cycle.

in Figure 19.1 shows the systems engineering "V" model. The timeline of the "V" model is synchronized with the timeline of the upper chart.

For cost management purposes, all the costs (negative values) and revenues (positive values) are added, and the cumulative cash flow is frequently reviewed and compared with the budgeted cash flow (i.e., predicted revenue minus budgeted costs). Three cumulative cash flow curves are presented in Figure 19.2. Let us assume that the three cumulative cash flow curves are for three alternative product programs. Alternative 1 incurs much more cost and also extends over a longer duration in the negative cash flow condition than Alternative 2. However, the product in

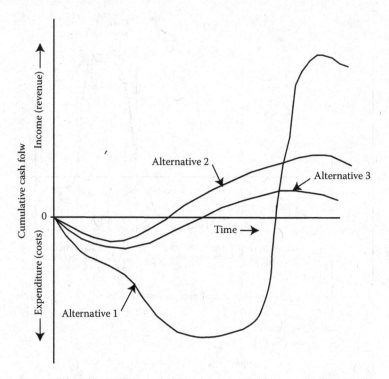

FIGURE 19.2   Cumulative cash flow curves for three alternative product programs.

Alternative 1 generates more revenue and at a much higher rate than Alternative 2. Alternative 3 does not incur much cost, but it generates the lowest amount of revenue. Understanding the nature of the cumulative cash flow curves (i.e., their levels and timings) is very important before committing to an alternative.

FIXED VERSUS VARIABLE COSTS

Many organizations organize their total costs into two major categories: fixed costs and variable costs. The fixed costs do not increase or decrease with the output quantity (i.e., production volumes) of products produced. The variable costs are a direct function of the output quantities (i.e., the variable costs increase with an increase in output quantity). The nonrecurring costs are generally treated as the fixed costs, and the recurring costs are the variable costs. The cost of any product output is the sum of the fixed and the variable costs. Manufacturers should seek to reduce both the fixed and the variable costs. However, decreasing the unit cost of an output through increasing product volumes is a much sought–after approach, as it spreads the fixed costs over a larger volume. Developing and/or using common components that can be shared across a larger number of products (or models, hence increasing their total volumes) can reduce the total cost of the components substantially.

Table 19.1 shows the effect of product volume on four products: A, B, C, and D. The product cost was computed by using the following simple formula:

$$\text{Product cost} = (\text{fixed costs/product volume}) + \text{variable cost per unit}$$

The table shows that the unit cost of Product A will decrease from \$3200.00 to \$3001.25 as the product volume is increased from 5,000 to 800,000 units. Similarly, the unit cost of Product D will decrease from \$2.00 to \$1.01 for product volumes of 5,000 to 800,000 units, respectively. This shows the importance and power of increasing the product volume in reducing the cost of products.

MAKE VERSUS BUY DECISIONS

Most product–producing organizations do not produce all the entities (i.e., systems, subsystems, or components) of the product within their organization. Many of the entities are purchased from other organizations (i.e., suppliers). Typically, standardized components that are common across many similar products are made by different organizations. Some examples of standardized components are fasteners (such as nuts, bolts, rivets, clips, and pins), electrical and electronic components (e.g., switches, resistors, transistors, and microprocessors), plumbing supplies (e.g., pipes, hoses, valves, and couplings), and so on. Some special components that require unique manufacturing processes and specialized systems, machines, or equipment are also purchased from suppliers with specialized production capabilities. For example, major automotive manufacturers typically purchase about 30%–70% of the components (or systems) in the automotive products from their suppliers. Aircraft companies also rely on suppliers to produce most of their components. For example, none of the commercial aircraft manufacturing companies produce jet engines, which contribute about 40%–50% of the cost of an airplane. Similarly, specialized systems such as electronic and electrical systems with components such as microprocessors, sensors, actuators, and printed circuit boards in most complex products are produced by suppliers.

The decision on whether to make or buy an entity depends on many considerations. Some important considerations are

1. Availability of in–house manufacturing capability and capacity (e.g., specialized equipment and personnel with unique backgrounds and skills needed to produce the required product volume)

TABLE 19.1
Effect of Product Volume on Product Cost

Product	Fixed Costs	Variable Costs/Unit	Product Cost ($)					
			Product Volume (Units)					
			5,000	20,000	40,000	100,000	300,000	800,000
A	\$1,000,000.00	\$3,000.00	\$3,200.00	\$3,050.00	\$3,025.00	\$3,010.00	\$3,003.33	\$3,001.25
B	\$200,000.00	\$500.00	\$540.00	\$510.00	\$505.00	\$502.00	\$500.67	\$500.25
C	\$50,000.00	\$10.00	\$20.00	\$12.50	\$11.25	\$10.50	\$10.17	\$10.06
D	\$5,000.00	\$1.00	\$2.00	\$1.25	\$1.13	\$1.05	\$1.02	\$1.01

2. Availability of reliable and low–cost suppliers that can deliver needed volumes of the entities meeting the required quality standards
3. Availability of capital required for internal production of the needed entities
4. Need to maintain confidentiality of the competitive information on future product designs or specialized knowledge on some unique processes needed to produce certain entities within the organization to retain competitive advantage

PARTS AND PLATFORM SHARING

One of the most important principles in reducing costs is sharing the costs with other products or vehicle lines by "commonizing." *Commonizing* is a term used in the auto industry for parts sharing; that is, the same part can be used in more than one vehicle model. Thus, the use of common parts can reduce the contribution of fixed costs to the total part cost by spreading the fixed costs over a larger volume of total number of vehicles produced using different body–styles, name plates, and brands of vehicles. It also involves using common standardized designs, parts, procedures, and equipment in manufacturing and assembly plants. It can be used in the following ways:

1. Parts sharing: Using the same standardized components in several different vehicles.
2. Platform sharing: Platform sharing means using the same body underside and chassis dimensions/design configuration and other system components to create different vehicles. This allows common production equipment (e.g., fixtures, conveyors, welding machines in plants, and assembly lines) and parts to be shared across several products. Thus, several vehicles with different body–styles can be assembled in the same assembly plants using the same workstations in the same sequence. (See Chapter 10 for definition of *vehicle platform*.)
3. Reduced engineering workload: Design, tooling, and testing costs can be substantially reduced by using the same design configuration and shared systems and components.

QUALITY COSTS

To ensure that the product being designed will meet the needs of its customers and satisfy them, the organization must perform a number of tasks, such as conducting a number of analyses and evaluations, implementing statistical process control, honoring warranty, and repairing or replacing failed components. The costs incurred for such tasks can be grouped into the following four categories (Campanella, 1990):

1. *Prevention Costs:* These costs are associated with the information gathered and analyses conducted to ensure that the right product is being designed and that the product will meet its customer needs (i.e., preventing the creation of a wrong or defective product). Some examples of the activities involved in this cost category are market research, benchmarking, product performance analyses, design reviews, supplier reviews and ratings, supplier quality planning, training, quality administration, and process validations. Bhise (2014) provides more information on quality issues and techniques.

2. *Appraisal Costs:* These costs are related to various appraisals or evaluations conducted to ensure that incoming components and materials and outgoing products will meet quality requirements. Examples of the activities involved in this cost category are purchasing appraisals, maintenance of laboratories with calibrated state–of–the–art testing equipment and trained staff, measurements and tests, inspections, and plant quality audits.

3. *Internal Failure Costs:* These costs are incurred at the manufacturer's facilities due to product failures during manufacturing, defects observed during testing, troubleshooting, and analyzing the failures, rejected and scrapped units (or components), rework, repairs, and so forth.

4. *External Failure Costs:* These costs are incurred after the product leaves the manufacturer's facilities and is sold to the customers. The costs are due to handling customer complaints, managing returned products, sending replacements, repairing failed products, product recalls, product litigations and liabilities, penalties, lost sales, and so forth.

MANUFACTURING COSTS

The manufacturing costs can be categorized into the following four broad categories:

1. *Costs of parts (components) and/or subassemblies purchased from the suppliers:* These costs include expenses incurred in purchasing components, standard fasteners, subassemblies, and so forth, from various suppliers.

2. *Costs of parts manufactured internally within the product manufacturer's plants:* These costs are associated with fixed costs for tooling, equipment, and facilities and variable costs associated with purchasing raw materials, expendable tools, processing and operating machines/equipment, inspection, direct labor, coolants, lubricants, utilities, and so on.

3. *Assembly costs:* These include assembly and inspection related to fixed and variable costs of equipment operation (e.g., fixed costs of fixtures and robots needed for assembly; variable costs to run the assembly robots and/or equipment), direct labor costs, and associated employee benefits.

4. *Overhead costs:* These costs include expenses related to indirect labor (e.g., administrative and plant maintenance personnel and costs of their benefits), employee training, utilities, insurance, property taxes, equipment dismantling, and so forth.

SAFETY COSTS

The safety–related costs can be categorized into the following four broad categories (also see Bhise [2014]):

1. *Accident prevention costs:* These costs represent amounts spent by the organization to avoid or prevent accidents and injuries (including injuries due to adverse health effects from longer–term exposures to unsafe conditions, e.g., cumulative trauma, noise and vibrations) from occurring. Accident prevention activities typically include safety analyses (e.g., conducting hazard analysis, conducting failure modes and effects analyses), incorporating engineering

changes (e.g., process and equipment improvements to reduce probability of accidents and injuries, adding safety devices), conducting safety evaluations/tests, safety reviews, providing safety training to employees, providing/installing and maintaining protection devices (e.g., hard hats, safety glasses, lock–out devices, anti–slip walking surfaces, and lifting devices to reduce back injuries).

2. *Costs due to accidents:* These costs include losses that an organization incurs due to accidents. The accidents can involve injuries (e.g., medical costs, temporary disability–related costs till an injured person returns to his/her regular job, costs due to permanent disability), loss of life, damage to facilities and equipment, and/or work stoppages. It should be noted that accident costs are almost always underestimated due to many unreported or unaccounted costs (e.g., loss of production or temporary work slow–downs due to accidents, retraining of replacement workers). In some cases, the incidental costs of accidents have been estimated to be four times as great as the directly accounted costs.

3. *Insurance costs:* These costs include expenditures to insure (i.e., insurance premiums and workers' compensation costs) against losses due to accidents and injuries, fatalities, and property damage (i.e., repairing or replacing damaged equipment).

4. *Product liability costs:* These are costs incurred in product liability cases resulting from injuries caused by the product due to defects in the products. These costs include costs to defend cases (e.g., fees charged by lawyers and experts) and compensation or settlement charges paid to the plaintiff, penalties, and fines.

PRODUCT TERMINATION COSTS

These costs are incurred after the decision is made to terminate the production of a product. These costs include

1. Costs of selling discontinued products at discounted prices or with sales incentives
2. Costs of lost sales of new products due to some customers purchasing the discontinued products at the discounted prices
3. Plant and equipment write–down costs
4. Plant shutdown, equipment removal and disposal costs
5. Environmental clean–up and site restoration costs
6. Materials recycling costs
7. Continual service, production, and distribution of spare parts for products in service till they are disposed of

TOTAL LIFE–CYCLE COSTS

These costs comprise the total of all the costs described in this section, from product conception to the end of production, disposal (or recycling) of all products from service, and facilities closing.

EFFECT OF TIME ON COSTS

As the costs are incurred over time, in determining all these costs, the effect of time due to factors such as interest rate (or discount rate), inflation rate, and fluctuations in currency exchange rates (if applicable) must be taken into account. Similarly, since the revenues are generated over the selling periods of the products, and payments are received over time, the effects of changes in interest rates, inflation, and currency exchange rates should also be considered.

Most complex product programs extend over many years. Therefore, cost computations need to consider the effects of interest and inflation.

With the annual compounding of the combined interest and inflation, the relationship between present and future value is (Blanchard and Fabrycky, 2011)

$$F = P(1+i)^n \text{ or } P = F\left[1/(1+i)^n\right]$$

where:

P = present value (or value at a time assumed to be the present)
i = combined annual interest and inflation rate

$$= i_r + i_f$$

i_r = annual interest rate
i_f = annual inflation rate
n = number of annual interest periods
F = future value after n periods

Using this formula, the value of $100 today will be $128 in 5 years at 5% combined annual interest and inflation rate (note: $128 = 100 (1+0.05)^5$). This means that $128 spent 5 years from now will be equivalent to $100 today, assuming a 5% rate of combined interest and inflation.

For a program extending over many periods, the present value of revenues minus the present value of costs can be computed for each period. The present values for each of the periods (assumed to be monthly, quarterly, or annually) can be summed over the entire duration of the program to obtain the present value of the cumulative cash flow. The present value is generally computed at the beginning of the product program to provide management with an estimate of cash flow over the life of the program.

PROGRAM FINANCIAL PLAN

EXAMPLE: AUTOMOTIVE PRODUCT PROGRAM CASH FLOW

This section presents a simplified cash flow analysis of an automotive product program. The analysis covers a 100 month period from 40 months before Job #1 to 60 months after Job #1. In the automotive industry, Job #1 represents the time at which the first production vehicle rolls out of the assembly plant (i.e., the vehicle is released for sale to the customer).

The assumptions used for the costs and revenue computations were as follows:
Program Milestones:

1. Program kick–off at −40 months (i.e., 40 months before Job #1).
2. Product development team formation begins at −39.5 months.
3. Strategic intent confirmation at −34 months.
4. Hard–points freeze at −29 months.
5. Feasibility sign–off at −27.5 months.
6. Program approval at −26 months.
7. Surface freeze at −24 months.
8. Appearance approval at −19 months.
9. Early prototype vehicles available for testing at −14 months.
10. Early production prototype vehicles available for testing at −9 months.
11. Final prototype vehicles available at −5 months.
12. Production begins at −3 months,
13. The vehicle is released in the public domain at Job #1 (0 months).
14. Production continued till 60 months after Job #1.

The costs estimates used for this illustration are provided in Table 19.2. The columns of the table are labeled in the top row from A through T. The columns are defined as follows:

TABLE 19.2

(PART 1) Costs, Revenue, and Present Value of Cash Flow in a Vehicle Program

A	B	C	D	E	F	G
Months from Job Development #1	Product Development Headcount	Product Development Manpower Costs	Services and Supplies Costs	Facilities and Tooling Costs	Product Development Costs Subtotal	Present Value of Development Costs Subtotal
−40	50	$4,00,000	$1,40,000		$5,40,000	$5,40,000
−39	100	$8,00,000	$2,80,000		$10,80,000	$10,77,307
−38	200	$16,00,000	$5,60,000		$21,60,000	$21,49,240
−37	500	$40,00,000	$14,00,000		$54,00,000	$53,59,702
−36	800	$64,00,000	$22,40,000		$86,40,000	$85,54,137
−35	1000	$80,00,000	$28,00,000		$1,08,00,000	$1,06,66,007
−34	1000	$80,00,000	$28,00,000		$1,08,00,000	$1,06,39,408
−33	1200	$96,00,000	$33,60,000		$1,29,60,000	$1,27,35,451
−32	1200	$96,00,000	$33,60,000		$1,29,60,000	$1,27,03,692
−31	1200	$96,00,000	$33,60,000		$1,29,60,000	$1,26,72,012
−30	1200	$96,00,000	$33,60,000		$1,29,60,000	$1,26,40,411
−29	1200	$96,00,000	$33,60,000		$1,29,60,000	$1,26,08,889
−28	1200	$96,00,000	$33,60,000		$1,29,60,000	$1,25,77,445
−27	1200	$96,00,000	$33,60,000		$1,29,60,000	$1,25,46,080

TABLE 19.2 (CONTINUED)
(PART 1) Costs, Revenue, and Present Value of Cash Flow in a Vehicle Program

A	B	C	D	E	F	G
Months from Job #1	Product Development Headcount	Product Development Manpower Costs	Services and Supplies Costs	Facilities and Tooling Costs	Product Development Costs Subtotal	Present Value of Development Costs Subtotal
−26	1200	$96,00,000	$33,60,000		$1,29,60,000	$1,25,14,793
−25	1200	$96,00,000	$33,60,000	$4,00,00,000	$5,29,60,000	$5,10,13,164
−24	1200	$96,00,000	$33,60,000	$4,00,00,000	$5,29,60,000	$5,08,85,949
−23	1200	$96,00,000	$33,60,000	$4,00,00,000	$5,29,60,000	$5,07,59,051
−22	1200	$96,00,000	$33,60,000	$4,00,00,000	$5,29,60,000	$5,06,32,470
−21	1200	$96,00,000	$33,60,000	$4,00,00,000	$5,29,60,000	$5,05,06,204
−20	1200	$96,00,000	$33,60,000	$4,00,00,000	$5,29,60,000	$5,03,80,254
−19	1200	$96,00,000	$33,60,000	$4,00,00,000	$5,29,60,000	$5,02,54,617
−18	1200	$96,00,000	$33,60,000	$4,00,00,000	$5,29,60,000	$5,01,29,294
−17	1200	$96,00,000	$33,60,000	$4,00,00,000	$5,29,60,000	$5,00,04,283
−16	1200	$96,00,000	$33,60,000	$4,00,00,000	$5,29,60,000	$4,98,79,584
−15	1200	$96,00,000	$33,60,000	$4,00,00,000	$5,29,60,000	$4,97,55,196
−14	1200	$96,00,000	$33,60,000	$4,00,00,000	$5,29,60,000	$4,96,31,119
−13	1200	$96,00,000	$33,60,000	$4,00,00,000	$5,29,60,000	$4,95,07,350
−12	1200	$96,00,000	$33,60,000	$4,00,00,000	$5,29,60,000	$4,93,83,890
−11	1200	$96,00,000	$33,60,000	$4,00,00,000	$5,29,60,000	$4,92,60,739
−10	1200	$96,00,000	$33,60,000	$4,00,00,000	$5,29,60,000	$4,91,37,894
−9	1200	$96,00,000	$33,60,000	$4,00,00,000	$5,29,60,000	$4,90,15,355
−8	1200	$96,00,000	$33,60,000	$4,00,00,000	$5,29,60,000	$4,88,93,123
−7	1200	$96,00,000	$33,60,000	$4,00,00,000	$5,29,60,000	$4,87,71,195
−6	1000	$80,00,000	$28,00,000	$4,00,00,000	$5,08,00,000	$4,66,65,374
−5	1000	$80,00,000	$28,00,000	$4,00,00,000	$5,08,00,000	$4,65,49,001
−4	800	$64,00,000	$22,40,000	$4,00,00,000	$4,86,40,000	$4,44,58,606
−3	600	$48,00,000	$16,80,000	$4,00,00,000	$4,64,80,000	$4,23,78,347
−2	400	$32,00,000	$11,20,000	$4,00,00,000	$4,43,20,000	$4,03,08,187
−1	300	$24,00,000	$8,40,000		$32,40,000	$29,39,369
0	200	$16,00,000	$5,60,000		$21,60,000	$19,54,693
1					$0	$0
2					$0	$0
3					$0	$0
4					$0	$0
5					$0	$0
6					$0	$0
7					$0	$0
8					$0	$0
9					$0	$0
10					$0	$0
11					$0	$0
12					$0	$0
12					$0	$0

TABLE 19.2

(PART 2) Costs, Revenue, and Present Value of Cash Flow in a Vehicle Program

A	H	I	J	K	L	M	N	O
Months from Job #1	Manufacturing Headcount	Manufacturing Manpower Costs	Number of Vehicles Produced	Parts, Materials, and overhead Costs	Manufacturing Costs Subtotal	Sales and Marketing Costs	Total Costs	Revenue from Vehicle Sales
−40							$5,40,000	
−39							$10,80,000	
−38							$21,60,000	
−37							$54,00,000	
−36							$86,40,000	
−35							$1,08,00,000	
−34							$1,08,00,000	
−33							$1,29,60,000	
−32							$1,29,60,000	
−31							$1,29,60,000	
−30							$1,29,60,000	
−29							$1,29,60,000	
−28							$1,29,60,000	
−27							$1,29,60,000	
−26							$1,29,60,000	
−25							$5,29,60,000	
−24							$5,29,60,000	
−23							$5,29,60,000	
−22							$5,29,60,000	
−21							$5,29,60,000	
−20							$5,29,60,000	
−19							$5,29,60,000	

...ue of Cash Flow in a Vehicle...

Period								
-18							$5,29,60,000	
-17							$5,29,60,000	
-16							$5,29,60,000	
-15							$5,29,60,000	
-14							$5,29,60,000	
-13							$5,29,60,000	
-12							$5,29,60,000	
-11							$5,29,60,000	
-10							$5,29,60,000	
-9							$5,29,60,000	
-8							$5,29,60,000	
-7							$5,29,60,000	
-6							$5,08,00,000	
-5							$5,08,00,000	
-4							$4,86,40,000	
-3	500	$60,00,000	1,000	$80,00,000	$1,40,00,000	$20,00,000	$6,24,80,000	$2,80,00,000
-2	600	$72,00,000	4,000	$3,20,00,000	$8,35,20,000	$80,00,000	$13,58,40,000	$11,20,00,000
-1	800	$96,00,000	8,000	$6,40,00,000	$7,68,40,000	$1,60,00,000	$9,60,80,000	$22,40,00,000
0	1000	$1,20,00,000	10,000	$8,00,00,000	$9,41,60,000	$2,00,00,000	$11,63,20,000	$28,00,00,000
1	1000	$1,20,00,000	24,000	$19,20,00,000	$20,40,00,000	$4,80,00,000	$25,20,00,000	$67,20,00,000
2	1000	$1,20,00,000	24,000	$19,20,00,000	$20,40,00,000	$4,80,00,000	$25,20,00,000	$67,20,00,000
3	1000	$1,20,00,000	24,000	$19,20,00,000	$20,40,00,000	$4,80,00,000	$25,20,00,000	$67,20,00,000
4	1000	$1,20,00,000	24,000	$19,20,00,000	$20,40,00,000	$4,80,00,000	$25,20,00,000	$67,20,00,000
5	1000	$1,20,00,000	24,000	$19,20,00,000	$20,40,00,000	$4,80,00,000	$25,20,00,000	$67,20,00,000
6	1000	$1,20,00,000	24,000	$19,20,00,000	$20,40,00,000	$4,80,00,000	$25,20,00,000	$67,20,00,000
7	1000	$1,20,00,000	25,000	$20,00,00,000	$21,20,00,000	$5,00,00,000	$26,20,00,000	$70,00,00,000
8	1000	$1,20,00,000	26,000	$20,80,00,000	$22,00,00,000	$5,20,00,000	$27,20,00,000	$72,80,00,000
9	1000	$1,20,00,000	27,000	$21,60,00,000	$22,80,00,000	$5,40,00,000	$28,20,00,000	$75,60,00,000
10	1000	$1,20,00,000	28,000	$22,40,00,000	$23,60,00,000	$5,60,00,000	$29,20,00,000	$78,40,00,000
11	1000	$1,20,00,000	29,000	$23,20,00,000	$24,40,00,000	$5,80,00,000	$30,20,00,000	$81,20,00,000
12	1000	$1,20,00,000	30,000	$24,00,00,000	$25,20,00,000	$6,00,00,000	$31,20,00,000	$84,00,00,000

TABLE 19.2

(PART 3) Costs, Revenue, and Present Value of Cash Flow in a Vehicle Program

A	P	Q	R	S	T
Months from Job#1	Present Value of Total Cost	Present Value of Revenue from Vehicle Sales	Present Value of Cumulative Total Cost	Present Value of Cumulative Total Revenue	Present Value Cash Flow
−40	$5,40,000	$0	$5,40,000	$0	−$5,40,000
−39	$10,77,307	$0	$16,17,307	$0	−$16,17,307
−38	$21,49,240	$0	$37,66,547	$0	−$37,66,547
−37	$53,59,702	$0	$91,26,249	$0	−$91,26,249
−36	$85,54,137	$0	$1,76,80,386	$0	−$1,76,80,386
−35	$1,06,66,007	$0	$2,83,46,393	$0	−$2,83,46,393
−34	$1,06,39,408	$0	$3,89,85,801	$0	−$3,89,85,801
−33	$1,27,35,451	$0	$5,17,21,252	$0	−$5,17,21,252
−32	$1,27,03,692	$0	$6,44,24,944	$0	−$6,44,24,944
−31	$1,26,72,012	$0	$7,70,96,956	$0	−$7,70,96,956
−30	$1,26,40,411	$0	$8,97,37,366	$0	−$8,97,37,366
−29	$1,26,08,889	$0	$10,23,46,255	$0	−$10,23,46,255
−28	$1,25,77,445	$0	$11,49,23,700	$0	−$11,49,23,700
−27	$1,25,46,080	$0	$12,74,69,780	$0	−$12,74,69,780
−26	$1,25,14,793	$0	$13,99,84,573	$0	−$13,99,84,573
−25	$5,10,13,164	$0	$19,09,97,736	$0	−$19,09,97,736
−24	$5,08,85,949	$0	$24,18,83,685	$0	−$24,18,83,685
−23	$5,07,59,051	$0	$29,26,42,736	$0	−$29,26,42,736
−22	$5,06,32,470	$0	$34,32,75,206	$0	−$34,32,75,206
−21	$5,05,06,204	$0	$39,37,81,410	$0	−$39,37,81,410
−20	$5,03,80,254	$0	$44,41,61,664	$0	−$44,41,61,664
−19	$5,02,54,617	$0	$49,44,16,281	$0	−$49,44,16,281
−18	$5,01,29,294	$0	$54,45,45,575	$0	−$54,45,45,575
−17	$5,00,04,283	$0	$59,45,49,858	$0	−$59,45,49,858
−16	$4,98,79,584	$0	$64,44,29,443	$0	−$64,44,29,443
−15	$4,97,55,196	$0	$69,41,84,639	$0	−$69,41,84,639
−14	$4,96,31,119	$0	$74,38,15,757	$0	−$74,38,15,757
−13	$4,95,07,350	$0	$79,33,23,108	$0	−$79,33,23,108
−12	$4,93,83,890	$0	$84,27,06,998	$0	−$84,27,06,998
−11	$4,92,60,739	$0	$89,19,67,737	$0	−$89,19,67,737
−10	$4,91,37,894	$0	$94,11,05,630	$0	−$94,11,05,630
−9	$4,90,15,355	$0	$99,01,20,986	$0	−$99,01,20,986
−8	$4,88,93,123	$0	$1,03,90,14,109	$0	−$1,03,90,14,109
−7	$4,87,71,195	$0	$1,08,77,85,303	$0	−$1,08,77,85,303
−6	$4,66,65,374	$0	$1,13,44,50,677	$0	−$1,13,44,50,677
−5	$4,65,49,001	$0	$1,18,09,99,678	$0	−$1,18,09,99,678

(Continued)

TABLE 19.2 (CONTINUED)

(PART 3) Costs, Revenue, and Present Value of Cash Flow in a Vehicle Program

A	P	Q	R	S	T
Months from Job#1	Present Value of Total Cost	Present Value of Revenue from Vehicle Sales	Present Value of Cumulative Total Cost	Present Value of Cumulative Total Revenue	Present Value Cash Flow
−4	$4,44,58,606	$0	$1,22,54,58,284	$0	−$1,22,54,58,284
−3	$5,69,66,418	$1,76,82,426	$1,28,24,24,702	$1,76,82,426	−$1,26,47,42,276
−2	$12,35,43,865	$6,98,56,498	$1,40,59,68,567	$8,75,38,924	−$1,31,84,29,643
−1	$8,71,64,998	$13,79,88,144	$1,49,31,33,565	$22,55,27,069	−$1,26,76,06,496
0	$10,52,63,824	$17,03,55,734	$1,59,83,97,389	$39,58,82,803	−$1,20,25,14,586
1	$22,74,78,789	$40,38,06,184	$1,82,58,76,178	$79,96,88,987	−$1,02,61,87,191
2	$22,69,11,510	$39,88,20,923	$2,05,27,87,688	$1,19,85,09,909	−$85,42,77,779
3	$22,63,45,646	$39,38,97,207	$2,27,91,33,334	$1,59,24,07,117	−$68,67,26,218
4	$22,57,81,193	$38,90,34,279	$2,50,49,14,528	$1,98,14,41,396	−$52,34,73,132
5	$22,52,18,148	$38,42,31,387	$2,73,01,32,675	$2,36,56,72,783	−$36,44,59,893
6	$22,46,56,507	$37,94,87,789	$2,95,47,89,182	$2,74,51,60,572	−$20,96,28,610
7	$23,29,88,975	$39,04,19,536	$3,18,77,78,157	$3,13,55,80,108	−$5,21,98,049
8	$24,12,78,488	$40,10,23,524	$3,42,90,56,645	$3,53,66,03,632	$10,75,46,987
9	$24,95,25,207	$41,13,06,178	$3,67,85,81,852	$3,94,79,09,810	$26,93,27,958
10	$25,77,29,296	$42,12,73,818	$3,93,63,11,148	$4,36,91,83,628	$43,28,72,480
11	$26,58,90,914	$43,09,32,653	$4,20,22,02,062	$4,80,01,16,281	$59,79,14,219
12	$27,40,10,223	$44,02,88,790	$4,47,62,12,285	$5,24,04,05,070	$76,41,92,785

TABLE 19.2

(PART 4) Costs, Revenue, and Present Value of Cash Flow in a Vehicle Program

A	B	C	D	E	F	G
Months from Job #1	Product Development Headcount	Product Development Manpower Costs	Services and Supplies Costs	Facilities and Tooling Costs	Product Development Costs Subtotal	Present Value of Development Costs Subtotal
13					$0	$0
14					$0	$0
15					$0	$0
16					$0	$0
17					$0	$0
18					$0	$0
19					$0	$0

(Continued)

TABLE 19.2 (CONTINUED)

(PART 4) Costs, Revenue, and Present Value of Cash Flow in a Vehicle Program

A	B	C	D	E	F	G
Months from Job #1	Product Development Headcount	Product Development Manpower Costs	Services and Supplies Costs	Facilities and Tooling Costs	Product Development Costs Subtotal	Present Value of Development Costs Subtotal
20					$0	$0
21					$0	$0
22					$0	$0
23					$0	$0
24					$0	$0
25						
26						
27						
28						
29						
30						
31						
32						
33						
34						
35						
36						
37						
38						
39						
40						
41						
42						
43						
44						
45						
46						
47						
48						
49						
50						
51						
52						
53						
54						
55						
56						
57						
58						
59						
60						

TABLE 19.2

(PART 5) Costs, Revenue, and Present Value of Cash Flow in a Vehicle Program

A	H	I	J	K	L	M	N	O
Months from Job #1	Manufacturing Headcount	Manufacturing Manpower Costs	Number of Vehicles Produced	Parts , Materials, and Overhead Costs	Manufacturing Costs Subtotal	Sales and Marketing Costs	Total Costs	Revenue from Vehicle Sales
13	1000	$1,20,00,000	30,000	$24,00,00,000	$25,20,00,000	$6,00,00,000	$31,20,00,000	$84,00,00,000
14	1000	$1,20,00,000	30,000	$24,00,00,000	$25,20,00,000	$6,00,00,000	$31,20,00,000	$84,00,00,000
15	1000	$1,20,00,000	30,000	$24,00,00,000	$25,20,00,000	$6,00,00,000	$31,20,00,000	$84,00,00,000
16	1000	$1,20,00,000	30,000	$24,00,00,000	$25,20,00,000	$6,00,00,000	$31,20,00,000	$84,00,00,000
17	1000	$1,20,00,000	30,000	$24,00,00,000	$25,20,00,000	$6,00,00,000	$31,20,00,000	$84,00,00,000
18	1000	$1,20,00,000	30,000	$24,00,00,000	$25,20,00,000	$6,00,00,000	$31,20,00,000	$84,00,00,000
19	1000	$1,20,00,000	30,000	$24,00,00,000	$25,20,00,000	$6,00,00,000	$31,20,00,000	$84,00,00,000
20	1000	$1,20,00,000	30,000	$24,00,00,000	$25,20,00,000	$6,00,00,000	$31,20,00,000	$84,00,00,000
21	1000	$1,20,00,000	30,000	$24,00,00,000	$25,20,00,000	$6,00,00,000	$31,20,00,000	$84,00,00,000
22	1000	$1,20,00,000	30,000	$24,00,00,000	$25,20,00,000	$6,00,00,000	$31,20,00,000	$84,00,00,000
23	1000	$1,20,00,000	30,000	$24,00,00,000	$25,20,00,000	$6,00,00,000	$31,20,00,000	$84,00,00,000
24	1000	$1,20,00,000	30,000	$24,00,00,000	$25,20,00,000	$6,00,00,000	$31,20,00,000	$84,00,00,000
25	1000	$1,20,00,000	30,000	$24,00,00,000	$25,20,00,000	$6,00,00,000	$31,20,00,000	$84,00,00,000
26	1000	$1,20,00,000	30,000	$24,00,00,000	$25,20,00,000	$6,00,00,000	$31,20,00,000	$84,00,00,000
27	1000	$1,20,00,000	30,000	$24,00,00,000	$25,20,00,000	$6,00,00,000	$31,20,00,000	$84,00,00,000
28	1000	$1,20,00,000	30,000	$24,00,00,000	$25,20,00,000	$6,00,00,000	$31,20,00,000	$84,00,00,000
29	1000	$1,20,00,000	30,000	$24,00,00,000	$25,20,00,000	$6,00,00,000	$31,20,00,000	$84,00,00,000
30	1000	$1,20,00,000	30,000	$24,00,00,000	$25,20,00,000	$6,00,00,000	$31,20,00,000	$84,00,00,000
31	1000	$1,20,00,000	30,000	$24,00,00,000	$25,20,00,000	$6,00,00,000	$31,20,00,000	$84,00,00,000
32	1000	$1,20,00,000	30,000	$24,00,00,000	$25,20,00,000	$6,00,00,000	$31,20,00,000	$84,00,00,000
33	1000	$1,20,00,000	30,000	$24,00,00,000	$25,20,00,000	$6,00,00,000	$31,20,00,000	$84,00,00,000
34	1000	$1,20,00,000	30,000	$24,00,00,000	$25,20,00,000	$6,00,00,000	$31,20,00,000	$84,00,00,000
35	1000	$1,20,00,000	30,000	$24,00,00,000	$25,20,00,000	$6,00,00,000	$31,20,00,000	$84,00,00,000
36	1000	$1,20,00,000	30,000	$24,00,00,000	$25,20,00,000	$6,00,00,000	$31,20,00,000	$84,00,00,000

(Continued)

TABLE 19.2 (CONTINUED)
(PART 5) Costs, Revenue, and Present Value of Cash Flow in a Vehicle Program

A	H	I	J	K	L	M	N	O
Months from Job #1	Manufacturing Headcount	Manufacturing Manpower Costs	Number of Vehicles Produced	Parts, Materials, and Overhead Costs	Manufacturing Costs Subtotal	Sales and Marketing Costs	Total Costs	Revenue from Vehicle Sales
37	1000	$1,20,00,000	29,000	$23,20,00,000	$24,40,00,000	$5,80,00,000	$30,20,00,000	$81,20,00,000
38	1000	$1,20,00,000	28,000	$22,40,00,000	$23,60,00,000	$5,60,00,000	$29,20,00,000	$78,40,00,000
39	1000	$1,20,00,000	27,000	$21,60,00,000	$22,80,00,000	$5,40,00,000	$28,20,00,000	$75,60,00,000
40	1000	$1,20,00,000	26,000	$20,80,00,000	$22,00,00,000	$5,20,00,000	$27,20,00,000	$72,80,00,000
41	1000	$1,20,00,000	25,000	$20,00,00,000	$21,20,00,000	$5,00,00,000	$26,20,00,000	$70,00,00,000
42	1000	$1,20,00,000	24,000	$19,20,00,000	$20,40,00,000	$4,80,00,000	$25,20,00,000	$67,20,00,000
43	1000	$1,20,00,000	24,000	$19,20,00,000	$20,40,00,000	$4,80,00,000	$25,20,00,000	$67,20,00,000
44	1000	$1,20,00,000	24,000	$19,20,00,000	$20,40,00,000	$4,80,00,000	$25,20,00,000	$67,20,00,000
45	1000	$1,20,00,000	24,000	$19,20,00,000	$20,40,00,000	$4,80,00,000	$25,20,00,000	$67,20,00,000
46	1000	$1,20,00,000	24,000	$19,20,00,000	$20,40,00,000	$4,80,00,000	$25,20,00,000	$67,20,00,000
47	1000	$1,20,00,000	24,000	$19,20,00,000	$20,40,00,000	$4,80,00,000	$25,20,00,000	$67,20,00,000
48	1000	$1,20,00,000	24,000	$19,20,00,000	$20,40,00,000	$4,80,00,000	$25,20,00,000	$67,20,00,000
49	1000	$1,20,00,000	24,000	$19,20,00,000	$20,40,00,000	$4,80,00,000	$25,20,00,000	$67,20,00,000
50	1000	$1,20,00,000	24,000	$19,20,00,000	$20,40,00,000	$4,80,00,000	$25,20,00,000	$67,20,00,000
51	1000	$1,20,00,000	24,000	$19,20,00,000	$20,40,00,000	$4,80,00,000	$25,20,00,000	$67,20,00,000
52	1000	$1,20,00,000	24,000	$19,20,00,000	$20,40,00,000	$4,80,00,000	$25,20,00,000	$67,20,00,000
53	1000	$1,20,00,000	24,000	$19,20,00,000	$20,40,00,000	$4,80,00,000	$25,20,00,000	$67,20,00,000
54	1000	$1,20,00,000	24,000	$19,20,00,000	$20,40,00,000	$4,80,00,000	$25,20,00,000	$67,20,00,000
55	1000	$1,20,00,000	24,000	$19,20,00,000	$20,40,00,000	$4,80,00,000	$25,20,00,000	$67,20,00,000
56	1000	$1,20,00,000	24,000	$19,20,00,000	$20,40,00,000	$4,80,00,000	$25,20,00,000	$67,20,00,000
57	1000	$1,20,00,000	24,000	$19,20,00,000	$20,40,00,000	$4,80,00,000	$25,20,00,000	$67,20,00,000
58	1000	$1,20,00,000	24,000	$19,20,00,000	$20,40,00,000	$4,80,00,000	$25,20,00,000	$67,20,00,000
59	1000	$1,20,00,000	24,000	$19,20,00,000	$20,40,00,000	$4,80,00,000	$25,20,00,000	$67,20,00,000
60	1000	$1,20,00,000	24,000	$19,20,00,000	$20,40,00,000	$4,80,00,000	$25,20,00,000	$67,20,00,000

TABLE 19.2
(PART 6) Costs, Revenue, and Present Value of Cash Flow in a Vehicle Program

A	P	Q	R	S	T
Months from Job #1	Present Value of Total Cost	Present Value of Revenue from Vehicle Sales	Present Value of Cumulative Total Cost	Present Value of Cumulative Total Revenue	Present Value Cash Flow
13	$27,33,26,906	$43,48,53,126	$4,74,95,39,191	$5,67,52,58,196	$92,57,19,005
14	$27,26,45,293	$42,94,84,569	$5,02,21,84,484	$6,10,47,42,765	$1,08,25,58,280
15	$27,19,65,379	$42,41,82,290	$5,29,41,49,864	$6,52,89,25,055	$1,23,47,75,191
16	$27,12,87,162	$41,89,45,472	$5,56,54,37,025	$6,94,78,70,526	$1,38,24,33,501
17	$27,06,10,635	$41,37,73,305	$5,83,60,47,660	$7,36,16,43,831	$1,52,55,96,171
18	$26,99,35,796	$40,86,64,993	$6,10,59,83,456	$7,77,03,08,824	$1,66,43,25,368
19	$26,92,62,639	$40,36,19,746	$6,37,52,46,095	$8,17,39,28,570	$1,79,86,82,475
20	$26,85,91,161	$39,86,36,786	$6,64,38,37,256	$8,57,25,65,356	$1,92,87,28,101
21	$26,79,21,358	$39,37,15,344	$6,91,17,58,613	$8,96,62,80,701	$2,05,45,22,087
22	$26,72,53,225	$38,88,54,661	$7,17,90,11,838	$9,35,51,35,362	$2,17,61,23,524
23	$26,65,86,758	$38,40,53,986	$7,44,55,98,595	$9,73,91,89,348	$2,29,35,90,753
24	$26,59,21,953	$37,93,12,579	$7,71,15,20,548	$10,11,85,01,927	$2,40,69,81,379
25	$26,52,58,806	$37,46,29,708	$7,97,67,79,354	$10,49,31,31,635	$2,51,63,52,281
26	$26,45,97,312	$37,00,04,650	$8,24,13,76,666	$10,86,31,36,284	$2,62,17,59,618
27	$26,39,37,469	$36,54,36,691	$8,50,53,14,135	$11,22,85,72,975	$2,72,32,58,840
28	$26,32,79,271	$36,09,25,127	$8,76,85,93,406	$11,58,94,98,102	$2,82,09,04,696
29	$26,26,22,714	$35,64,69,261	$9,03,12,16,120	$11,94,59,67,363	$2,91,47,51,243
30	$26,19,67,794	$35,20,68,406	$9,29,31,83,914	$12,29,80,35,769	$3,00,48,51,855
31	$26,13,14,508	$34,77,21,882	$9,55,44,98,422	$12,64,57,57,652	$3,09,12,59,229
32	$26,06,62,851	$34,34,29,020	$9,81,51,61,273	$12,98,91,86,671	$3,17,40,25,398
33	$26,00,12,819	$33,91,89,155	$10,07,51,74,092	$13,32,83,75,827	$3,25,32,01,734
34	$25,93,64,408	$33,50,01,635	$10,33,45,38,500	$13,66,33,77,461	$3,32,88,38,961
35	$25,87,17,614	$33,08,65,812	$10,59,32,56,114	$13,99,42,43,274	$3,40,09,87,160
36	$25,80,72,433	$32,67,81,049	$10,85,13,28,547	$14,32,10,24,323	$3,46,96,95,776
37	$24,91,77,936	$31,19,88,491	$11,10,05,06,482	$14,63,30,12,814	$3,53,25,06,332
38	$24,03,26,195	$29,75,11,375	$11,34,08,32,677	$14,93,05,24,189	$3,58,96,91,512
39	$23,15,17,053	$28,33,44,167	$11,57,23,49,731	$15,21,38,68,356	$3,64,15,18,626
40	$22,27,50,353	$26,94,81,421	$11,79,51,00,083	$15,48,33,49,777	$3,68,82,49,693
41	$21,40,25,937	$25,59,17,779	$12,00,91,26,020	$15,73,92,67,555	$3,73,01,41,535
42	$20,53,43,649	$24,26,47,968	$12,21,44,69,669	$15,98,19,15,523	$3,76,74,45,854
43	$20,48,31,570	$23,96,52,314	$12,41,93,01,239	$16,22,15,67,837	$3,80,22,66,598
44	$20,43,20,768	$23,66,93,643	$12,62,36,22,007	$16,45,82,61,480	$3,83,46,39,473
45	$20,38,11,240	$23,37,71,500	$12,82,74,33,247	$16,69,20,32,980	$3,86,45,99,733
46	$20,33,02,982	$23,08,85,432	$13,03,07,36,229	$16,92,29,18,412	$3,89,21,82,182
47	$20,27,95,992	$22,80,34,994	$13,23,35,32,221	$17,15,09,53,406	$3,91,74,21,184
48	$20,22,90,267	$22,52,19,747	$13,43,58,22,488	$17,37,61,73,153	$3,94,03,50,665

TABLE 19.2 (CONTINUED)
(PART 6) Costs, Revenue, and Present Value of Cash Flow in a Vehicle Program

A	P	Q	R	S	T
		Present Value of			
Months from Job #1	Present Value of Total Cost	Revenue from Vehicle Sales	Present Value of Cumulative Total Cost	Present Value of Cumulative Total Revenue	Present Value Cash Flow
49	$20,17,85,802	$22,24,39,257	$13,63,76,08,290	$17,59,86,12,410	$3,96,10,04,119
50	$20,12,82,596	$21,96,93,093	$13,83,88,90,886	$17,81,83,05,503	$3,97,94,14,617
51	$20,07,80,644	$21,69,80,833	$14,03,96,71,530	$18,03,52,86,335	$3,99,56,14,805
52	$20,02,79,944	$21,43,02,057	$14,23,99,51,475	$18,24,95,88,392	$4,00,96,36,918
53	$19,97,80,493	$21,16,56,352	$14,43,97,31,968	$18,46,12,44,745	$4,02,15,12,777
54	$19,92,82,287	$20,90,43,311	$14,63,90,14,255	$18,67,02,88,056	$4,03,12,73,801
55	$19,87,85,324	$20,64,62,529	$14,83,77,99,579	$18,87,67,50,585	$4,03,89,51,006
56	$19,82,89,600	$20,39,13,609	$15,03,60,89,179	$19,08,06,64,195	$4,04,45,75,016
57	$19,77,95,112	$20,13,96,157	$15,23,38,84,291	$19,28,20,60,352	$4,04,81,76,061
58	$19,73,01,858	$19,89,09,785	$15,43,11,86,149	$19,48,09,70,137	$4,04,97,83,988
59	$19,68,09,833	$19,64,54,109	$15,62,79,95,982	$19,67,74,24,246	$4,04,94,28,264
60	$19,63,19,035	$19,40,28,749	$15,82,43,15,017	$19,87,14,52,995	$4,04,71,37,978

Note: Discount rate for present value calculations = 3.
Selling price ($) per vehicle = 28,000.

A = Time in months from Job #1.

B = Product development headcount (i.e., number of professionals assigned to the program on a full–time basis).

C = Product development manpower costs (value in column B times average monthly salary and benefits paid to the product development professionals). Note: Head count cost (i.e., salary plus benefits) is assumed to be $8000/month.

D = Services and suppliers costs (i.e., monthly costs of services or materials provided by noncompany personnel such as contractors, suppliers, vendors, outside test labs).

E = Facilities and tooling costs (i.e., cost of designing and building vehicle–producing facilities [plants], tools, and equipment).

F = Product development costs subtotal = sum of values in columns C, D, and E in each row.

G = Present value of cost in column F computed at the beginning of the program (at −40 months) using interest rate (note: annual interest rate of 3% is used in this example by setting its value at the bottom of the table [second last row under column F]). Note: The present value is computed based on 1 month intervals and monthly interest rate on 0.25% (i.e., 3% annual rate divided by 12).

H = Manufacturing headcount (number of manufacturing personnel assigned to the program on a full–time basis).

I = Manufacturing personnel cost (assumed to be $30/hour; manufacturing plant is assumed to operate 8 hours per shift, 2 shifts/day, and 25 days/month).

J = Number of vehicles produced in the month corresponding to the month specified in the row.

K = Parts, materials, and overhead costs (e.g., parts and systems purchased from suppliers to assemble the vehicles; assumed to be $8000/vehicle).

L = Manufacturing costs subtotal (sum of values in columns I and K in each row).

M = Sales and marketing costs for the vehicles produced and shipped to the dealers (shown in column J; assumed to be $2000/vehicle).

N = Total costs (sum of values in columns F, L, and M).

O = Revenue from vehicle sales (vehicle selling price is assumed to be $28,000 [see input value in the last row of the table under column D]). (Note: It is assumed that the dealer pays the manufacturer as soon as the vehicle is shipped.)

P = Present value of total cost (in column N).

Q = Present value of revenue from vehicle sales (in column O).

R = Present value of cumulative total costs.

S = Present value of cumulative total revenue.

T = Present value of cash flow (value in column S minus value in column R).

Table 19.2 shows that the point of maximum cumulative expenditure occurred at 2 months before Job #1. The maximum cumulative expenditure in the program was $1.36 billion (see present value of cash flow [column T] in Table 19.2). The cash flow (profit) at 24 and 60 months after Job #1 was $2.41 and $4.05 billion, respectively.

Figure 19.3 presents costs incurred during the product development phases from −40 months to Job #1. The figure shows four separate traces: (a) product development manpower costs (column C in Table 19.2), (b) services and supplies costs (column D in Table 19.2), (c) facilities and tooling costs (column E in Table 19.2), and (d) total product development costs (column F in Table 19.2).

Figure 19.4 presents the cumulative present value curves of total costs (column R), revenue (column S), and cash flow (column T) for the vehicle program shown in Table 19.2. The cumulative cash curve was obtained by summing all the costs (negative values) and revenues from product sales (positive values). It should be noted that the cost and the revenue values in Figure 19.4 and Table 19.2 were considered with a 3% discount rate. However, a multiplier of 5 was used here to the discount rate in the present value computation of revenue to generate more conservative present values of the revenue.

The spreadsheet used to compute the costs is provided on the website of this book (see file called "Program Cost Flow by Months").

A financial analysis such as this one should be conducted jointly by the product planning, financial, and marketing personnel in the vehicle development team and included in the business plan of the new vehicle. The financial analysis should be

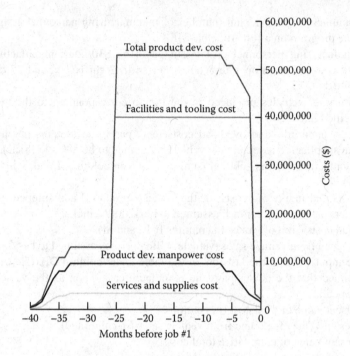

FIGURE 19.3   Costs incurred during product development.

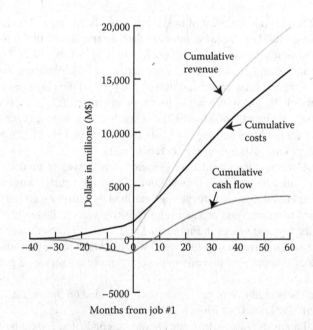

FIGURE 19.4   Cumulative present value curves of costs, revenue, and cash flow (profit) from 40 months before to 60 months after Job #1.

presented to the senior management of the company along with the vehicle concept to seek formal approval of the program.

The key points the management needs to understand from the financial analysis are: (a) the estimate of the maximum amount that the company needs to provide in undertaking the vehicle program is $1.36 billion, (b) the vehicle program can potentially pay back $2.41 billion after 24 months of vehicle production, and (c) after 36 months of production, the program can potentially pay back $3.37 billion. The outputs of the financial analysis should be incorporated in the business plan. The business plan is covered in more detail in Chapter 5.

Table 19.3 presents output of another spreadsheet program, which performs the vehicle program financial analysis at quarterly intervals. The Excel spreadsheet program is also available on the website of this book (see file called "Program Cost Flow by Quarters").

CHALLENGES IN ESTIMATING COSTS AND REVENUES

Estimating product development and production costs is generally conducted using a combination of past experience (i.e., historic data from past vehicle programs with similar vehicle configurations), a thorough understanding of tasks needed to perform to develop the proposed vehicle, coordinated efforts between different specialty disciplines, and consideration of the challenges involved in developing and incorporating new technological advances in the vehicle. The time estimates can suffer from a number of over– or under–estimating errors and associated risks, such as insufficient time allocated for verification of functionality of the new design and cost overruns due to performing too many design analyses.

Similarly, projecting future sales volumes of the proposed vehicle is also very challenging because of uncertainty over achieving the required levels of vehicle attributes and other new vehicle introductions by the competitors (which may affect sales volumes of the proposed vehicle). Since the total revenue is the product of sales volume times the vehicle purchase price, the uncertainty in predicting the purchase prices of the vehicles (which in turn, involves estimating take rates of various optional features) also affects the accuracy of revenue projections.

The next section describes two approaches used in estimating future product prices.

PRODUCT PRICING APPROACHES

TRADITIONAL COSTS–PLUS APPROACH

The traditional approach in determining the product price is to add all the costs per unit (cost of producing and selling a vehicle) and the required profit per unit to come up with the price for the unit. This approach generally does not provide strong incentives to reduce costs, as the profits for the manufacturer are assured. The approach also assumes that customers are willing to pay the price (i.e., it is the producer's market—the producer sets the price without regard to the customers). This approach worked well in the past, when customers had a very limited number of choices in the market for selecting a product.

TABLE 19.3
(PART 1) Vehicle Program Cost Flow Analysis by Quarters

No.	Description	2015			
		Q1	Q2	Q3	Q4
1	Number of salaried employees	100	300	500	600
2	Average monthly salary	$5,500	$5,500	$5,500	$5,500
3	Total salaried employee cost	$1,650,000	$4,950,000	$8,250,000	$9,900,000
4	Number of hourly employees	30	200	400	600
5	Average hourly pay	$25.00	$25.00	$25.00	$25.00
6	Total hourly employee cost	$396,000	$2,640,000	$5,280,000	$7,920,000
7	Costs of benefits (%)	28	28	28	28
8	Total employee cost	$2,618,880	$9,715,200	$17,318,400	$22,809,600
9	Tooling and equipment costs	$0	$50,000	$150,000	$500,000
10	Facilities (plants and buildings) cost	$25,000	$100,000	$2,000,000	$5,000,000
11	Operational and maintenance costs	$10,000	$20,000	$50,000	$50,000
12	Cost of raw materials and supplier–produced parts/ vehicle				
13	Average marketing, advertising, and sales–related costs/ vehicle				
14	Average plant to dealer transportation cost/ vehicle				
15	Total costs per quarter	$2,653,880	$9,885,200	$19,518,400	$28,359,600
	Number of quarters from beginning of program	1	2	3	4
	Present value (PV) of total cost per quarter (at end of each quarter)	$2,608,236	$9,548,093	$18,528,530	$26,458,330
16	Cumulative total costs (in PV)	$2,653,880	$12,539,080	$32,057,480	$60,417,080
17	Number of vehicles produced				
18	Number of vehicles sold				
19	Average selling price/ vehicle				
20	Total revenue per quarter	$0	$0	$0	$0

TABLE 19.3 (CONTINUED)
(PART 1) Vehicle Program Cost Flow Analysis by Quarters

No.	Description	2015			
		Q1	Q2	Q3	Q4
	Present value (PV) of total revenue per quarter (at end of each quarter)	$0	$0	$0	$0
21	Total cumulative revenue (in PV)	$0	$0	$0	$0
22	Vehicles in inventory (not sold)	0	0	0	0
23	Total cumulative net cash flow (in PV)	−$2,653,880	−$12,539,080	−$32,057,480	−$60,417,080

Note: Annual interest rate for present value (PV) calculations = 7%.

MARKET PRICE–MINUS PROFIT APPROACH

In this approach, the producer determines the lowest price based on the prices of other similar products sold in the markets, then subtracts his or her dealer margin and expected profit, and the balance is considered as the target cost for producing the product. The target cost is then divided and assigned to each entity in the product. All internal and external suppliers are asked to meet their respective target costs by improving the product design and their manufacturing processes and operations.

For example, in determining the price of a low–cost vehicle for the U.S. market, Hussain and Randive (2011) surveyed the prices of low–cost vehicles sold in the U.S. market. They found that the lowest price of small economy vehicles sold in the U.S. market during 2010–2011 was about $10,000. Thus, they set the target manufacturer's retail price of $8000 (20% below the lowest–price vehicle sold in the U.S. market). Assuming the dealer margin of 10% ($800) and the manufacturer's profit of $200 (2.78% of factory cost), they set the target cost at $7000/vehicle, and then proceeded to develop a target cost for each vehicle system (see Figure 19.5). This assumes that they challenged their suppliers to deliver the systems at the target costs. This approach was also used during the development of the Tata Nano, the lowest–cost vehicle, sold for about INR 100,000 ($2000) in India (Hussain and Randive, 2011).

OTHER COST MANAGEMENT SOFTWARE APPLICATIONS

Many different software applications are available to perform product life–cycle costing and to create various reports (e.g., by systems, program phases, and months; comparisons of actual versus budgeted costs). Many of the applications are integrated with other functions such as management information systems, product planning, and supply–chain management. The software systems are also used for

TABLE 19.3

(PART 2) Vehicle Program Cost Flow Analysis by Quarters

No.	Description	2016			
		Q1	Q2	Q3	Q4
1	Number of salaried employees	600	500	400	400
2	Average monthly salary	$5,650	$5,650	$5,650	$5,650
3	Total salaried employee cost	$10,170,000	$8,475,000	$6,780,000	$6,780,000
4	Number of hourly employees	800	1,200	1,500	1,800
5	Average hourly pay	$26.00	$26.00	$26.00	$26.00
6	Total hourly employee cost	$10,982,400	$16,473,600	$20,592,000	$24,710,400
7	Costs of benefits (%)	29	29	29	29
8	Total employee cost	$27,286,596	$32,183,694	$35,309,880	$40,622,616
9	Tooling and equipment costs	$2,000,000	$5,000,000	$200,000	$50,000
10	Facilities (plants and buildings) cost	$5,000,000	$1,000,000		
11	Operational and maintenance costs	$50,000	$50,000	$60,000	$75,000
12	Cost of raw materials and supplier–produced parts/vehicle			$8,000	$8,000
13	Average marketing, advertising, and sales–related costs/vehicle			$300	$500
14	Average plant to dealer transportation cost/vehicle			$700	$700
15	Total costs per quarter	$34,336,596	$38,233,694	$40,069,880	$114,347,616
	Number of quarters from beginning of program	5	6	7	8
	Present value (PV) of total cost per quarter (at end of each quarter)	$31,483,655	$34,454,008	$35,487,640	$99,529,488
16	Cumulative total costs (in PV)	$94,753,676	$132,987,370	$173,057,250	$287,404,866

TABLE 19.3 (CONTINUED)

(PART 2) Vehicle Program Cost Flow Analysis by Quarters

No.	Description	2016 Q1	Q2	Q3	Q4
17	Number of vehicles produced			500	8,000
18	Number of vehicles sold				8,000
19	Average selling price/vehicle			$22,000	$22,000
20	Total revenue per quarter	$0	$0	$0	$176,000,000
	Present value (PV) of total revenue per quarter (at end of each quarter)	$0	$0	$0	$155,873,305
21	Total cumulative revenue (in PV)	$0	$0	$0	$155,873,305
22	Vehicles in inventory (not sold)	0	0	500	500
23	Total cumulative net cash flow (in PV)	−$94,753,676	−$132,987,370	−$173,057,250	−$131,531,561

Note: Annual interest rate for present value (PV) calculations = 7%.

production scheduling, component ordering, inventory control, product control, shop floor management, cost accounting, and so forth. Some examples of such software systems are: manufacturing resource planning (MRP) and enterprise resource planning (ERP). The software systems are available from a number of developers (e.g., SAP, Oracle, Microsoft, EPICOR, and Sage).

TRADE−OFFS AND RISKS

Programs and projects involving the development of complex automotive products encounter a number of developmental problems and challenges. Many problems involve trade−offs between different attribute requirements and trade−offs between a number of design and manufacturing issues. The costs and timings are directly affected by how the trade−off issues are resolved. The design teams deal with these issues constantly during various design stages. Many of these problems are not sufficiently known in the early stages; hence, the budgets prepared during the early stages need to be constantly reviewed, and some changes in target costs and timings may need to be incorporated in subsequent budgets and milestones of the program.

TABLE 19.3

(PART 3) Vehicle Program Cost Flow Analysis by Quarters

No.	Description	2017			
		Q1	Q2	Q3	Q4
1	Number of salaried employees	300	200	200	200
2	Average monthly salary	$5,900	$5,900	$5,900	$5,900
3	Total salaried employee cost	$5,310,000	$3,540,000	$3,540,000	$3,540,000
4	Number of hourly employees	1,800	1,800	1,800	1,800
5	Average hourly pay	$27.00	$27.00	$27.00	$27.00
6	Total hourly employee cost	$25,660,800	$25,660,800	$25,660,800	$25,660,800
7	Costs of benefits (%)	30	30	30	30
8	Total employee cost	$40,262,040	$37,961,040	$37,961,040	$37,961,040
9	Tooling and equipment costs	$50,000	$50,000	$50,000	$50,000
10	Facilities (plants and buildings) cost				
11	Operational and maintenance costs	$75,000	$75,000	$75,000	$75,000
12	Cost of raw materials and supplier–produced parts/vehicle	$8,000	$8,000	$7,500	$7,500
13	Average marketing, advertising, and sales–related costs/vehicle	$1,000	$1,000	$800	$800
14	Average plant to dealer transportation cost/vehicle	$700	$700	$700	$700
	Number of quarters from beginning of program	9	10	11	12
	Present value (PV) of total cost per quarter (at end of each quarter)	$159,015,460	$178,811,236	$180,197,711	$177,098,487
16	Cumulative total costs (in PV)	$473,291,906	$685,977,946	$904,063,986	$1,122,150,026
17	Number of vehicles produced	15,000	18,000	20,000	20,000

TABLE 19.3 (CONTINUED)
(PART 3) Vehicle Program Cost Flow Analysis by Quarters

No.	Description	2017			
		Q1	Q2	Q3	Q4
18	Number of vehicles sold	12,000	20,000	20,000	20,000
19	Average selling price/vehicle	$22,000	$22,000	$22,000	$23,000
20	Total revenue per quarter	$264,000,000	$440,000,000	$440,000,000	$460,000,000
	Present value (PV) of total revenue per quarter (at end of each quarter)	$233,809,957	$389,683,261	$389,683,261	$407,396,137
21	Total cumulative revenue (in PV)	$389,683,261	$779,366,523	$1,169,049,784	$1,576,445,921
22	Vehicles in inventory (not sold)	3,500	1,500	1,500	1,500
23	Total cumulative net cash flow (in PV)	−$83,608,645	$93,388,577	$264,985,798	$454,295,895

Note: Annual interest rate for present value (PV) calculations = 7%.

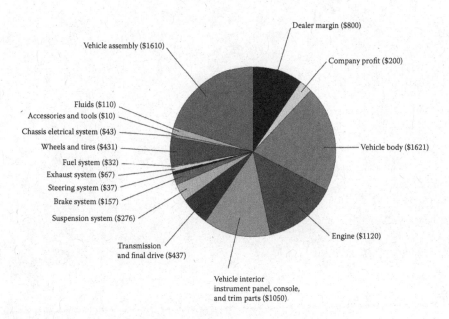

FIGURE 19.5 Low−cost vehicle target cost breakdown.

CONCLUDING REMARKS

Bringing the product to the market at the right time and at the right price is very important. Therefore, costs and timings are important parameters used by program and project managers to evaluate and control their progress. Both these parameters affect the profitability of the organization and its competitive position in the market. Since the initial estimates of these parameters made during the early planning stages are usually not very accurate, they need to be adjusted to account for problems and challenges encountered during the program. Cost overruns and timing delays are universally hated by the management. On the other hand, completion of the program before its planned end–date or under its budget is a reason for celebration of this great accomplishment and deserves special recognition for the program teams.

REFERENCES

Bhise, V. D. 2014. *Designing Complex Products with Systems Engineering Processes and Techniques*. Boca Raton, FL: CRC Press.

Blanchard, B. S. and W. J. Fabrycky. 2011. *Systems Engineering and Analysis*. 5th edn. Upper Saddle River, NJ: Prentice Hall.

Campanella, J. 1990. *Principles of Quality Costs*. 2nd edn. Milwaukee, WI: ASQC Quality Press.

Hussain, T. and S. Randive. 2010. *Defining a Low Cost Vehicle for the U.S. Market*. Published by the Institute for Advanced Vehicle Systems, College of Engineering and Computer Science, the University of Michigan–Dearborn, Dearborn, MI. Website: www.engin. umd.umich.edu/IAVS/books/A_Low_Cost_Vehicle_Concept_for_the_U.S._Market. pdf (Accessed: September 13, 2015).

20 Vehicle Package Engineering Tools

INTRODUCTION

Vehicle packaging is a key activity during the development of a new vehicle. Vehicle packaging involves creating vehicle drawings or computer-aided design (CAD) models that show locations and spaces occupied by the vehicle systems, occupants, and items brought into the vehicle. Thus, all design and engineering activities need to constantly work with the CAD models to ensure that their designs of all vehicle entities (e.g., systems, subsystems, or components) fit within the vehicle space defined by the exterior and interior vehicle surfaces.

While it is possible to work using reduced (i.e., smaller than actual, minified) scale drawings of the new vehicle and its systems, full-size drawings are often created to get a better idea of the actual space available to incorporate various vehicle systems and their components within the vehicle. The package engineers do not only create full-size scale drawings and project full-size images of the CAD models on large screens; they also create full-size mock-ups and bucks (i.e., physical models) to get a good understanding of the "feel" of the spaces required to design and fit all systems and their interfaces to ensure that a functional vehicle can be created.

The outputs of the vehicle packaging activity are thus required by all the disciplines and activities involved in the design process to visualize the product and understand the space constraints within which all the vehicle systems need to be configured.

VEHICLE PACKAGING BACKGROUND

WHAT IS VEHICLE PACKAGING?

Packaging is a term used in the automobile industry to describe the activities involved in locating various vehicle systems (e.g., body system, chassis system, powertrain system, climate control system, and fuel system) and their components and occupants in the vehicle space. Thus, it is about space allocation for various vehicle systems (i.e., hardware), accommodating people (i.e., the driver and the passengers), and providing storage spaces for various items (e.g., suitcases, boxes, golf bags, sunglasses, cell phones, and beverage containers) that people bring into their vehicles.

The term *packaging* is used in the industry because the task of package engineering is essentially bringing in systems and components produced by others (e.g., manufacturing departments and suppliers) and fitting them into the vehicle space so that they will function properly to satisfy the customers and users of the vehicle.

355

WHAT IS PACKAGED IN A VEHICLE?

The entities that the package engineers need to package in the vehicle space are

1. Occupants (driver and passengers)
2. Vehicle framework (i.e., vehicle chassis and body), which holds all the vehicle systems and provides for exterior dimensions and shape
3. All vehicle systems required for controlling the vehicle (e.g., powertrain, suspension system, steering and braking systems, electrical system, and controls and displays)
4. All vehicle systems developed for the comfort, convenience, safety, communication, and entertainment of the vehicle occupants
5. Storage spaces or compartments for items brought and stored in the vehicle (e.g., luggage, cargo, and other items such as papers, maps, beverage containers, gloves, owner's manual, purses, sunglasses, cell phones, garage door openers, CDs, glasses, coins, clip boards, pens, and ice scrapers)

VEHICLE PACKAGING ORGANIZATIONS

Vehicle packaging is a function of one or more departments (groups of people in the vehicle development process) responsible for the overall allocation of space for occupants and vehicle systems by creating drawings. It is an engineering and CAD activity. The main, or the overall, vehicle packaging function is performed in close collaboration with the styling and appearance department, in which industrial designers (specialized in automotive design) create the exterior and interior surfaces of the vehicle by developing shapes and selecting surfacing materials with visual characteristics (e.g., color and texture) and tactile feel (e.g., smoothness and compressibility). The main packaging department is thus located close to the design studios so that constant communication is facilitated while developing the basic architecture (i.e., overall size of the vehicle envelope, proportions of different major compartments, wheelbase, and front and rear overhangs) of the vehicle.

The actual tasks conducted by vehicle package engineers vary in different organizations. Some do pure drawing (i.e., drafting or CAD modeling activities), whereas others perform various degrees of engineering analyses (e.g., computations of forces and stresses, heat transfer, electrical loads, and aerodynamic flows) to support decisions related to the selection of functions performed by different systems and their performance requirements and characteristics (e.g., capacities configurations, interfaces with other systems, dimensions, and selection of materials).

SPECIALIZATION WITHIN VEHICLE PACKAGE ENGINEERING

There are two specializations within vehicle package engineering work:

1. *Occupant (People:Driver and Passengers) Packaging*: This specialization includes knowledge of design practices and procedures in (a) determining driver positioning: driver location (seating reference point [SgRP]) and

driver seating posture; location of primary controls, i.e., steering column, steering wheel, gear shifter, and pedals, (b) packaging of instrument panel, console, and parking brake, (c) positioning and postures of passengers (design and locations of seats), (d) entry and egress evaluations involving items such as body openings, seats, doors, grab handles (or handle bars), door opening handles, and steps), (e) clearances required to orient and position various components (i.e., people, hardware, and service tools), and (e) field of view analyses (window openings, mirrors, and obscurations). The vehicle ergonomic department usually works closely with the occupant packaging department to ensure that various design tools, ergonomics guidelines, and data are available for accommodating a large portion of the vehicle user population during all packaging activities.

2. *Mechanical (Hardware) Packaging*: This specialization includes knowledge of design practices and procedures in (a) body structure space planning (estimation of shapes, sizes, and cross sections of various body and chassis parts), (b) powertrain and fuel systems packaging (estimating and positioning of envelopes of engines, transmissions, and final drives, including pipes, hoses, wiring, and fittings), (c) chassis system packaging (spaces for wheels, tires, suspensions, and steering and braking systems), (d) packaging of other mechanical and electrical systems (e.g., instrument panel, doors, console, door locks, power window mechanisms, latches, hinges, wiring harness, lamps, heating, and air-conditioning systems), and (e) luggage/cargo area and storage spaces. Various engineering offices (e.g., body engineering, powertrain engineering, climate control engineering, electrical engineering, fuel systems engineering, and manufacturing engineering) work closely with the mechanical packaging engineers to ensure functional, manufacturing, and assembly feasibility during various hardware packaging activities.

VEHICLE PACKAGING PERSONNEL

The technical background of people involved in vehicle packaging varies between different organizations. The employees in vehicle package engineering typically include drafters (or draftsmen, who used to draw 2-D drawings; called *engineering designers*) and CAD modelers/designers and engineers (primarily mechanical engineers), who develop 3-D models that are useful in product visualization. With the integration of CAD and computer-aided engineering (CAE) software applications, more companies are now assigning these tasks to engineers with master's degrees.

Vehicle package engineers need to understand (a) functioning, spatial needs, and interfaces of various vehicle systems, (b) customer wants and needs, (c) customer characteristics and ergonomic design considerations, (d) design (styling) trends and wants, (e) engineering requirements and standards (system design standards and applicable government requirements), and (e) manufacturing and assembly processes and requirements. Thus, they can create a vehicle package that maintains compatibility between the basic vehicle architecture, functioning of systems, and manufacturing and assembly requirements.

The package engineers are responsible for the development, design, verification, and delivery of digital data that represent various vehicle systems and component designs. They are part of a team that develops and manages the requirements for delivering mechanical package compatibility. Their key job responsibilities are to drive mechanical package compatibility by working in conjunction with design engineers, core functional engineering departments, suppliers, CAD, manufacturing, and studio personnel. The package engineers are required to lead the resolution of issues and trade-off discussions to ensure system designs that deliver digital mechanical package compatibility.

Typically, package engineers are required to have a bachelor's degree in mechanical engineering, with a preference for a master's degree in a related engineering field, and work experience and skills as follows:

1. Automotive/aerospace experience in product development and/or manufacturing
2. Demonstrated ability to proactively lead cross-functional teams working to resolve issues
3. Proficiency in technical problem-solving skills to resolve mechanical package issues
4. Demonstrated negotiation skills in teamworking to broker design discussions and design solutions (i.e., studying alternative designs and process trade-offs between different vehicle attributes)
5. Proficiency in CAD skills: training and certifications in CAD, dimensional engineering, and digital buck generation, and working knowledge of CAD systems (e.g., CATIA, Unigraphics)
6. Presentation skills required for management reviews (oral and written communications)
7. Ability to guide and work with physical model builders (in wood, metal, upholstery, and plastic workshops) to create and verify interior and exterior bucks used for product reviews and market surveys
8. Knowledge of the systems engineering process and techniques
9. Project management skills and data-driven approach to managing compatibility deliverables
10. Self-starting, with ability to prioritize and manage multiple tasks with minimal direction
11. Ability to learn and apply new techniques/skills quickly (on-the-job training)
12. Proficiency in standard PC skills (Outlook, Excel, Word, PowerPoint) and accessing databases on corporate standards, lessons learned, bill of materials, and so forth

PACKAGE ENGINEERING AND ERGONOMICS

Many of the occupant packaging tools and data on driver characteristics used by the vehicle packaging engineers have been developed and updated by ergonomics engineers working in the automotive industries, research institutes, or universities.

Figure 20.1 presents a Venn diagram illustrating overlapping and nonoverlapping technical areas in vehicle package engineering and ergonomics (also called *human factors engineering*).

Ergonomics engineers work with vehicle package engineers in reviewing the vehicle design and packaging issues and provide the latest information on occupant packaging procedures, driver anthropometric dimensions, and biomechanical and driver interface design considerations. The design considerations involved here are shown in Figure 20.1 as the intersection of the two disciplines in the Venn diagram, which include location and positioning (i.e., posturing) of the driver and occupants, selection and locations of controls and displays, design of seats to allow comfortable accommodation in the postures preferred by the driver and passengers, easy entry into and exit from the vehicle, and evaluation of visual fields available to the driver and passengers through the window openings and other vision systems (e.g., mirror systems). Bhise (2012) provides details of issues and tasks performed by ergonomics engineers in the automotive design process.

The left side of the Venn diagram shows other tasks performed by vehicle package engineers in other functional areas, such as body engineering, powertrain engineering, chassis engineering, and electrical engineering. The right side of the Venn diagram shows the tasks performed by ergonomics engineers to ensure that other driver interfaces (i.e., controls and displays, seats) of other systems, such as those mounted in the instrument panel, doors, header, and center console, can be operated by the driver easily and without excessive driver workload or distraction.

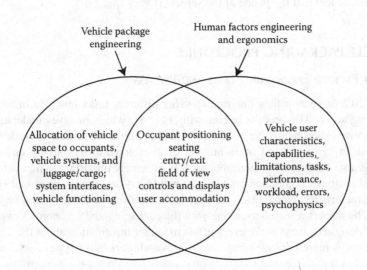

FIGURE 20.1 Venn diagram showing overlap between vehicle package engineering and ergonomics activities.

PRINCIPLES USED IN VEHICLE PACKAGING

The packaging engineering work is performed considering many principles. The three basic principles underlying package engineering work are

1. *Customer satisfaction*: This involves striving to achieve product quality by simultaneous consideration of all vehicle attributes and ensuring that all decisions involving trade-offs between all attributes are made to achieve high levels of customer satisfaction.
2. *Systems consideration*: The vehicle should be designed as a system with a careful balance between "forms" and "functions." It should be noted that "form" here refers to styling and appearance attributes, and "function" refers to the operational performance of each of the vehicle systems. Thus, the package engineer's task is not just to package a collection of components; each vehicle system must be located to meet its allocated functions and interface requirements and at the same time meet the needs of the design/styling department. The functionality of each system must be considered to ensure that the systems and their lower-level entities (i.e., subsystems, sub-subsystems, and so on, until the lowest component level) fit within the vehicle space and meet their respective requirements, which are cascaded down from all vehicle-level attributes (see Chapters 8 and 9 for more details).
3. *People maximum, machine minimum*: This principle refers to minimizing the space taken up by the hardware (machine) so as to provide larger usable space for the occupants (people). The goal should be to allocate the maximum amount of space within the occupant compartment and trunk/cargo areas to achieve the feeling of interior "spaciousness" (i.e., the customers should feel that the inside of the vehicle is very spacious).

VEHICLE PACKAGING PROCEDURE

VEHICLE PACKAGE ENGINEERING TASKS AND PROCESS

Figure 20.2 presents a flow diagram showing different tasks involved in occupant packaging work. The process begins with Task #1, which involves understanding the customers, benchmarking competition, and defining the vehicle to be designed. This task involves inputs from a number of disciplines to thoroughly understand the program objectives and assumptions. It is extremely important to first define the intended customer population, that is, who would buy and use the proposed vehicle. The characteristics, capabilities, desires, and needs of these users must be understood. The market researchers, along with the packaging and ergonomics engineers and the designers, must make every effort to gather information about the intended population. A representative sample of owners and users of the type of vehicle from its intended market segment (e.g., luxury small four-door car, economy two-door hatchback, or mid-size luxury sports utility vehicle [SUV]) should be invited and shown early product concepts. They should be extensively interviewed and asked to

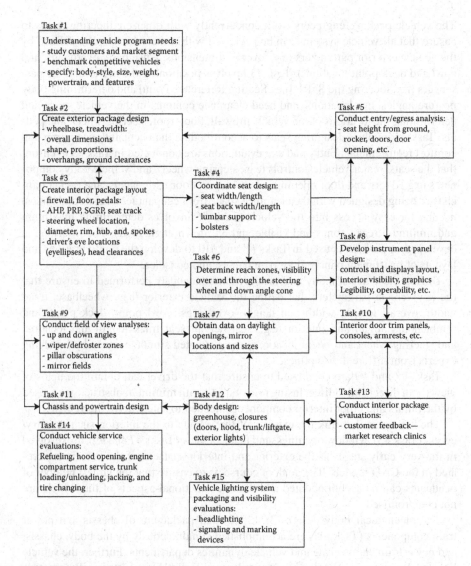

Task #1
Understanding vehicle program needs:
- study customers and market segment
- benchmark competitive vehicles
- specify: body-style, size, weight,
  powertrain, and features

Task #2
Create exterior package design
- wheelbase, treadwidth:
- overall dimensions
- shape, proportions
- overhangs, ground clearances

Task #5
Conduct entry/egress analysis:
- seat height from ground,
  rocker, doors, door
  opening, etc.

Task #3
Create interior package layout
- firewall, floor, pedals:
- AHP, PRP, SGRP, seat track
- steering wheel location,
  diameter, rim, hub, and, spokes
- driver's eye locations
  (eyellipses), head clearances

Task #4
Coordinate seat design:
- seat width/length
- seat back width/length
- lumbar support
- bolsters

Task #8
Develop instrument panel
design:
controls and displays layout,
interior visibility, graphics
Legibility, operability, etc.

Task #6
Determine reach zones, visibility
over and through the steering
wheel and down angle cone

Task #9
Conduct field of view analyses:
- up and down angles
- wiper/defroster zones
- pillar obscurations
- mirror fields

Task #7
Obtain data on daylight
openings, mirror
locations and sizes

Task #10
Interior door trim panels,
consoles, armrests, etc.

Task #11
Chassis and powertrain design

Task #12
Body design:
greenhouse, closures
(doors, hood, trunk/liftgate,
exterior lights)

Task #13
Conduct interior package
evaluations:
- customer feedback—
  market research clinics

Task #14
Conduct vehicle service
evaluations:
Refueling, hood opening, engine
compartment service, trunk
loading/unloading, jacking, and
tire changing

Task #15
Vehicle lighting system
packaging and visibility
evaluations:
- headlighting
- signaling and marking
  devices

FIGURE 20.2 Vehicle packaging engineering tasks and process.

respond to a number of questions related to how well they like or dislike the product
concepts and characteristics of the products, their preferences for alternative vehicle
designs, their habits related to vehicle uses, and so forth. Their relevant anthropometric dimensions can also be measured to create a database for evaluation of various
vehicle dimensions. Quality function deployment (QFD) is an excellent tool, and it
can be used at this early stage to translate the customer needs of the vehicle being
designed into functional (engineering) specifications for the vehicle (see Chapter 18;
Besterfield et al., 2003; Bhise, 2012).

The exterior design, as shown in Task #2, usually leads the design process. The
exterior shape of the vehicle is developed by creating a number of design alternatives.

The vehicle package engineers work concurrently with other engineering teams to ensure that all vehicle systems can be packaged within the vehicle space defined by the vehicle exterior parameters (e.g., overall dimensions, wheelbase, overhangs, and cowl and deck point locations). Task #3 involves positioning the driver and the passengers (i.e., locating the SgRP [i.e., Seating Reference Point] and determining body posture angles, eye locations, and head clearance contour) in the vehicle space and determining the locations of the vehicle firewall, floor, roof, steering wheel, and pedals. Task #4 involves designing seats to accommodate the occupant positioning and comfort requirements. Entry and exit evaluations are conducted in Task #5 to ensure that the seats, major vehicle controls (e.g., steering wheel), and vehicle body components (e.g., doors and door openings, door hinges, door trim panels, and rocker panels) are being designed with the clearance space (for occupant feet, heads and torsos) needed for easy access into the vehicle. Task #6 involves determining maximum and minimum reach zones and visible and 35° down angle areas. The information developed in Task #6 is used in Tasks #8 and #10 to develop the configurations and layouts of instrument panels, door trim panels, and consoles.

During these tasks, many analyses are simultaneously performed to ensure that the key vehicle parameters that define the vehicle exterior (e.g., wheelbase, tread width, overall length, width and height, overhangs, cowl point, deck point, and tumblehome angle) and interior (e.g., seat height, seat track length and location, and steering wheel and pedal locations) are evaluated simultaneously by involving experts from different disciplines.

Tasks #7 and #9 are conducted to ensure that the driver can obtain the fields of view (both direct and indirect [using mirrors] and with minimum obstructions caused by the pillars and other interior components) needed to safely drive the vehicle.

The key areas that link the exterior of the vehicle to the interior, such as entry/egress (Task #5), window openings, and fields of view (Tasks #7 and #9), are resolved in the very early stages as the exterior and interior surfaces of the vehicle are created in the CAD models. The goal, of course, is to ensure that the largest number of occupants can be accommodated and that the functional aspects of the vehicle are not compromised.

The mechanical body design (Task #12) and packaging of chassis and power train components (Task #11) are accomplished simultaneously by the body, chassis, and powertrain engineering and vehicle dynamics departments. Further, the vehicle lighting design and packaging of exterior lamps (Task #15) and interior illuminating light sources (e.g., interior dome, map/reading and convenience lamps, and illumination of lighted graphics and components in Task #8) are conducted to ensure that the vehicle can be used safely and comfortably during nighttime.

A number of special evaluations are also conducted to verify that the drivers and the passengers can enter the vehicle and exit from the vehicle comfortably (Task #5) and that various interior vehicle features and their dimensions related to space (e.g., headroom, legroom, shoulder room, and hip room) are acceptable to the customers by conducting market research clinics (Task #13). Various evaluation methods used in the entire vehicle development process are summarized in Chapter 21.

Many of these tasks are performed simultaneously through constant communication and reviews by engineers and experts from various engineering, marketing, and management personnel to ensure that all important customer needs and trade-offs between various design (styling and appearance), engineering, and manufacturing requirements are considered to produce a superior vehicle package by comparison with other leading benchmarked products.

The next sections of this chapter provide details related to the dimensions and positioning procedures involved in Tasks #2 through #10. Additional information on many of the tasks can also be obtained from Bhise (2012).

STANDARD PRACTICES USED IN VEHICLE PACKAGING

Most automotive companies and their suppliers use vehicle packaging standards and practices developed by SAE International, Inc. (formerly called the Society of Automotive Engineers, Inc. [SAE]). The SAE standards are developed by the SAE Technical Committees. The committee members are professionals working in the automotive industry, government (regulatory and research departments), and universities (faculty members and researchers). The SAE standards are voluntary unless they are adopted in the U.S. Federal Motor Vehicle Safety Standards (FMVSS) or other government standards (National Highway Traffic Safety Administration [NHTSA], 2015). The SAE J1100 standard provides definitions of many exterior and interior dimensions and reference points used in the industry (SAE, 2009).

All linear vehicle- and occupant-related SAE dimensions are generally measured in millimeters. The prefixes L, H, and W denote length- (longitudinal—from front to rear of the vehicle), height- (vertical—up/down), and width- (lateral—left/right) related dimensions, respectively. All angles are designated by the prefix A and are measured in degrees. The numbers following the prefix define specific dimensions; for example, H30-1 specifies the height of the driver's SgRP from the driver's right heel point on the undepressed accelerator pedal measured in millimeters. The number 1 following H30 denotes the first seat row. (See SAE standard J1100 in the SAE Handbook [SAE, 2009] for more details on the nomenclature and dimensions.) The coordinate system for vehicle design and the x,y,z coordinates of locations of components in vehicle space are specified in the SAE 1100 and J182 standards.

MECHANICAL PACKAGING

Mechanical packaging involves the incorporation of geometric representations of all vehicle hardware into the CAD model of the vehicle. The mechanical packaging work begins with the creation of vehicle axes and the overall vehicle envelope, defined by the following three SAE dimensions: (1) L103: overall length, (2) W103: overall width, and (3) H101: overall height. Figure 20.3 presents the overall vehicle envelope.

The vehicle axes X, Y, and Z are defined and drawn as shown in Figure 20.4 (refer to SAE standard J 182 [SAE, 2009]).

The 3-D Cartesian coordinate system used to define locations of various points in the vehicle space is generally defined as follows:

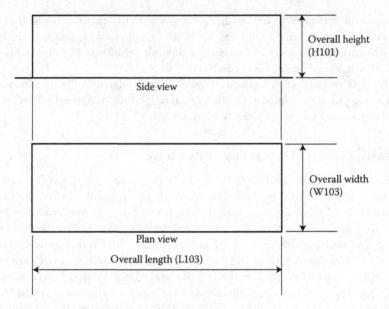

FIGURE 20.3   Overall vehicle envelope.

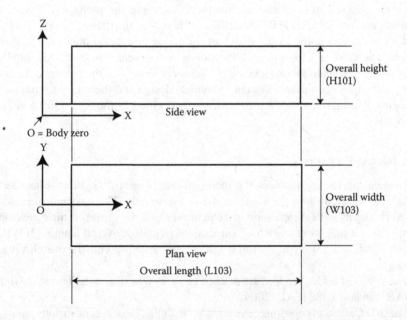

FIGURE 20.4   Overall vehicle envelope with vehicle coordinate axes X, Y, Z and origin O.

1. The positive direction of the longitudinal X-axis is pointing from the front to the rear of the vehicle.
2. The positive direction of the vertical Z-axis is pointing from the ground up.
3. The positive direction of the lateral Y-axis is pointing from the left side of the vehicle to the right side.
4. The origin of the coordinate system (called the *body zero*) is located forward of the front bumper (to make all x-coordinate values positive), below the ground level (to make all z-coordinate values positive), and at the midpoint of the vehicle width.

Figure 20.5 shows the locations of the wheels (wheel and tire diameters, wheelbase, front and rear tread widths, and overhangs). It also shows the cowl and deck points, which indicate the highest points on the hood and deck-lid, respectively, at the vehicle centerline. The cowl and deck points represent the intersections of the windshield and hood surfaces, and the backlite and trunk surfaces, respectively, in the vertical plane passing through the vehicle centerline. These two points are used by package engineers to indicate the height of the space needed to package the engine under the hood (cowl point) and luggage space under the rear deck-lid (deck point).

Since the engine is typically the largest and most important functional system, the engine envelope is created and located within the vehicle space (see Figure 20.6). The engine envelope defines the overall 3-D space with all required clearances for the engine (along with all other attached accessories, such as the alternator, steering

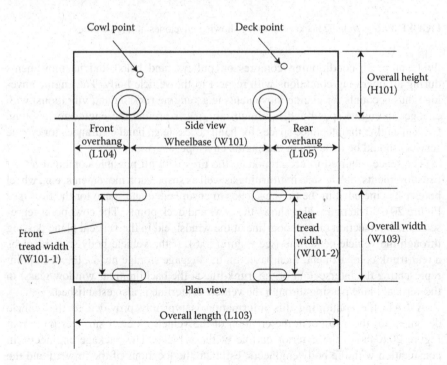

FIGURE 20.5 Vehicle envelope with locations of wheels and cowl and deck points.

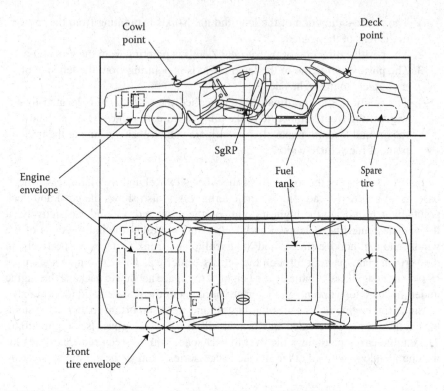

FIGURE 20.6   Vehicle package drawing showing envelopes of large entities.

fluid pump, air-conditioning compressor, pulleys, and belts) and its movements during all operating conditions with respect to the vehicle body. The engine envelope thus accounts for engine movements (e.g., engine rocking and vibrations) with changes in engine speed and minimum air spaces around the engine for cooling. Critical engine maintenance spaces for hand access (e.g., hand clearances for engine service) should be included in the engine envelope.

The space required to accommodate the tires with all possible combinations of tire movements (due to steering/turning as well as suspension movements, e.g., wheel bounces) is included in the tire envelopes to ensure adequate space for the tires (see Figure 20.6). The figure also shows the cowl and deck points. The cowl point represents the intersection of the hood-line at the windshield in the vertical plane passing through the vehicle centerline (see Figure 20.6). If the vehicle body-style involves a rear trunk compartment, then based on the luggage storage needs, the deck point representing the intersection of the trunk-line at the backlite (rear window glass) in the vertical plane passing through the vehicle centerline is also established.

A CAD file containing this information is typically provided to the vehicle designers for them to begin the creation of the vehicle exterior surfaces (note that Figure 20.6 shows the exterior outline of the vehicle). The package engineers, in consultation with the body engineers, estimate the locations of the firewall and the vehicle floor by taking into account the ground clearance and space required to create the floor board and floor cross members. The occupant compartment space is thus established between the firewall and the back side of the rear seats (the space rearward of the rear seat generally defines the boundary between the occupant compartment and the luggage compartment).

The mechanical packaging engineers maintain constant communication with the designers (to understand their exterior and interior styling needs) and begin the incorporation of the following hardware, which requires large spaces:

1. Vehicle body details (i.e., cross sections) of major body components (e.g., body frame, pillars, cross members, rocker panel, roof rails, roof headers, front and rear fascia with lighting equipment [headlamps and tail lamps])
2. Powertrain system (engine, transmission, and final drive)
3. Suspension systems (front and rear suspensions)
4. Wheels and tires (front and rear tire envelopes and spare tire)
5. Steering system (steering column, power-assist mechanism and linkages)
6. Fuel system (fuel tank, filler pipe, fuel system module, and fuel lines)
7. Engine cooling system (radiator, hoses, and coolant pump)
8. Climate control system (heat exchangers, air mixing chamber, blower, compressor, and air ducts)
9. Electrical system (battery, alternator, electronic processing units, and wiring harnesses)
10. Pedal box, linkages, and braking system master cylinder

Occupant Packaging

The occupant compartment package is created by using the following basic steps. The steps are described in more detail in later sections of this chapter (see "Driver Package Development Steps and Calculations" section).

1. *Locate vehicle floor line* (in side view: top of the carpet on the floor panel taking into account ground clearance and space required for cross members of the floor; see Figure 20.7).
2. *Locate firewall* taking into account space requirements of the engine compartment entities and pedal placement (see vertical dotted line in Figure 20.7).
3. *Determine the driver's seating location* by establishing the location of the SgRP. It is located at (a) H30-1 vertical distance above the vehicle floor (see Figure 20.7), (b) X_{95} horizontal distance rearward from the ball of the driver's foot (see Figure 20.8) (BOF; also called pedal reference point [PRP]) on the accelerator pedal, and (c) W20-1 lateral distance from the vehicle centerline.

   The driver's BOF on the accelerator pedal is located 203 mm from the driver's accelerator heel point (AHP) along the pedal plane line in the side view (refer to SAE procedures specified in SAE J826, J1517, and J4004 standards). The accelerator pedal angle is computed by using a quadratic equation as a function of H30-1 (refer to SAE J1516).

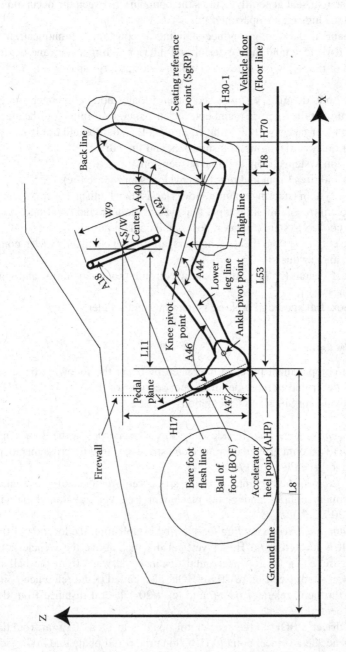

FIGURE 20.7   Interior package side view with reference points and dimensions.

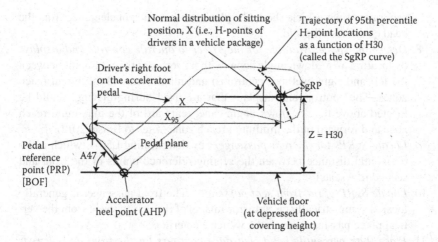

FIGURE 20.8 Distribution of horizontal location on H-points and 95th percentile H-point location.

It should be noted that H30-1 (the height of the driver's SgRP) above the vehicle floor is an important dimension in establishing the height of the cab (which in turn, is determined by the body-style of the vehicle). The top of the vehicle floor is defined by the driver's right heel contact point on the carpet (or rubber mat) on the vehicle floor. The driver's heel point is defined by the location of the driver's right foot (i.e., heel) on the top of the depressed carpet (due to the weight of the foot) on the vehicle floor when the foot is on the undepressed accelerator pedal.

4. *Determine seat track length* by computing $X_{2.5}$ and $X_{97.5}$ to determine horizontal seat track length (TL1) using SAE J1517 or J4004. Note that $X_{2.5}$ and $X_{97.5}$ are the 2.5 and 97.5 percentile horizontal locations of driver seating positions as measured by the hip point locations of drivers from the driver's BOF on the accelerator pedal.

5. *Determine drivers' eye locations* by drawing 95th percentile eyellipses using the SAE J941 procedure. The equations for determining the centroids of the left and right ellipsoids (with respect to the SgRP), the lengths of the three axes of the ellipsoids, and the forward tilt angle of the ellipsoids are also provided in SAE J941.

6. *Locate steering wheel* by determining the W9, W7, L11, and H17 dimensions (see Figure 20.7). The steering wheel should be rearward of the maximum reach distance (obtained from SAE J287) and forward of the minimum comfortable reach distance (Bhise, 2012). Also, evaluate obscurations caused by the steering wheel (using SAE J1050) and thigh clearance (for the driver's thigh during entry and exit) between the bottom of the steering wheel and the seat cushion.

7. *Locate 99th percentile head clearance contour* using the SAE J1052 procedure. The head clearance contour is an ellipsoidal surface that defines 99% of the tops, sides, and backs of heads of drivers. The roof liner, header, and

roof rails of the vehicle should be placed with sufficient clearance from the head clearance ellipsoid.

8. *Determine space available for the driver to operate controls and displays* by locating a 35° down angle cone (with its vertex at the mid-point between the left and right eyellipse centroids) and maximum and minimum reach zones. The controls and displays that are used during driving should be located above the 35° down angle cane, rearward of the maximum reach zone and forward of the minimum reach zone (refer to Bhise, 2012).

9. *Locate SgRPs for the rear passengers* by determining L50-1 (which is the horizontal distance between the seating reference points of the front and second row seats).

10. *Locate SgRP of the front seat passenger*. The front passenger is generally located symmetrically (as a mirror image) of the driver's seat from the vertical plane passing through the vehicle centerline.

11. *Place 99th percentile head clearance contours for the front and all rear passengers*. The roof liner, rear header, and roof rails of the vehicle should be placed with sufficient clearance from the head clearance ellipsoid.

12. *Location of seats*. The seats are located based on the SgRPs of the occupants. The SgRP location of each seat must be placed at the corresponding SgRP location shown in vehicle drawings or a CAD model.

13. *Location of the instrument panel*. The instrument panel is located based on the maximum and minimum reach and visibility requirements (visibility through the steering wheel and visibility over the top of the steering wheel, the instrument panel top, and the hood) (refer to Bhise, 2012).

14. *Evaluate the interior package*. Evaluations are generally conducted to assess the acceptability of seating positions, locations of various interior items, and clearances by asking customers (representative subjects) to sit in a full-size vehicle package buck (or a programmable vehicle buck or a virtual reality simulator) configured to the proposed vehicle dimensions and to provide ratings on a number of package dimensions (see next section and Chapter 21 for details).

CAD Models and Package Bucks

A CAD model of the vehicle being designed is created by the package engineers to illustrate the outputs of all mechanical and occupant packaging steps. The model helps in understanding the 3-D aspects of the exterior and interior of the vehicle. The model also shows how various entities within the vehicle are located, configured, and interfaced with other entities within the vehicle space.

The CAD model can be viewed on computer screens or projected on a large screen to get a better idea of the true size of the vehicle. The vehicle model can also be rotated and viewed from different viewing locations and orientations. However, CAD models, even when projected in virtual reality simulators such as computer-aided virtual environments (CAVEs), do not give a realistic sense of the true size and spatial aspects of the vehicle. Therefore, the package engineers create physical models (or bucks) to illustrate the vehicle design and to conduct design reviews, evaluations, and verification of the locations of the occupants and various entities in the vehicle space.

A full-size package buck is a mock-up created (using wood, fiberglass, an aluminum frame structure, etc.) to allow assessment of the interior space, locations of primary controls, entry/egress, visibility, loading/unloading, and storage compartments. The interior buck typically includes seats, steering column and steering wheel, pedals, and surfaces to represent the instrument panel, door trim panels, center console, headliner and daylight (window) openings. Programmable vehicle models (PVMs) can be also used. PVMs are computer-controlled adjustable bucks that can quickly adjust package dimensions to input parameters. Additional information on package bucks and PVMs is presented in Chapter 13.

INTERIOR PACKAGE REFERENCE POINTS AND SEAT TRACK–RELATED DIMENSIONS

The reference points used for the location of the driver and their relevant dimensions are

1. *AHP*: This is the heel point of the driver's right shoe that is on the depressed floor covering (carpet) on the vehicle floor when the driver's foot is in contact with the undepressed accelerator (gas) pedal (see Figure 20.7). SAE standard J1100 defines it as "A point on the shoe located at the intersection of the heel of shoe and the depressed floor covering, when the shoe tool (specified in SAE J826 or J4002) is properly positioned (i.e., the ball of foot contacting the lateral centerline of the undepressed accelerator pedal, while the bottom of shoe is maintained on the pedal plane)."

2. *Pedal Plane Angle (A47)*: This is defined as the angle of the accelerator pedal plane in the side view measured in degrees from the horizontal (see Figure 20.7). The pedal plane is not the plane of the accelerator pedal but the plane representing the bottom of the manikin's shoe, defined in SAE J826 or J4002. (A47 can be computed using the equations provided in SAE J1516 or J4004, or it can be measured using the manikin tools described in SAE J 826 or J4002 (see Step 2 in "Driver Package Development Steps and Calculations" later in this chapter).)

3. *BOF*: This is the point on the top portion of the driver's foot that is normally in contact with the accelerator pedal. The BOF is located 200 mm from the AHP measured along the pedal plane (SAE J4004, SAE 2009).

4. *PRP*: This is on the accelerator pedal lateral centerline where the ball of foot (BOF) contacts the pedal when the shoe is properly positioned (i.e., heel of shoe at AHP and bottom of shoe on pedal plane). SAE standard J4004 provides a procedure for locating the PRP for curved and flat accelerator pedals using the SAE J4002 shoe tool. If the pedal plane is based on SAE standards J826 and J1516, the BOF point should be taken as the PRP.

5. *SgRP*: This is the location of a special hip point (H-point) designated by the vehicle manufacturer as a key reference point to define the seating location for each designated seating position. Thus, there is a unique SgRP for each designated seating position (e.g., the driver's seating position, the

front passenger's seating position, and the left rear passenger's seating position). An H-point simulates the hip joint (in the side view as a hinge point) between the torso and the thighs, and thus, it provides a reference for locating a seating position. In the plan view, the H-point is located on the centerline of the occupant.

The driver's SgRP is specified as follows:

a. It is designated by the vehicle manufacturer.

b. It is located near or at the rearmost point of the seat track travel.

c. The SAE (in standards J1517 or J4004) recommends that the SgRP should be placed at the 95th percentile location of the H-point distribution obtained by the seat position model (called the *SgRP curve*) at an H-point height (H30-1) from the AHP specified by the vehicle manufacturer.

The driver's SgRP is the most important and basic reference point in defining the driver package. The driver's SgRP must be established early in the vehicle program and should not be changed later during the vehicle development process, because

i. It determines the locations of the driver and the seat track in the vehicle package.

ii. All driver-related design and evaluation analyses are conducted with respect to this point, that is, location of eyes, interior and exterior visibility fields, specifications of spaces (e.g., headroom, legroom, and shoulder room), reach zones, locations of controls and displays, and door openings (for entry/exit). Thus, any change in the SgRP location will require recalculation of other reference points and analyses.

The occupant positioning tools available are

H-Point Location Model: The original H-point location model was developed by Philippart et al. (1984) based on measurements of preferred sitting locations of a large number of drivers in actual vehicles with different package parameters. The sitting position of each driver was defined as the location of the driver's H-point. The H-point location was determined by the horizontal seat track position selected by the driver at the seat height (measured by H30; see Figure 20.8) in the vehicle. (Note: The H30 dimension for the driver's seating height is designated as H30-1 in newer versions of the SAE standards.) For any given vehicle, the H-point of a population of drivers can be represented by their distribution of horizontal locations. Figure 20.8 shows the distribution of the horizontal location (X) of the H-points. The 95th percentile value of the distribution is generally selected as the location of the SgRP, as shown in Figure 20.8. The SgRP is defined as the point located at X_{95} horizontal distance from the BOF point and H30 vertical distance from the AHP. The trajectory of X_{95} locations as a function of H30 is called the *SgRP curve* (see Figure 20.8). The equation of the SgRP curve (provided in SAE standards J1516 and J4004a) is provided in a later section of this chapter (see Step 1 in "Driver Package Development Steps and Calculations").

H-Point Location Fixtures: The SgRP can be located in a physical property (i.e., an actual vehicle or a package buck) by placing the SAE H-point machine (HPM) specified in SAE standard J826 or the H-point device (HPD) specified in SAE standard J4002. The HPM and HPD are 3-D fixtures, and they can be placed in a seat at any designated seating location to measure or verify the location of the SgRP at that location.

The HPM is referred to in the auto industry and by some seat manufacturers as the *OSCAR*. Since the seat is compressible and flexible, the HPM is placed on the seat and used as a development and verification tool by seat manufacturers and vehicle manufacturers. The tool is used to determine whether the SgRP of the seat that is built and installed in an actual vehicle falls within the manufacturing tolerances from the design SgRP location (shown in the vehicle CAD model or drawings). The description and procedure for location of the HPM are provided in SAE standard J826. The HPD is designed with a three-segmental back pan to account for the effect of the shape of the seat back (especially in the lumbar region). SAE standard 4002 provides drawings, detailed specifications, and procedures for the use of the HPD.

The SAE HPM and HPD (HPM in SAE standard J826; HPD in SAE standard 4002) are designed such that when they are placed in a seat they deflect the seat similarly to the way a real person will deflect the seat. Each device weighs 76 kg (167 lb, which is the 50th percentile U.S. male weight) and has the torso contour of a 50th percentile U.S. male. The devices use 95th percentile legs (10th and 50th percentile leg lengths are also available).

6. *Seat Track Length*: This is defined as the horizontal distance between the foremost and rearmost locations of the H-point of a seated driver. To accommodate 95% of the driver population with a 50:50 male-to-female ratio, the foremost and the rearmost points can be defined by determining 2.5 and 97.5 percentile H-point locations from the BOF on the accelerator pedal. The computation procedures for determining different percentile values are specified in SAE standards J1517 and J4004. SAE standard J1517 was replaced by SAE standard J4004 standard, and the SAE recommends that J4004 should be used to determine the seat track length and the accommodation levels for the U.S. driving population. It should be noted that since the introduction of SAE standards J4002, 4003, and 4004, the package engineering community within various automotive companies is slowly transitioning from the old (J826, J1516, and J1517) procedures to the revised (J4002, 4003, and J4004) procedures. Therefore, relevant information from both the procedures is provided here.

The original seat position location model developed by Philippart et al. (1984) was included in SAE standard J1517. SAE standard J1517 was thus developed by measuring the actual seated positions of large numbers of drivers in vehicles with different H30 values (after they had driven the vehicles and adjusted the seat location to their preferred position) (Philippart et al., 1984). Therefore, the H-point location model is based on "functional" anthropometric data (i.e., real drivers seated in actual vehicles in their preferred driving posture). SAE standard J1517, entitled "Driver-Selected Seat Position," provides statistical prediction equations for seven percentile values ranging from 2.5 to 97.5 of H-point locations in the vehicle space. The 2.5 and 97.5 percentile H-point location prediction equations are generally used to establish seat track travel to accommodate 95% of drivers. The equations are quadratic functions of H30 for Class A vehicles (passenger cars and light trucks) and linear functions of H30 for Class B vehicles (medium and heavy commercial trucks). The Class A vehicle equations are based on a 50:50 male-to-female ratio. The Class B vehicle driver-selected seat position lines are specified in SAE J1517 for 50:50, 75:25, and 90:10 to 95:5 male-to-female ratios.

SAE standard J4004 presents an H-point location procedure based on more recent work by Flannagan et al. (1996, 1998) for Class A vehicles. The recommended seat track lengths to accommodate different percentages of drivers and the reference location (X_{ref}) for placement of the seat track and the PRP are provided in SAE standard J4004. X_{ref} is defined as a linear function of H30, steering wheel location (L6), and type of transmission (with or without a clutch pedal). (SAE standard J4004 suggests that until the year 2017, the BOF and AHP determined according to SAE standard J1517 may be used in lieu of the pedal reference point cited in J4004. However, SAE standard J4004 should be used to determine the seat track length and the accommodation levels for the U.S. driving population.)

According to SAE standard J4004, the seat track length should be 206, 240, and 271 mm to accommodate 90%, 95%, and 97.5% of drivers, respectively.

The equations illustrating these two procedures are provided in a later section of this chapter (see page 381).

INTERIOR DIMENSIONS

A number of interior package dimensions shown in Figure 20.7 are described in this section. The dimensions are defined using the nomenclature specified in SAE standard J1100.

1. *Location of SgRP from AHP*: The horizontal and vertical distances between the AHP and the SgRP are defined as L53 and H30, respectively (see Figure 20.7).
2. *Posture Angles*: The driver's posture is defined by the angles of different body segments (defined by lines of the body segments, such as torso line, thigh line, and lower leg line) of the HPM or the HPD. The angles shown in Figure 20.7 are defined as follows:
   a. Torso angle (A40): This is the angle between the torso line (also called the backline) and the vertical. It is also called the *seatback angle* or the *back angle*.
   b. Hip angle (A42): This is the angle between the thigh line and the torso line.
   c. Knee angle (A44): This is the angle between the thigh line and the lower leg line. It is measured on the right leg with the right foot positioned on the accelerator pedal.
   d. Ankle angle (A46): This is the angle between the (lower) leg line and the bare foot flesh line, measured on the right leg with the right foot on the accelerator pedal.

   e.   Pedal plane angle (A47): This is the angle between the accelerator pedal plane and the horizontal.

3. *Steering Wheel Location*: The center of the steering wheel is specified by locating its center by dimensions L11 and H17 in the side view. L11 is the horizontal distance of the steering wheel center from the AHP. H17 is the vertical distance of the steering wheel center from the AHP. The steering wheel center is located on the top plane of the steering wheel rim (see Figure 20.7). The lateral distance between the center of the steering wheel and the vehicle centerline is defined as W7. The diameter of the steering wheel is defined as W9. The angle of the steering wheel plane with respect to the vertical is defined as A18 (see Figure 20.7).

4. *Entrance Height (H11)*: This is the vertical distance from the driver's SgRP to the upper trimmed body opening (see Figure 20.9). The trimmed body opening is defined as the vehicle body opening with all plastic trim (covering) components installed. This dimension is used to evaluate head clearance as the driver enters the vehicle and slides over the seat during entry and egress.

5. *Belt Height (H25)*: This is the vertical distance between the driver's SgRP and the bottom of the side window daylight opening (DLO) at the SgRP X-plane (i.e., the plane perpendicular to the longitudinal X-axis and passing through the SgRP) (see Figure 20.9). The belt height is important to determine the driver's visibility to the sides. It is especially important in tall vehicles, such as heavy trucks and buses, to evaluate whether the driver can see vehicles in the adjacent lanes, especially on the right-hand side (of a left-hand drive vehicle). The belt height is also an important exterior styling characteristic: many luxury sedans have high belt height from the ground as compared with their overall vehicle height.

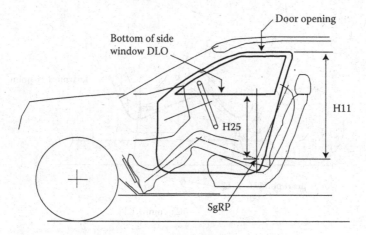

FIGURE 20.9   Entrance height (H11) and belt height (H25).

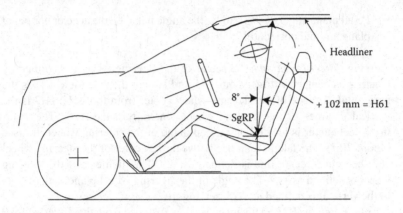

FIGURE 20.10   Effective head room (H61).

6. *Effective Headroom (H61)*: This is the distance along a line 8° rear of the
vertical from the SgRP to the headlining, plus 102 mm (to account for the
SgRP to bottom of buttocks distance) (see Figure 20.10). It is one of the
commonly reported interior dimensions and is usually included in vehicle
brochures and websites.

7. *Leg Room (L33)*: This is the maximum distance along a line from the ankle
pivot center to the farthest (rearmost) H-point in the travel path (i.e., seat
adjusted to the rearmost position of its travel) plus 254 mm (to account for
the ankle point to accelerator pedal distance), measured with the right foot
on the undepressed accelerator pedal (see Figure 20.11). It is also one of the
commonly reported interior dimensions and is usually included in vehicle
brochures and websites.

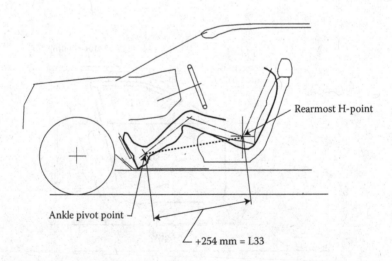

FIGURE 20.11   Leg room (L33).

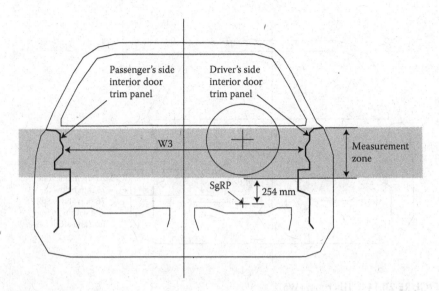

FIGURE 20.12   Shoulder room (W3).

8. *Shoulder Room (W3)* (Minimum cross-car width at beltline zone): This is the minimum cross-car distance between the trimmed doors within the measurement zone. The measurement zone lies between the beltline and 254 mm above the SgRP, in the X-plane passing through the SgRP (see Figure 20.12). It is also one of the commonly reported interior dimensions and is usually included in vehicle brochures and websites.

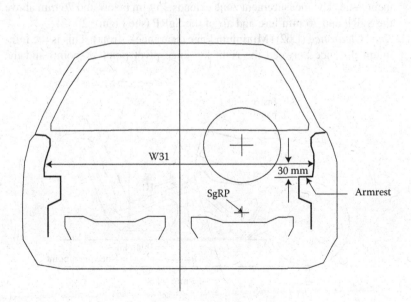

FIGURE 20.13   Elbow room (W31).

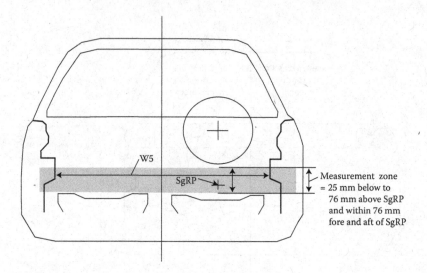

FIGURE 20.14   Hip room (W5).

9. *Elbow Room (W31)* (Cross-car width at armrest): This is the cross-car distance between the trimmed doors, measured in the X-plane passing through the SgRP, at a height 30 mm above the highest point on the flat surface of the armrest. If no armrest is provided, it is measured at 180 mm above the SgRP (see Figure 20.13).

10. *Hip Room (W5)* (Minimum cross-car width at SgRP zone): This is the minimum cross-car distance between the trimmed doors within the measurement zone. The measurement zone extends 25 mm below and 76 mm above the SgRP and 76 mm fore and aft of the SgRP (see Figure 20.14).

11. *Knee Clearance* (L62) (Minimum knee clearance—front): This is the minimum distance between the right leg knee pivot point (K-point) and the

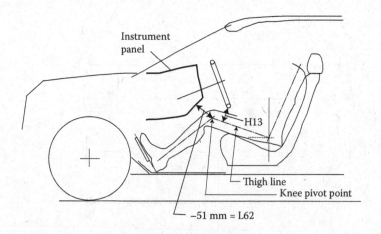

FIGURE 20.15   Knee clearance (L62) and thigh room (H13).

nearest interference, minus 51 mm (to account for the knee point to front of the knee distance) measured in the side view, on the same Y-plane as the K-point, with the heel of the shoe at floor reference point/heel point (FRP) (see Figure 20.15).

12. *Thigh Room (H13)* (Steering wheel to thigh line): This is the minimum distance from the bottom of the steering wheel rim to the thigh line (see Figure 20.15).

DRIVER PACKAGE DEVELOPMENT STEPS AND CALCULATIONS

This section covers 14 basic steps involved in (a) positioning the driver, (b) determining the seat track length, (c) positioning eyellipses, (d) positioning head clearance envelopes, (e) determining maximum and minimum reach envelopes, (f) positioning the steering wheel, and (g) determining the couple distance between the front and second rows of seats.

1. *Determine H30* = height of the SgRP from the AHP. (It is also called H30-1.)

   The H30 value is usually selected by the package engineer based on the type of vehicle to be designed. The H30 dimension is one of the basic dimensions used in the SAE standards to define Class A vehicles (passenger cars and light trucks) and Class B vehicles (medium and heavy trucks). The values of H30 for Class A vehicles range between 127 and 405 mm.

   It should be noted that smaller values of H30 will allow lower roof height (measured from the vehicle floor) and will require longer horizontal space (dimension L53 and X_{95}) to accommodate the driver—like sitting in a sports car. Conversely, if a large value of H30 is selected, a taller cab height and shorter horizontal space (dimension L53 and X_{95}) will be required to accommodate the driver. Class B vehicles (medium and heavy trucks) will have large values of H30 (typically 350 mm and above), so that less horizontal cab space is used to accommodate the driver, and thus, longer longitudinal space is available for the cargo area.

   The BOF to SgRP horizontal distance is usually determined by computing the X_{95} value (which defines the location at which 95% of drivers will have their hip point rearward of the BOF point). The X_{95} value is computed using the following equation, given in SAE J1517. (This equation is called the SgRP curve in SAE J4004.)

   $$X_{95} = 913.7 + 0.672316z - 0.00195530z^2$$

   where z = H30 (mm)

2. Determine pedal plane angle (A47)

   The value of the pedal plane angle in degrees from horizontal is obtained using the following equation from SAE standard J 1516.

   $$A47 = 78.96 - 0.15z - 0.0173z^2$$

where $z = \text{H30}$ (cm) (Note: the z-value is in centimeters for this equation only.)

In SAE standard J4004, the pedal plane angle is defined as alpha (α), where

$\alpha = 77 - 0.08$ (H30) degrees from horizontal. (Note: H30 is specified in millimeters.)

3. *The vertical height (H) between BOF and AHP* can be computed as follows:

$$H = 203 \times \sin(A47)$$

It should be noted that the distance between AHP and BOF is specified as 203 mm in SAE standard J1517 and 200 mm in SAE standard J4004.

4. *The horizontal length (L) between BOF and AHP* can be computed as follows:

$$L = 203 \times \cos(A47)$$

5. *The horizontal distance between AHP and SgRP* (defined as L53) can be computed as follows:

$$L53 = X_{95} - L$$

6. *The seat track length* is defined by the total horizontal distance of the fore and aft movement of the H-point (for a seat that does not have vertical movement of the H-point). The foremost and rearmost H-points on the seat track are defined by the vehicle manufacturer.

To accommodate 95% of drivers (with 50% males and 50% females), the foremost point is located at $X_{2.5}$ horizontal distance rearward of the BOF and the rearmost point is located at $X_{97.5}$ horizontal distance rearward of the BOF. SAE standard J1517 provides the following equations to determine values of $X_{2.5}$ and $X_{97.5}$:

$$X_{2.5} = 687.1 + 0.895336z - 0.00210494z^2$$

$$X_{97.5} = 936.6 + 0.613879z - 0.00186247z^2$$

Where
$z = \text{H30}$ (mm)

$$TL23 = X_{95} - X_{2.5}$$

= horizontal distance between the SgRP and the foremost H-point

$$TL2 = X_{97.5} - X_{95}$$

= horizontal distance between the SgRP and the rearmost H-point
total seat track length to accommodate 95% of the drivers = TL1
where

$$TL1 = TL23 + TL2 = X_{97.5} - X_{2.5}$$

If SAE standard J4004 is used to locate the seat track, then the x distance of the H-point reference point aft the PRP is computed as follows:

$$X_{ref} = 718 - 0.24(H30) + 0.41(L6) - 18.2t$$

where:
L6 is the horizontal distance from the PRP to the steering wheel center (see Figure 20.16)

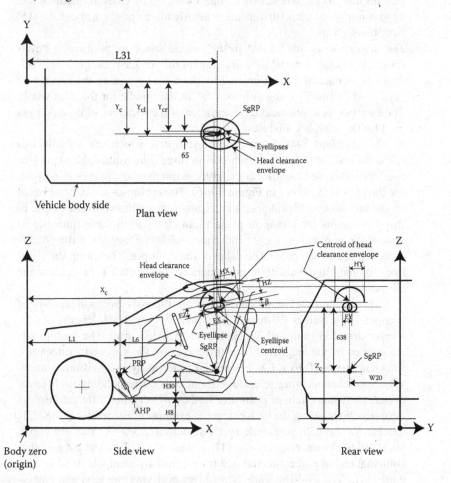

FIGURE 20.16 Location of eyellipses and head clearance envelope.

t is the transmission type (*t* = 1 if clutch pedal is present [manual transmission], and *t* = 0 if clutch pedal is not provided [automatic transmission])

The foremost and rearmost points on the seat track are obtained from data presented in SAE J4004. The value of TL1 for 95% accommodation is specified in the standard as 240 mm.

7. *The seatback angle* (also called the *torso angle*) is defined by dimension A40 (measured in degrees with respect to the vertical). With the reclinable seatback feature, a driver can adjust the angle to his/her preferred seatback angle. The seatback angle in the 1960s and 1970s was defined as 24° or 25° by many manufacturers for bench seats (which had a fixed seatback angle). But with the adjustable seatback feature (i.e., reclinable seats) introduced in later vehicles, most drivers preferred to sit more upright, with seatback angles of about 18°–22° in passenger cars and about 15°–18° degrees in pickups and SUVs. The seatback angles selected by Class B (medium and large commercial truck) drivers are generally more upright at about 10°–15° from the vertical.

8. *The driver's eyes* are located in the vehicle space by positioning "eyellipses" in the CAD model (or a drawing) of the vehicle package. The "eyellipse" is a concocted word created by the SAE by joining the two words "eye" and "ellipse" (using only one "e" in the middle for the joint word). The eyellipse is a statistical representation of the locations of drivers' eyes used for the visibility analyses.

   SAE standard J941 defines these eyellipses, which are actually two ellipsoidal surfaces (one for each eye) in three dimensions (they look like two footballs fused together at the average interocular distance of 65 mm; see the plan view shown in Figure 20.16). The eyellipses are defined based on the tangent cutoff principle; that is, any tangent drawn to the ellipse in two dimensions (or a tangent plane to an ellipsoid in three dimensions) divides the population of eyes above and below the tangent in the proportions defined by the percentile value of the eyellipse. Therefore, the sightlines for visibility evaluations are generally constructed as tangents to the ellipsoids.

   SAE standard 941 has defined four eyellipsoids by combinations of two percentile values (95th and 99th) and two seat track lengths (TL23, shorter than 133 mm, and TL23, greater than 133 mm). The eyellipsoids are defined by the lengths of their three axes (in the x, y, and z directions, shown in Figure 20.16 as EX, EY, and EZ). Generally, most visibility analyses are conducted using selected pairs of eyes corresponding to different drivers (e.g., short, tall, or nearest or farthest driver from different displays or obstructions) using the 95th percentile eyellipses. The values of EX, EY, and EZ for the 95th percentile eyellipse with TL23 > 133 mm are 206.4, 60.3, and 93.4 mm, respectively. (The values of EX, EY, and EZ for other combinations of percentile and seat track travel are available in SAE standard J941.) The eyellipses are located by specifying the x, y, and z coordinates of their centroids. The ellipsoids are also tilted downward in the forward direction by $\beta = 12°$ (i.e., the horizontal axes of the ellipsoids are rotated counterclockwise by 12°; see Figure 20.16).

The coordinates of the left and right eyellipse centroids ([X_c, Y_{cl}, Z_c] and [X_c, Y_{cr}, Z_c], respectively) with respect to the body zero are defined in SAE standard J941 as follows (see Figure 20.16):

$$X_c = L1 + 664 + 0.587(L6) - 0.178(H30) - 12.5t$$
$$Y_{cl} = W20 - 32.5$$
$$Y_{cr} = W20 + 32.5$$
$$Z_c = 638 + H30 + H8$$

where:
(L1, W1, H1) = coordinates of the PRP (or BOF)
L6 = horizontal distance between the BOF (or PRP) and the steering wheel center
$t = 0$ for vehicle equipped with automatic transmission and $t = 1$ for vehicle with clutch pedal (manual transmission) (Note: The SgRP coordinates with respect to the body zero are (L31, −W20, H8 + H30). L1 = L31−X_{95} and Y_c = −W20.) (It is the lateral distance of the driver centerline from the X-axis. See Figure 20.16.)

9. *Locations of Tall and Short Driver Eyes*: The tall and short drivers' eyes on the 95th percentile eyellipse are located at 46.7 mm (half of EZ = 93.4 mm) above and below the eyellipse centroid. By taking into account that the eyellipses are tilted 12° forward, the height of 46.7 mm can be adjusted to 46.7/ Cos 12 or 47.74 mm.

10. *The head clearance envelopes* are defined in SAE standard J1052 (see Figure 20.16). They were developed to provide clearance for the driver's hair on the top, front, and sides of the head. They are defined as ellipsoidal surfaces (upper half of the ellipsoid, i.e., above the centroid only) in three dimensions. The size of the ellipsoid is defined by specified dimensions of three axes of the ellipsoid from its centroid. The dimensions are shown in Figure 20.16 as HX, HY, and HZ (Note: these are half-axis lengths—measured from the centroid). The values of HX, HY, and HZ for the 99th percentile head clearance ellipsoid are 246.04, 166.79, and 151 mm, respectively, for seat track lengths (TL23) over 133 mm. To accommodate drivers (or passengers) who require more head clearance space (e.g., who wear caps or hats), additional clearance spaces should be provided over the 99th percentile head clearance contours.

The head clearance envelopes are also defined as tangent cutoff ellipsoids, and clearances from vehicle surfaces such as the roof, header, or roof rails can be measured by determining the amount of movement (in the three directions defined by the vehicle coordinate system) of the head clearance envelope needed to touch different interior surfaces. The centroid of the head clearance contour is located at (x_h, y_h, z_h) distance from the mid-eye centroid (i.e., the midpoint of the left and right eyellipse centroids). For seat track travel (TL23) greater than 133 mm, the values of (x_h, y_h, z_h) coordinates in millimeters are (90.6, 0.0, 52.6).

SAE standard J1052 provides four head clearance ellipsoids for the combinations of two percentile values (95th and 99th) and two seat track lengths (TL23 below 133 mm and above 133 mm). In addition, to accommodate horizontal head shift of occupants seated in the outboard (toward the side glass) locations, the standard provides an additional lateral shift of 23 mm of the ellipsoid on the outboard side. The ellipsoids are also tilted downward in the counterclockwise direction by 12° (see Figure 20.16).

11. *The maximum reach* data are provided in SAE standard J287. The reach distances are based on the controls reach studies conducted by the SAE (Hammond and Roe, 1972; Hammond et al., 1975). In these studies, each subject was asked to sit in an automotive buck at his/her preferred seating position (by adjusting the seat controls) with respect to the steering wheel and the pedals. The subject was then asked to grasp each knob (1 in. in diameter—like the old push-pull head lamp switch knob) with three fingers and slide the knob (mounted at the end of horizontally sliding bar) as far forward as he/she could reach at each of the vertical and lateral bar locations (see Figure 20.17). The experimenters were looking for the maximum rather than the preferred reach distances. SAE standard J287 provides

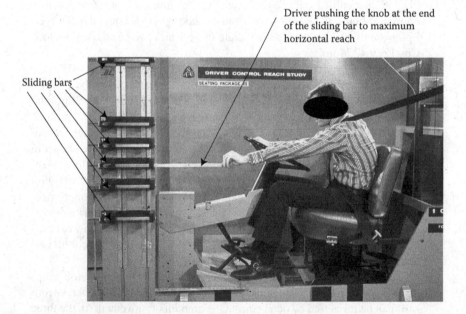

FIGURE 20.17   Maximum hand reach study buck. (The buck shown in the above picture was configured to represent a heavy truck package.)

tables that present horizontal distances forward from an "HR" (hand reach) reference plane at combinations of different lateral and vertical locations with respect to the driver's SgRP.

The HR reference plane is a vertical plane located perpendicular to the longitudinal axis (X-axis) of the vehicle (see Figure 20.18). The horizontal location of the HR plane from the AHP is established by computing the value (in millimeters) of a variable that is also called HR. The value of HR is computed by using the following equation:

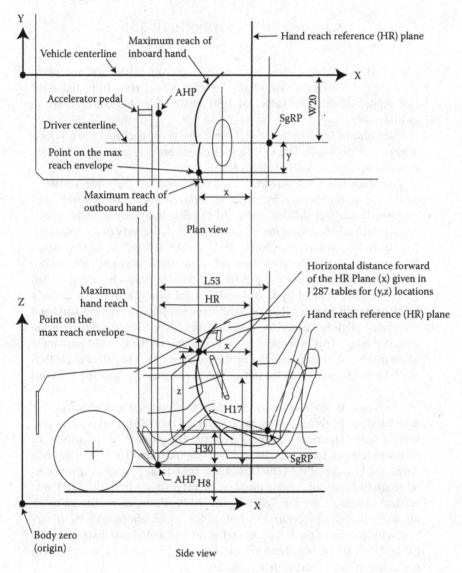

FIGURE 20.18 Plan and side views showing the HR plane and horizontal distances forward of the HR plane provided in tables of SAE Standard J287.

$$HR = \left[786 - 99G\right]$$

where G = general package factor

If the computed value of HR is greater than L53, then the HR plane is located at the SgRP.

The G-value is computed using the following formula in SAE standard J287 FEB2007 (SAE, 2009):

$$G = 0.00327(H30) + 0.00285(H17) - 3.21$$

where H17 = height of the center of the steering wheel (on the plane placed on the driver's side of the steering wheel rim) from the AHP (see Figure 20.18). The values of H30 and H17 in this equation are in millimeters.

The value of G varies from −1.3 (for a sports car package) to +1.3 (for a heavy truck package). The G-values of passenger cars are typically negative. Pickup trucks have G-values close to zero.

The reach tables are provided for combinations of the three variables: (a) type of restraints used by the driver (unrestrained = lap belt only, and restrained = lap and shoulder belt), (b) G-value (range of G-values specified for each table), and (c) male-to-female population mix (three males-to-females ratios are specified: 50:50, 75:25, and 90:10). Figure 20.18 presents side and plan views showing the reach contours. The reach contours are defined such that 95% of the drivers will be able to reach each specified reach location, defined by lateral and vertical distance from the driver's SgRP. It should be noted that the hand reach envelopes of the inboard hand are placed slightly farther forward than the hand reach envelopes of the outboard hand. This is because of the restriction on the forward movement of the driver's outboard shoulder by the shoulder belt. The driver could thus reach farther forward with the inboard hand as compared with the outboard hand.

The reach contours actually generate two complex surfaces, one for each hand, in the three dimensions. SAE standard J287 provides tables presenting horizontal reach distances from the HR plane for combinations of males-to-females ratio, type of seat belt (shoulder and lap belt or lap belt only), and G-value of the vehicle package. Each table provides the horizontal reach distance for combinations of lateral distance from the SgRP and vertical distance from the SgRP. Figure 20.19 illustrates maximum reach distances (i.e., 5th percentile reach distance, which allows 95% of drivers to reach) forward of the HR plane and at vertical and lateral distances from the SgRP for the outboard and inboard hand (i.e., for the left and right hand, respectively, in a left-hand drive vehicle).

To account for differences in reach distances obtained by an extended finger (e.g., reaching a push button with an extended single finger) or full

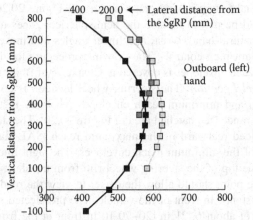

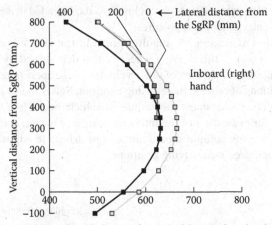

FIGURE 20.19 Maximum horizontal hand reach distances. (Plotted from data in Table 4 of SAE J287 FEB2007 [Society of Automotive Engineers, Inc., *SAE Handbook*. Warrendale, PA, Society of Automotive Engineers, 2009].)

grasp (all fingers grasping a control, such as a gear-shift knob on a floor-shift vehicle), 50 mm is added or subtracted, respectively, from the value obtained from the tables provided in SAE standard J287.

12. *The minimum reach* is the shortest distance at which the hand controls can be placed from the driver. It is defined by the closest reach distance of a short driver seated at the foremost point on the seat track (i.e., her H-point located at the forwardmost point of the seat track) to reach the hand controls. The minimum reach zones are defined as two hemispherical zones with their centers at the short driver's elbow points touching the seat back

and radius equal to upper hand grasp length (see Figure 20.20). The hand controls should be placed outside the hemispherical zones to avoid awkward and unnatural hand, wrist, and finger angles and movements when reaching and grasping controls. The drawing procedure for the minimum comfortable reach envelopes is covered in Chapter 5 of Bhise (2012).

13. *Steering Wheel Location.* The steering wheel location is constrained by the maximum and minimum reach envelopes, visibility of the roadway, and thigh clearance (see hatched area in Figure 20.20). The steering wheel should be placed rearward of the maximum reach (SAE standard J287) and forward of the minimum reach envelopes. The sight line (or the visibility) over the top of the steering wheel rim from the short driver's (5th percentile) eye point should allow the driver to view the road surface at a preselected distance in front of the vehicle. The preselected ground intercept distance of about 6–21 m (20–70 ft) in front of the front bumper is generally considered acceptable. The thigh clearance between the bottom of the steering wheel and the top of the seat should allow accommodation of at least the 95th percentile thigh thickness of the intended driver population during the entry and egress postures.

In addition to meeting the requirements illustrated in Figure 20.20, the nominal location of the steering wheel is also determined by benchmarking steering wheel locations of other vehicles (e.g., superimposing steering wheel locations of other vehicles using common SgRP and/or BOF) and by using subjective assessment techniques in vehicle bucks (see Chapters 11, 21, and 22 for evaluation of the interior package). Further, use of a tilt and telescopic steering column should allow most drivers to adjust the steering wheel to their preferred driving position.

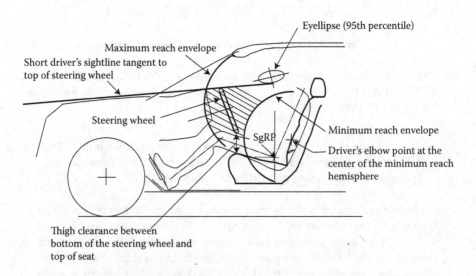

FIGURE 20.20   Considerations related to location of the steering wheel.

14. *SgRP Couple Distance (L50)*. This is the longitudinal distance between the SgRPs of adjacent seating rows. The longitudinal distances between rows of seats are defined as follows:

L50-1 = SgRP couple distance between the front seat row and the second seat row

L50-2 = SgRP couple distance between the second seat row and the third seat row

The couple distances are determined by considering the knee and foot clearance space needs of the passengers in the rear seat from their anthropometric data, benchmarking of existing vehicles, and evaluations of physical properties by customers in package evaluation clinics.

ENTRY AND EXIT CONSIDERATIONS

The driver and occupants should able to enter and exit from the vehicle quickly and comfortably, without any awkward postures or undue physical effort, which may involve excessive bending, turning, twisting, stretching, leaning, and/or hitting of body parts on surrounding vehicle components. In this section, we will cover many problems that drivers and passengers experience during entry into and exit from vehicles and relate these to different vehicle package dimensions.

Drivers experience different problems while entering and exiting vehicles with different body-styles, from low sports cars to sedans to SUVs to pickups and heavy trucks. Assuming that a vehicle or a physical buck is available, the best method to uncover these problems would be to ask a number of male and female drivers with different anthropometric characteristics (e.g., tall, short, slim, or obese) to get in and get out of the vehicles after they have adjusted the seat and steering wheel to their preferred driving positions, and observe (or video record and replay in slow motion) and ask them to describe the difficulties that they encountered. Such exercises are usually performed by package and ergonomics engineers during the evaluation of a physical buck of a new vehicle. Comparisons are also made with different benchmarked vehicles to understand the differences in vehicle dimensional characteristics and assist features (e.g., door handles, grab handles, steps, and hidden rockers) used during entry/exit performance and the difficulty/ease ratings provided by the drivers and passengers to determine whether a given vehicle package will be acceptable or will need improvements.

PROBLEMS DURING ENTRY AND EXIT

The problems that drivers experience while entering or exiting from a passenger car depend on their gender and anthropometric characteristics:

1. *Drivers with short legs* (predominantly women) will complain that
   a. The seat and step-up (top of the rocker panel) are too high. The rocker panel is the lower part on the side of the vehicle body under the doors (which in effect creates the lower part of the of the door frame), over which the occupant's feet move during entry and exit (see Figure 20.21).

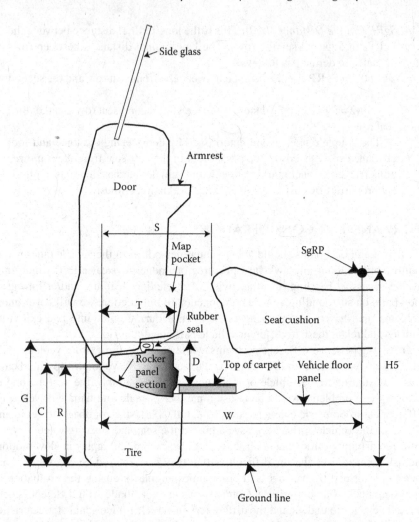

FIGURE 20.21   Cross section of the vehicle at the driver's SgRP (in the rear view) showing dimensions relevant for entry and exit.

b.   The step-over (rocker panel section) is too wide. The lateral distance to the outer edge of the rocker panel from the driver centerline is too great for the driver to move his/her legs during entry from the ground to inside the vehicle (see Dimension W in Figure 20.21).

c.   The clearance between the driver's knees and the lower part of the instrument panel and/or the steering column is insufficient. This problem occurs because the short driver needs to move her seat farther forward to reach the pedals with her shorter feet.

2. *Older, obese, mobility-challenged drivers* will complain that

a.   The seat is either too high or too low (see dimension H5 in Figure 20.21). This indicates that the driver had difficulty climbing up into the seat (e.g., strain in the knees while climbing up in a truck) or sitting down into the seat due to larger muscular forces needed in the leg and back muscles to move the driver's body on to the seat during entry. Similarly, during exit, a low-mounted seat requires greater muscular force in rising from the seat during exit (e.g., from a sports car).

   b. The upper part of the body door opening (entrance height defined as H11; see Figure 20.9) is too low. The person will experience difficulty in bending or moving his/her head under the lower edge of the upper body opening.
   c. The step-over is too wide.
   d. The thigh clearance (between the bottom edge of the steering wheel and the top surface of the seat) is insufficient (see Figure 20.15).
   e. The steering wheel-to-stomach clearance (between the lower part of the steering wheel and the driver's stomach) is insufficient.
   f. The door does not open wide enough (i.e., the space between the inner door trim panel of the opened door and the vehicle body-side is insufficient).

3. *Drivers with a tall torso* will complain that
   a. The upper body opening (entrance height, H11) is too low.
   b. The A-pillar (front roof pillar) is too close to their head when they lean or bend the torso forward.
   c. The seat bolsters (i.e., the raised sides of the seat cushion) are too high.
   d. The head clearance is insufficient.

4. *Drivers with long legs* will complain that
   a. The seat track does not extend sufficiently rearward (the seat track is too short and placed father forward in the vehicle).
   b. The front edge of the B-pillar (the roof pillar between the front side window and the rear side window) is too far forward. The seatback in this case is moved rearward of the front edge of the B-pillar, requiring the driver to brush past the B-pillar to get into the seat.
   c. The lower rear edge of the cowl side is too far rearward. (Note: the cowl panel is the body panel behind the rear edge of the engine hood and forward of the lower edge of the windshield.) The legroom for the driver is not sufficient to move his legs from the ground to inside the vehicle. This problem usually results in the driver's shoes hitting the door opening edge under the cowl side (look for shoe scuff-marks on the trim parts on the lower and forward part of the door opening).
   d. The door does not open wide enough (i.e., the space between the opened door and the vehicle body is insufficient).

Thus, the design of the door opening size and shape—created by the rocker panel (on the bottom side), the B-pillar (on the rear side), the roof rail (the vehicle body component above the doors and mounted on the sides of the vehicle roof), the A-pillar (the rear edge of the front roof pillar), the lower part of the cowl and the knee bolster, the door opening angles, the positioning of the seat and seat dimensions, the steering wheel (diameter and location), and the door grab and opening handles—all affect the driver's ease during entry and exit.

VEHICLE FEATURES AND DIMENSIONS RELATED TO ENTRY AND EXIT

The vehicle features and vehicle dimensions to which vehicle designers and engineers should pay attention to facilitate ease during entry/egress are as follows:

Door Handles

1. *Height of the Outside Door Handle*: The short 5th percentile woman should be able to grasp the door handle without raising her hand over her standing shoulder height, and the tall 95th percentile male should be able to grasp the handle without bending down (i.e., not below his standing wrist height).
2. *Longitudinal Location of the Outside Door Handle*: The handle should be placed as close to the rear edge of the hinged door as possible to avoid the lower right corner of the driver's door hitting the driver's shin during door opening.
3. *Inside Door Handle Location*: While the driver is closing the door (once he/she has entered the vehicle and is seated in the driver's seat), the inside door grasp (or grab) handle should not require him/her to adopt a "chicken winging"-type wrist posture. This means that the inside door handle should be placed (a) forward of the minimum reach zone, (b) rearward of the maximum reach zone, (c) not below the door armrest height, and (d) not above the seated shoulder height. For exit, the inside door opening handle location should also meet these location requirements.
4. *Handle Grasps*: The grasp area clearances should be checked to ensure that the outside door handles and the inside grasp handles (or pull cups) can allow the insertion of four fingers of the 95th percentile male's palm (considering palm width, finger widths, and finger thickness). Further, to facilitate gloved-hand operation, additional clearances would be needed, depending on the type of winter gloves used. Additional clearances (at least 15 mm) should be provided to avoid scratches on nearby surfaces due to finger rings and long fingernails.

Lateral Section at the SgRP and Foot Movement Areas

The following vehicle dimensions, shown in Figure 20.21, are important for ease during entry and exit:

1. Vertical height of the SgRP from the ground (H5)
2. Lateral distance of the SgRP from the outside edge of the rocker (W)
3. Lateral distance of the outside of the seat cushion to the outside of the rocker (S)
4. Lateral overlap thickness of the lower part of the door (T)
5. Vertical distance from the top of the rocker to the ground (G)

6. Vertical distance from the top of the floor to the top of the rocker (D)
7. Height of the outer edge of the rocker panel (R)
8. Curb clearance of doors at design weight (C)

To improve the ease of the driver's entry and exit, the magnitudes of the above dimensions (separately and in combination) need to be considered during the early stages of the vehicle design.

The H5 dimension should allow drivers to easily slide in and out of their seat without climbing up into the seat or sitting down into it. Thus, H5 should be about 50 mm below the buttock height of most of the users (considering the up/down adjustment of the seat cushion). A top of the seat to ground distance of about 500–650 mm is generally considered to facilitate easy ingress and egress for the U.S. population.

Dimension W should be as short as possible (see Figure 20.21). This means that Dimension S, lateral distance from the outer edge of the seat to the outer edge of the rocker, should be short enough to allow the driver's foot to be placed close to the vehicle and on the ground during entry/exit. The lateral distance from the outer edge of the rocker panel to the SgRP (i.e., Distance W) should be about 420–480 mm to accommodate most drivers. The height of the lower door edge C (at maximum vehicle weight) should be sufficient to clear the curb height, so that the door swings over the curb and does not hit most curbs (see Figure 20.21).

The distance of the top of the rocker from the ground (Dimension G) and of the rocker top from the vehicle floor (Dimension D) should be as small as possible to reduce foot lifting during entry and exit. These dimensions are dependent on ground and curb clearances.

The width dimension T, which is the lateral dimension from the outer edge of the rocker to the lower edge of the inner door trim panel (which is generally the lower inward protruding edge of the map pocket), should be as small as possible. The smaller Dimension T, the more foot passage space will be available during entry and exit. This is especially important when the door cannot be opened wide due to restricted space at the side, such as in garages and when parking close to another vehicle (side-by-side).

Body Opening Clearances from SgRP Locations

A number of protruding points outlining the entry/exit space defined by the door openings and the instrument panel should be measured (in the side view) from the SgRPs of the front and rear occupant positions. Larger distances of these points from the respective SgRPs will allow more room during entry and exit.

For the driver's door opening, the following points should be checked: (a) the top point on the door opening (defines the entrance height [H11] for the head clearance), (b) the rearmost point at approximately the middle of the A-pillar (defines the head swing clearance to A-pillar during leaning and torso bending), (c) minimum knee clearance point to the lower portion of the instrument panel, (d) points on the front edge of the door opening under the cowl area for foot clearance, (e) points defining the top of the rocker, and (f) points on the forward edge of the B-pillar.

Similarly, (a) points around the rear door openings on the rear edge of the B-pillar, (b) points on the back side of the front seat back, (c) points on the lower edge of the roof rail, and (d) points on the top of the rocker define clearances for rear passenger entry and egress.

Package engineers generally compare the dimensions to the above points and the SgRP locations of the vehicle being designed with corresponding dimensions of other benchmarked vehicles. Additional information on entry/exit can be obtained from Bhise (2012).

DRIVER FIELD OF VIEW

Field of view analyses link the vehicle's interior design to its exterior design. The interior package provides the driver's eye locations, interior mirror, and other interior objects (e.g., the instrument panel, headrests, and the steering wheel) that can cause obstructions in the driver's field of view. The vehicle exterior defines the daylight openings and exterior mirrors. Thus, the interior and exterior designs must be developed in close coordination to ensure that drivers can see all the fields needed to drive their vehicles safely. Important design considerations and visibility-related issues are described in the following subsections.

VISIBILITY OF AND OVER THE HOOD

1. The visibility of the road surface (i.e., the closest longitudinal forward distance from the front bumper at which the road surface is visible, also called the *ground intercept distance*) is of critical concern to many drivers. The problem is worse for short drivers. In general, most drivers want and like to see the end of the hood, the vehicle corners (extremities), and the road at a close distance. As more aerodynamic vehicle designs with low front ends were introduced in the United States (after the mid-1980s), many drivers who were accustomed to long hoods with their visible front edges and corners complained about not being able to see the ends of their hoods.

2. The view of the hood allows better perception of the vehicle heading with respect to the roadway, providing a feeling of ease in lane maintenance (lateral location of the vehicle in the lane) and while parking. (Note: racing cars have a wide painted strip over the hood at the driver centerline to provide highly visible vehicle-heading cues in the driver's peripheral vision while looking straight ahead.)

3. Heavy trucks with long hoods experience a greater obstruction of the road ahead due to the hood. The problem can be severe if the obstruction is large enough to hide a small vehicle (e.g., bike rider, sports car) located in front of the hood. The problem often occurs when a truck with a long hood is behind a small vehicle while waiting at an intersection.

COMMAND SEATING POSITION

1. The "command seating position" provides the feeling of "sitting high" in the vehicle. It is opposite to the feeling of "sitting in a well" or "sitting too low" in the vehicle (experienced by drivers in many sports cars).
2. For command seating position, provide (a) higher SgRP location from the ground, (b) low cowl point, (c) low beltline, (d) visibility of the hood, and (e) greater visibility of the roadway (shorter ground intercept distance from the front bumper). It should be noted that the command sitting position is one of the key positive attributes of an SUV.
3. This command sitting feeling is also appreciated by short female drivers (with 2.5 or 5 percentile female sitting eye height).

SHORT DRIVER PROBLEMS

Short drivers are drivers with shorter (5th percentile and below) sitting eye heights and/or shorter (5th percentile and below) leg lengths. The visibility problems encountered by such short drivers are

1. The road can be obscured by the steering wheel (top part of the rim) and instrument panel (or instrument cluster binnacle, causing a smaller down angle [e.g., angle A61–1, defined in SAE J1100, SAE 2009]).
2. The driver may be unable to see any part of the hood (no visibility of the end of the hood). (Note: providing a raised hood ornament near the front of the hood can provide useful information in maintaining vehicle heading. Similarly, providing visibility of the corners of the hood or ends of the front fenders [via placement of "flag poles" as provided on some trucks] can improve ease in parking and lane maintenance).
3. The forward direct view may be obscured by the side view mirrors. (The upper edge of the side view mirrors should be placed at least 20 mm below the 5th percentile female's eye point.)
4. The closest distance at which a driver can see the road (over the hood) is much longer for the short driver than the road visibility distances for other drivers.
5. Shorter drivers will experience reduced rear visibility problems during backing up (especially with a higher deck point and taller rear headrests). (Note: one check that many package engineers consider is whether a short driver can see a 1 m high target [simulating a toddler] in the rear view while backing up in the direct rear view with the driver's head turned rearward and also while looking in the inside mirror.) A rear view camera system helps in eliminating this problem (see Chapter 6).
6. Since short drivers (with shorter leg lengths) sit farther forward in the seat track, the driver's side A-pillar will create larger obscurations (large obstruction angle) in the forward field of view for short drivers as compared with taller drivers.

7. Short drivers require larger head turn angles to view side view mirrors due to their farther-forward seating position as compared with taller drivers. This problem is more severe for short drivers, particularly older short females who have arthritis (which reduces the range of head turn angles).

TALL DRIVER PROBLEMS

Tall drivers are drivers with greater (95th percentile and above) sitting eye heights and/or longer (95th percentile and above) leg lengths. The visibility problems encountered by such tall drivers are

1. External objects placed at higher locations placed above the upper sightline at up angle (defined by dimension A60-1; see SAE J1100 standard [SAE, 2009]) may be obstructed from the view of tall drivers. In such situations, a tall driver may have to duck his head down to view high-mounted objects such as overhead traffic signals at intersections. The visibility near the top portion of the windshield is further limited by the shade bands and/or blackout paint applied around the edges of the windshields.
2. The inside rearview mirror may block the tall driver's direct forward field. Therefore, the lower edge of the inside mirror should be placed at least 20 mm above the tall driver's eye point (i.e., 95th percentile eye height).
3. The tall driver also sits farther from the mirrors due to a more rearward sitting position. Thus, the mirrors provide smaller fields of view to tall drivers as compared with the mirror fields of other drivers.
4. The tall driver may have more side visibility problems because of (a) farther-forward B-pillar obscurations in direct side viewing (because the tall driver sits farther rearward in the seat track than shorter drivers) and (b) farther-forward peripheral awareness zones (i.e., the peripheral zones [Bhise, 2012] do not extend as far rearward as for shorter drivers) while using the side view mirrors.

SUN VISOR DESIGN ISSUES

1. The sun visor drop-down height and length should be designed to prevent the incidence of direct sunlight on the driver's eyes from different sun angles and sun glare from the windshield and driver's side window.
2. The sun visor dropped-down position should be adjustable, and it should be capable of dropping down to accommodate the short driver's needs.
3. If the sun visor hinge mechanism becomes loose, the sun visor may accidentally swing and drop down and cause obstruction in the forward field of view. This obstruction will be more severe for taller drivers.

WIPER AND DEFROSTER REQUIREMENTS

The SAE standards J902 and J903 (SAE, 2009) (also FMVSS 103 and 104; NHTSA, 2015) provide requirements on how to establish areas in the driver's forward field that must be defogged (or defrosted) and wiped by the wipers, respectively. The requirements specify the sizes of these areas and the percentages of each area to be covered (cleaned) by the defoggers and wipers. The areas are specified by establishing four tangent planes to the 95th percentile eyellipses.

The areas to be covered by the wiper sweep pattern are defined as areas A, B, and C (SAE J903; SAE, 2009). The wiper sweep area must be designed so that at least 80% of area A, at least 94% of area B, and at least 99% of area C are wiped by the wipers. The areas A, B, and C are defined by drawing up, down, left, and right tangent planes to the eyellipses, as shown in Figure 20.22. The area A is bounded by the upper tangent plane at 10° up angle, the bottom tangent plane at 5° down angle (see side view in Figure 20.22), the left tangent plane at 18° angle to the left eyellipse, and the right tangent plane at 56° angle to the right eyellipse (see plan view in Figure 20.22). Similarly, the angles defining area B are 5° up, 3° down, 14° left, and 53° right. The angles defining area C are 5° up, 1° down, 10° left, and 15° right.

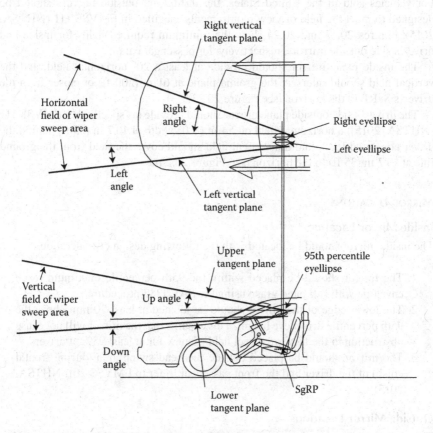

FIGURE 20.22 Plan and side views showing the four tangent planes that define the wiping areas to be covered by the wipers.

Obscurations Caused by A-Pillars

The left and right front roof pillars (the A-pillars), depending upon their size and the shape of their cross sections at different heights with respect to the driver's eye location, can cause binocular obstructions in the driver's direct forward field of view. The obstructions can hide targets such as pedestrians and other vehicles during certain situations. For example, during an approach and left turn through an intersection, a pedestrian crossing the street on the driver's left side and vehicles approaching from the driver's right side can be partially or completely obscured by the left and right A-pillars, respectively.

Vehicle body designers must conduct visibility analyses of such situations and minimize the obstructions caused by the pillars. Appendix C of SAE standard J1050 (SAE, 2009) provides a procedure to measure the visual obstruction caused by the A-pillar [see also Bhise (2012)].

Mirror Field of View Requirements

For vehicles sold in the United States, the inside and outside mirrors should be designed to meet the field of view requirements specified in FMVSS 111 (NHTSA, 2015). Figures 20.23 and 20.24 show the minimum required fields for inside and driver's side outside mirrors, respectively, for passenger cars.

The inside plane mirror should provide at least a 20° horizontal field, and the vertical field should intersect the ground plane at 61 m (200 ft) or closer from the driver's SgRP to the horizon (see Figure 20.23).

The driver's side outside plane mirror should provide (as specified in FMVSS 111 [NHTSA, 2015]) a horizontal field of 2.4 m (8 ft) width at 10.7 m (35 ft) behind the driver at ground level, and the vertical field should cover the field from the ground line at 10.7 m (35 ft) to the horizon (see Figure 20.24).

Mirror Locations

Inside Mirror Location

The inside mirror should be located with the following design considerations:

1. The mirror should be placed within the 95th percentile maximum reach envelope with full hand grasp using the SAE J287 procedure.
2. The lower edge of the mirror should be located at least 20 mm above the 95th percentile driver eye height. This ensures that the mirror will not cause obstruction in the forward direct field of view for at least 95% of drivers.
3. The mirror should be placed outside the head swing area (during frontal crash) of the driver and the front passenger (refer to FMVSS 201; NHTSA, 2015).

Outside Mirror Locations

The driver's side outside mirrors should be located with the following design considerations:

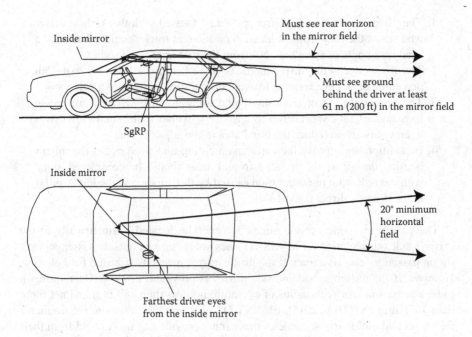

FIGURE 20.23 Inside mirror field required for passenger cars.

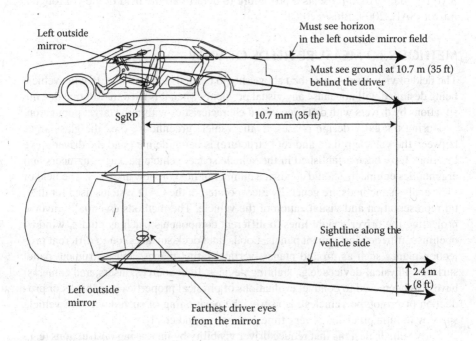

FIGURE 20.24 Driver's side mirror field required for passenger cars.

1. The driver's side outside mirror should be located such that a short driver who sits at the forwardmost location on the seat track should not require a head turn angle of more than 60° from the forward line of sight.
2. The upper edge of the mirror should be placed at least 20 mm below the 5th percentile driver eye location to avoid obscuration in the direct side view.
3. The mirror-aiming mechanism should allow for a horizontal aim range large enough for a short driver to see part of his/her vehicle and a tall driver to aim outward to reduce the blind area in the adjacent lane.
4. In addition, to improve the aerodynamic drag and wind noise, the mirror housing design needs (reduced frontal area) should be considered along with the reduction in obscuration caused by the mirror and the left A-pillar in the driver's direct field of view.

The passenger's side outside mirror is generally located symmetrically to the driver's side outside mirror. FMVSS 111 does not require an outside passenger mirror on passenger cars or a truck if the inside mirror meets its required field of view. However, if the passenger side outside mirror is provided, FMVSS 111 requires it to be a convex mirror with radius of curvature not less than 889 mm and not more than 1651 mm (NHTSA, 2015). FMVSS 111 also provides alternate requirements for trucks and multipurpose vehicles that cannot provide any useful field from their inside mirrors due to blockage by cargo or passenger areas.

Procedure for Determining Driver's Field of View through Mirrors

SAE Standard J1050 presents a procedure to determine the field of view through a mirror (SAE, 2009; Bhise, 2012).

METHODS TO MEASURE FIELDS OF VIEW

The field of view issues described above should be analyzed to ensure that the vehicle being designed will not cause any visual problems when it is used in different driving situations by drivers with differing visual characteristics within the target population.

During the early design phases, as the vehicle greenhouse (i.e., the glass areas between the vehicle pillars and roof structure) is being defined and the driver's eye locations have been established in the vehicle space, vehicle package engineers and ergonomics engineers should conduct a number of field of view analyses. The field of view analysis methods are generally incorporated in the CAD systems used for digital representation and visualization of the vehicle. The methods essentially involve projecting the driver's sight lines to different components (such as pillars, window openings, mirrors, instrument panels, hoods, and deck surfaces) on to different projection planes such as ground planes, vertical target planes, and instrument panel surfaces. Physical devices (e.g., sighting devices, light sources, lasers, and cameras) have also been used to conduct evaluations of physical properties (e.g., bucks or production or prototype vehicles). However, the positioning of such devices in vehicle space with high precision is very time consuming and costly.

Early vehicle designs that reduce driver visibility by increasing obstructions (e.g., due to larger pillars, headrests, or high beltlines) should be investigated fully using

CAD procedures. Questionable problems can be further evaluated by creating full-size bucks or even drivable mock-ups for market research clinics or human factors field tests. Such problems, if not fixed early, would be extremely time consuming and expensive to change during the later stages of vehicle development.

POLAR PLOTS

Creating a series of polar plots to conduct different field of view analyses is a very effective method for visualizing and measuring fields of view issues (McIssac and Bhise, 1995). A polar plot is especially useful for ergonomic analyses, as it allows direct measurements of angular fields, angular location of different objects, angular sizes of different objects, and angular amplitudes of eye movements and head movements required to view different objects. It also allows incorporation of views from both eyes, and thus, facilitates the evaluation of monocular, ambinocular, and binocular fields and obscurations. It also simplifies the 3-D analysis by reducing it to a 2-D analysis (i.e., space defined by azimuth angles and elevation angles).

   A polar plot involves plotting the visual field from the driver's (one or both) eye points in angular coordinates. It is equivalent to projecting the driver's view on a spherical surface with the driver's eyes at the center of the sphere. The driver's eye point is considered as the origin from which sight lines originate. Each sight line aimed at a target point can be located by determining its azimuth angle in degrees (θ) and its elevation angle in degrees (Φ) with respect to the eye point (as the origin) of a coordinate system. If Point P is defined by (x,y,z) as its Cartesian coordinates (with the eye point as the origin), then its polar (angular) coordinates (θ, Φ) can be computed as follows:

$$\theta = \tan^{-1}\left(y/x\right)$$
$$\Phi = \tan^{-1}\left[z/\left(x^2 + y^2\right)^{0.5}\right]$$

   It should be noted that this polar plotting method does not use the distances from an eye point to any object points for the analysis. Bhise (2012) presents the procedure for creating polar plots and also provides an illustration of a polar plot.

   Another advantage of the polar plot is that the polar coordinates of objects included in the plot provide angular locations that can be used to directly measure a driver's line-of-sight locations (i.e., combined eye movements and head turns) from the straight-ahead location (which is the origin of the polar plot and thus has the polar location of [0,0]). Similarly, the size of the object shown in the polar plot can be directly measured to determine the angular sizes of monocular and binocular obstructions. Thus, the angular locations of the pillars, up and down angles, and the binocular obscurations caused by each of the pillars can also be measured directly from the polar plot.

   The roadway and many other external objects on the roadway can also be included in polar plots. These external objects help in understanding many visibility issues in terms of what objects can be seen through the window openings and what objects are fully or partially obstructed by vehicle components. Through extensive photographic measurements of objects in the driver's view, Ford Motor Company (1973) developed targets that encompass the regions on the roadways where different objects appeared in the photographic data. These Ford targets represent different external objects, such as overhead signs, side-mounted signs, traffic signals, vehicles approaching from intersecting roadways, vehicles in adjacent lanes, vehicles ahead, and vehicles behind the driver's car. The targets can be placed in polar plots to evaluate fields of view from the vehicles. McIssac and Bhise (1995) describe the use of the targets in polar plots. The paper also describes the use of polar plots in determining indirect visual fields from plane and convex mirrors by plotting virtual images of objects seen in the mirrors and the outlines of the mirrors.

OTHER PACKAGING ISSUES AND VEHICLE DIMENSIONS

Other package and ergonomics design–related issues, such as opening the hood and servicing the engine, opening the trunk (or liftgate), and loading and unloading items are covered in Bhise (2012). An Excel-based spreadsheet program is provided on the publisher's website for the readers to better understand the various inputs and calculate the resulting package dimensions. The program can be also used to set up different driver packages, analyze an existing package, or conduct sensitivity analyses by changing combinations of different input parameters and studying the resulting driver packages.

CONCLUDING REMARKS

Vehicle packaging requires inputs from customers, design studio professionals, and vehicle systems engineering departments to ensure that acceptable trade-offs between occupant space and mechanical equipment spaces are made. The driver and occupants must be positioned so as to provide them with seating comfort, ease in entry and exit, and field of view for safe driving and use of controls and displays. A high percentage of occupants (e.g., 95%) must be accommodated in the vehicle space. Many SAE occupant packaging tools are used in the industry to ensure that 95% of the occupants are accommodated. In addition, many anthropometric databases are available (Bhise, 2012). Important SAE standards and key government requirements were reviewed in this chapter. In addition, package engineers perform a number of analyses and studies to verify that the vehicle package will satisfy drivers by meeting their needs. The package verification studies are covered in Chapters 11, 21, and 22. The customer needs and their translation into attribute requirements and cascading of the attribute requirements to vehicle systems are covered in Chapters 2, 3, and 9.

REFERENCES

Besterfield, D. H., C. Besterfield-Michna, G. H. Besterfield, and M. Besterfield-Scare. 2003. *Total Quality Management*. Third Edition. Upper Saddle River, NJ: Prentice Hall.

Bhise, V. D. 2012. *Ergonomics in the Automotive Design Process.* Boca Raton, FL: CRC Press. ISBN 978-1-4398-4210-2.

Flannagan, C. C., L. W. Schneider, and M. A. Manary. 1996. *Development of a Seating Accommodation Model.* SAE Technical Paper no. 960479. Warrendale, PA: Society of Automotive Engineers.

Flannagan, C. C., L. W. Schneider, and M. A. Manary. 1998. *An Improved Seating Accommodation Model with Application to Different User Populations.* SAE Technical Paper no. 980651. Warrendale, PA: Society of Automotive Engineers.

Ford Motor Company. 1973. Field of View from Automotive Vehicles. Report prepared under direction of L. M. Forbes, Report no. SP-381. Presented at the Automobile Engineering Meeting held in Detroit, Michigan. Published by the Society of Automotive Engineers, Inc., Warrendale, PA.

Hammond, D. C., D. E. Mauer, and L. Razgunas. 1975. *Controls Reach: The Hand Reach of Drivers.* SAE Paper no. 750357. Warrendale, PA: Society of Automotive Engineers.

Hammond, D. C. and R. W. Roe. 1972. *SAE Controls Reach Study.* SAE Paper no. 720199. Warrendale, PA: Society of Automotive Engineers.

McIssac, E. J. and V. D. Bhise. 1995. *Automotive Field of View Analysis Using Polar Plots.* Paper no. 950602. Warrendale, PA: Society of Automotive Engineers.

NHTSA. 2015. Federal Motor Vehicle Safety Standards. Website: http://www.nhtsa.gov/cars/rules/import/FMVSS/ (Accessed: July 5, 2015)

Philippart, N. L., R. W. Roe, A. J. Arnold, and T. J. Kuechenmeister. 1984. *Driver Selected Seat Position Model.* SAE Technical Paper no. 840508. Warrendale, PA: Society of Automotive Engineers.

Society of Automotive Engineers, Inc. 2009. *SAE Handbook.* Warrendale, PA: Society of Automotive Engineers.

21 Vehicle Evaluation Methods

INTRODUCTION

The objective of this chapter is to provide the reader with an understanding of what is meant by product evaluation and how evaluations are conducted to verify that the product will meet its stated requirements and to validate that the right product is developed.

Vehicle evaluation generally involves some form of testing (e.g., laboratory or field testing). Testing is an activity undertaken using well-established procedures to obtain detailed measurements and data on one or more characteristics and/or performance measures of the product and its systems, subsystems, and/or components. The collected test data are analyzed to determine whether the product and its systems, subsystems, and/or components meet their stated requirements specified during the design process.

The vehicle evaluations are conducted for both verification and validation purposes. Verification is the process of confirming that the product, its systems, and its components meet their respective requirements. The aim of the verification is to ensure that the tested item is built right, that is, it meets its requirements. Validation is the process of determining whether the product functions and possesses the characteristics expected by its customers when used in its intended uses and environments. The aim of the validation process is to ensure that the right product is designed and that the product will be liked by its intended customers.

OVERVIEW OF PRODUCT EVALUATION METHODS

During different stages in the life cycle of the product, it must be evaluated to ensure that it meets requirements (verification) and that the right product is designed (validation).

Tests conducted for evaluations should measure both "performance" (the tested entity performs, i.e., functions, such that it meets its stated requirements) and "preference" (i.e., customers find the product to be acceptable and prefer using it as compared with other similar products). Performance tests are generally based on objective measurements, and preference tests are based on subjective judgments of customers and/or experts.

Many different methods are used for vehicle evaluations. The evaluations can involve physical tests using measurement instruments or human subjects (e.g., customers or experts). The evaluations can be conducted at various levels of the product, that is, evaluations on components, subsystems, systems, or the whole product.

Some examples of evaluation methods used in the automotive industry are

1. Checklists, scorecards, and design reviews (using drawings, 3-D models, fly-throughs, benchmarking)
2. Customer clinics (in static or dynamic situations)
3. Modeling, prototyping, and simulations
4. Laboratory and bench tests
5. Field tests and drive evaluations
6. Field experience (customer complaints, warranty/repairs, accidents)

The basic methods for data collection underlying these evaluation methods are observation, communication, and experimentation. The concepts underlying these methods are covered in the next section.

TYPES OF DATA COLLECTION AND MEASUREMENT METHODS

Table 21.1 provides a summary of methods categorized by combinations of types of data collection methods and types of measurements.

The left-hand column of Table 21.1 shows that the data can be collected using methods of observation, communication, and experimentation. In the observation method, it is assumed that observations can be made about either a piece of equipment being tested or a subject performing a task. And the observations can be made by an experimenter or the data can be recorded (e.g., using a camera) for later observations by an experimenter. In the communication method, the subject (or the experimenter) can be asked to report on the state of the product or problems experienced during the performance of a task and/or asked to provide ratings on his/her impressions of the task. In the experimentation method, the test situations are designed by deliberate changes in combinations of certain independent variables (e.g., configurations of the product), and the responses are obtained by using combinations of methods of observation and/or communication. For further information on many available methods of data collection and their advantages and disadvantages, the reader should refer to Bhise (2014), Chapanis (1959), and Zikmund and Babin (2009).

Types of measurement can be categorized as objective or subjective, as shown in Table 21.1. Objective measures can be defined here as measurements that are not affected by the evaluator, the experimenter, or the subject performing the tasks in a product evaluation situation. Objective measures are generally obtained by the use of physical instruments or by unbiased and trained experimenters. On the other hand, subjective measurements are generally based on the subject's perception and experience during or after performing one or more tasks with the product being evaluated. Objective measures are generally preferred, because they are more precise and unbiased. However, there are many vehicle attributes that cannot be measured without using human subjects as the "measuring instruments." After the users have experienced the vehicle, they are better able to express their perceived impressions about the vehicle and its characteristics (e.g., quality, comfort, convenience, ride, and handling) by the use of methods of communication.

TABLE 21.1
Evaluation Methods Based on Combinations of Data Collection Methods and Type of Measurements

Type of Data Collection Method	Type of Measurements	
	ᵃObjective Measurements	Subjective Measurements
Observation	Events related to the state of the product or behavior of product users are observed. Data recorded with instruments or observed by a trained experimenter. Observations of the product during uses (e.g., recordings of performance measures, e.g., outputs, stresses, energy consumption, customer/user behavior, task performance, e.g., durations, errors, difficulties, conflicts, near-accidents).	Customer-volunteered responses (without communicating), for example, verbatim comments. Checklist completed by observers. Observers and/or recorders gather information about customer-initiated events. The gathered information is categorized, summarized, and analyzed with judgments made by expert evaluators.
Communication	Experimenter-reported objective measures (e.g., outcomes, output levels, response times, speeds, events—displayed by instruments).	Subject-reported detections and identifications of events. Responses in checklists or rating forms. Reporting of problems, difficulties, and errors during operation of equipment.
Experimentation	Measurements with instrumentation. Performance measurements or behavioral measurements—for different experimental conditions.	Data obtained from test subjects (e.g., ratings, behavioral measurements of difficulties, errors) are analyzed to determine differences between tested products with combinations of characteristics.

The following section provides additional information on the methods of data collection.

METHODS OF DATA COLLECTION AND ANALYSIS

OBSERVATIONAL METHODS

In observational methods, information is gathered by direct or indirect observations of the product and/or subjects during the use of the product. The observations can be used to evaluate product attributes such as understandability of controls and displays (easy- or difficult-to-understand operability), styling and appearance (like or dislike, by observing the subject's expressions), and adjusting seating position (easy- or difficult-to-adjust seat controls). An observer (or data collector) can directly observe or listen, or a video camera can be set up and its recordings can be played back at a later time. The observer needs to be trained to identify and classify different types of predetermined behaviors, events, problems, or errors that a subject commits or the state of the product (e.g., smoking, leaking, vibrating, making noise, or operating under a wrong mode) during the observation period. The observer can also record durations of different types of events (e.g., using a stopwatch), number of attempts made to perform an operation, number and sequence of controls used, number of glances made, state of the vehicle (e.g., speed or lateral control behavior), and so forth.

Some events, such as accidents, are rare, and they cannot be measured through direct observation, due to the excessive amount of direct observation time that would be needed for sufficient accident data to be collected. However, information about such events can be obtained through reports of near-accidents (i.e., situations in which accidents almost occurred but were averted) and indirect observations (e.g., through witnesses or from material evidence) gathered after such events. The information gathered through indirect observations may not be very reliable for a number of reasons (e.g., the witness may be guessing or even deliberately falsifying; or objects associated with the event of interest may have been displaced or removed).

COMMUNICATION METHODS

Communication methods involve asking the customer (or the user) to provide information about his or her impressions and experiences with the product. The most common technique involves a personal interview in which an interviewer asks the customer a series of questions. The questions can be asked prior to usage of the product, during usage, or after usage. The user can be asked questions that will require him or her to (a) describe the product or impressions of the product and its attributes (e.g., appearance, usability), (b) describe the problems experienced while using the product (e.g., difficulty in reading a display), (c) categorize the product using a nominal scale (e.g., acceptable or unacceptable; comfortable or uncomfortable; liked or disliked), (d) rate the product on one or more scales describing the magnitude of its characteristics and/or overall impressions (e.g.. using rating scales for workload ratings, comfort ratings, difficulty ratings, ratings on appearance and styling), or (e) compare the products presented in pairs based on a given attribute (e.g., ease of use, comfort, quality feel during operation of a control, using techniques such as the analytical hierarchy process covered in Chapter 17).

Commonly used communication methods in product evaluations include (1) rating on scales: using rating scales with numbers and/or adjectives (e.g., acceptance ratings and semantic differential scales), (2) paired comparison–based scales (e.g., using Thurstone's method of paired comparisons and analytical hierarchy process), which will be described later in a section entitled Paired Comparison=Based Methods in this chapter.

In addition, many tools used in fields such as industrial engineering, quality engineering and design for six sigma, and safety engineering can be used in evaluations. Some examples of such tools are process charts, task analysis, arrow diagrams, interface diagrams, matrix diagrams, quality function deployment (QFD), Pugh analysis, failure mode and effects analysis (FMEA), and fault tree analysis (FTA) (refer to Chapters 17 and 18 and Bhise, 2014). These tools rely heavily on information obtained through the methods of communication from the users/customers and members of the multifunctional design teams. Additional information on many of these tools can be obtained from Besterfield et al. (2003), Creveling and Slutsky (2003), Yang and El-Haik (2003), and Bhise (2012, 2014).

Experimental Methods

The purpose of experimental research is to allow the investigator to control the research situations (e.g., creating different vehicle concepts or test conditions) so that causal relationships between independent variables that define the vehicle characteristics (e.g., interface configuration, type of control, type of display, operating forces) and the response variable (i.e., the variable used to evaluate) can be determined. An experiment includes a series of controlled observations (or measurements of response variables) undertaken in artificial test situations with deliberate manipulations of combinations of independent variables to answer one or more hypotheses related to the effect of (or differences due to) the independent variables. Thus, in an experiment, one or more variables (called *independent variables*) are manipulated, and their effect on another variable (called the *dependent* or *response variable*) is measured, while all other variables that may confound the relationship(s) are eliminated or controlled.

The importance of experimental methods is that (a) they help identify the best combination of independent variables and their levels to be used in designing the product (i.e., the vehicle or its entities) and thus provide the most desired effect on the users, and (b) when the competitors' products are included in the experiment along with the manufacturer's product, the superior product can be determined. To ensure that this method provides valid information, the researcher designing the experiment needs to make sure that the experimental situation is not missing any critical factor related to the performance of the product or the task being studied. Additional information on experimental methods can be obtained from Kolarik (1995), Besterfield et al. (2003), or other textbooks on the design of experiments (commonly referred as DOE) or statistical experiment design.

EVALUATIONS DURING VEHICLE DEVELOPMENT

During the entire vehicle development process, a number of evaluations are conducted to ensure that the vehicle being designed will meet the needs of the customers and other corporate and government requirements. All design requirements–related issues covered throughout this book need to be systematically evaluated by carefully following evaluation plans (usually included in the systems engineering management plan; see Chapter 12). The results of the evaluations are generally reviewed in the vehicle development process at different milestones in meetings with members of various design and management teams.

Physical Tests with Measurement Instruments

Most engineering tests are conducted on physical entities in laboratories or field conditions (i.e., under actual driving situations) using well-calibrated measurement instruments with very low (i.e., acceptable) levels of measurement error. The goal is to ensure that components, subsystems, systems, and the product perform their functions (i.e., meet their respective requirements) under all foreseeable conditions (e.g., under worst case conditions involving high operating loads and extreme environmental conditions, and under normal operating conditions over 100,000 miles and 10 years of driving). The variables to be measured must be valid, that is, they represent the vehicle attributes that are critical-to-customer satisfaction (CTS) or critical-to-quality (CTQ). The test procedures should be predeveloped and proven (i.e., accepted by the technical community and incorporated in the company's engineering test procedure manual) to allow standardization and comparison of data with past tests and products.

Some examples of such tests are

1. Measurements of vehicle body for bending stiffness, torsional stiffness, deflections under loading, deformation under collision conditions (e.g., frontal, side, rear, and rollover accidents), water leaks, corrosion protection, vibrations, noise (squeaks and rattles), and so forth.
2. Engine tests using dynamometers for measurements of output power (torque vs. speed), fuel consumption, greenhouse gas emissions, oil consumption, vibrations, and noise under a range of operating speeds and temperatures.
3. Measurements of strength and operating characteristics of chassis components (suspensions, steering, and braking systems) such as suspension linkages, springs, dampers, and brake pads.
4. Measurements of lighting equipment (e.g., headlamps, tail lamps, and stop lamps) for light output (photometric requirements), abrasion/haze, vibration tests, corrosion protection, weathering effects on optical materials, impact integrity (e.g., ability to sustain stone damage), and so forth.
5. Interior materials testing for strength, wear resistance, abrasion resistance, compressibility, softness/hardness, color, gloss/reflectance, chemical resistance of materials (when they come into contact with water, salt, body fluids [human and pets], oils, gasoline, foods [milk, ketchup], etc.), and endurance under different loads and temperature ranges.
6. Durability tests to evaluate operability over extended durations and a large number of cycles of operations under various operating and environmental conditions.

Many standards and test procedures are available to product design engineers for verification of requirements on whole vehicles, vehicle systems, and their subsystems and components (SAE, 2009; NHTSA, 2015). Vehicle-level validation tests are covered in Chapter 14.

MARKET RESEARCH METHODS

The market research methods commonly employed in the auto industry to evaluate customer response to various attributes of the whole vehicle as well as its systems are described in the following subsections.

Mail Surveys

Mail surveys are administered when responses on one or more product issues are to be collected from a large number of respondents using a small budget. The advantage of the mail survey is that the questionnaires can be sent to a large number of respondents within a short period of time and are very inexpensive (only postage costs). The disadvantage is that the sponsoring agency has no control over who actually completes the questionnaire and how much of the questionnaire is completed. The return rate the questionnaire is usually very low (e.g., about 1%–5%). Thus, the reliability of responses obtained from mail surveys is generally very low or questionable.

Internet Surveys

Internet surveys have many characteristics that are similar to mail surveys, and they suffer from the same type of disadvantages. However, they can be administered without incurring postage charges and with a very short delivery and response time.

Personal Interviews

Personal interview is the most commonly used method to obtain feedback from customers on automotive product design issues. In this method, each participant is carefully screened to ensure that he or she comes from the population of the market segment targeted for the product being developed. Each participant is usually selected from a database of registered owners of certain models of vehicles in a selected market segment. Each participant is invited to attend the clinic at a preselected time and place to spend about one-and-half or two hours. The participant is paid for his/her participation time.

When the participant arrives at the clinic site, his/her identification and vehicle ownership are verified by checking his/her driver's license and vehicle registration/ownership papers. The participant is then asked to fill out a questionnaire on demographic information (e.g., profession, makes and models of vehicles owned, education, income, and vehicle use history). In some market research clinics, additional information from measurements of the participant's anthropometric dimensions and reactions to lists of product features (shown using pictures, videos, or models) is also obtained. Then, the participant, usually led by an interviewer, is provided with background information and instructions on the market research clinic and then asked to view or even to use the product (e.g., to sit in a vehicle buck or to drive a prototype vehicle) and then respond to a structured set of questions. A large number of participants (75–300, depending on the objective of the clinic) are usually interviewed, and the answers to questions are summarized and analyzed using statistical analysis techniques. Chapters 11 and 22 provide examples of questions asked during market research clinics, data collection, and summary.

Focus Group Sessions

The focus group session is also a very commonly used market research technique in the automotive industry. It is primarily used to obtain qualitative and fact-finding information on product characteristics. Here, a session leader (or a moderator) provides background information on one or more product issues and creates discussions among the group members to gather information on their views, reactions, likes, dislikes, concerns, and so forth. Typically, each group includes about 8–12 individuals. The individuals are carefully selected to represent a certain type of individual (e.g., based on their gender, age, educational level, profession, or vehicle ownership) in the market segment to promote discussions on the selected product issues.

For example, to understand the problems that customers encounter in using an electronic climate control system in a luxury car, the vehicle manufacturer invited eight elderly couples who owned the vehicle. The moderator showed them pictures of the climate control in various modes of operation and asked them about their understanding of what the climate control was doing and the problems they had during setting and operation of the climate control system. The couples described a number of problems. For example, one older male said that he never sets the climate control because he cannot read the labels. His wife said that she helps him set the climate control. But she needs to first open her purse and get her reading glasses; and then she needs to lean her head closer to the climate control to read the labels. Thus, the climate control designer realized that the legibility of the labels of the climate control must be improved to accommodate the needs of elderly customers.

ERGONOMIC EVALUATIONS

An automotive product is used by a number of users in a number of different usage situations. To ensure that the vehicle being designed will meet the needs of its customers, ergonomic engineers conduct evaluations of all ergonomically associated vehicle features under all possible usages. A usage can be defined in terms of each task that needs to be performed by a user to meet a certain objective. A task may have many steps or subtasks. For example, the task of getting into a vehicle would involve a user performing a series of subtasks, such as (a) unlocking the door, (b) opening the door, (c) entering the vehicle and sitting in the driver's seat, and (d) closing the door. Ergonomic evaluations are conducted for a number of purposes, such as (a) to determine whether the users will be able to use the vehicle or its features, (b) to determine whether the vehicle has any unacceptable features that will generate customer complaints after its introduction, (c) to compare the user preferences for a vehicle or its features with other similar vehicles, and (d) to determine whether the product will be perceived by the users to be the best in the industry.

The evaluations can be conducted by collecting data in a number of situations. Some examples of data collection situations are

1. A product (a vehicle or one or more of its systems, chunks [portion of the vehicle], or features) is shown to a user, and the user's responses (e.g., facial expressions, verbal comments) are noted (or recorded). (This situation occurs when a concept vehicle is displayed in an auto show.)
2. A product is shown to a user, and then, responses to questions asked by an interviewer are recorded. (This situation occurs in a market research clinic).
3. A customer is asked to use a product, and then, responses to a number of questions asked in a questionnaire or asked by an interviewer are recorded. (This situation can occur in a drive evaluation.)

4. A user is asked to use a number of products, and the user's performance in completing a set of tasks on each of the products is measured. (This situation can occur in a performance measurement study using a set of vehicles [or alternate designs of a vehicle system] in test drives).

5. A user is asked to use a number of products and then asked to rate the products based on a number of criteria (e.g., preference, usability, accommodation, and effort). (This situation occurs in field evaluations using a number of vehicles—the manufacturer's test vehicle and other competitive vehicles).

6. A sample of drivers are provided with instrumented vehicles that record vehicle outputs and video data of driver behavior and performance as the participants drive where they wish, as they wish, for weeks or months. This is probably the only valid method to discover what drivers actually do over time in the real world. (This situation occurs in naturalistic driving behavior measurement studies; Lee et al., 2007.)

These examples illustrate that an ergonomics engineer can evaluate a vehicle or its features by using a number of data collection methods and measurements.

Bhise (2014) presents detailed descriptions of the ergonomics methods used in the automotive industry. The methods can be summarized as

1. Databases on human characteristics and capabilities
2. Anthropometric and biomechanical human models
3. Checklists and score cards
4. Task analysis
5. Human performance evaluation models
6. Laboratory, simulator, and field studies
7. Human performance measurement methods

The application of human factors tools requires the implementer to be knowledgeable about human factors issues and principles and research studies in the subject area. These tools cannot be easily mastered without sufficient experience with the variables related to the users, the product, the users' tasks, and the product usage situations. The user's performance in using a product can be affected by a number of factors, such as familiarity with past models of similar products, adaptability to new situations, improved performance with practice (learning), or deliberate changes in behavior to please or displease the evaluator (or experimenter). Thus, many variables can affect the user's behavior, performance, and preferences. A trained human factors professional will generally take the necessary precautions to ensure that biases are not introduced during the application of the human factors methods.

Databases on Human Characteristics and Capabilities

A number of human factors engineering handbooks, textbooks, and standards provide data on various human characteristics and capabilities (e.g., anthropometric, biomechanical, and information processing characteristics) for various populations (by gender, age groups, occupations, and national origin) (Garrett, 1971; Van Cott and Kinkade, 1972; Jurgens et al., 1990; Pheasant and Haslegrave, 2006; Kroemer et al., 1994; Konz and Johnson, 2004; Sanders and McCormick, 1993; McDowell et al., 2008; Bridger, 2008; Bhise, 2012; Sanders, 1983; Card et al., 1983; SAE, 2009; Wickens et al., 1998; Woodson, 1992). In addition, many research reports and journals related to human factors provide useful data from many studies.

These databases provide information on the distributions and percentile values of various characteristics and capabilities that are needed to design products to fit most people in selected product user populations. Such data are needed for designing products. For example, while designing an automotive product, the designer must ensure that a short female can reach and operate the pedals and see over the steering wheel, and a tall male can fit inside the cockpit with sufficient headroom, legroom, hip room, shoulder room, and elbow room. The designers also need to know the level of familiarity of the users with the operation of the controls and displays, the performance characteristics of the product, and the characteristics of the operational environment. Further, the decision-making and product-operating capabilities of the users must be known to the product designers.

Anthropometric and Biomechanical Human Models

A number of two-dimensional (2-D) and three-dimensional (3-D) anthropometric and biomechanical models are presented in the literature, and many are available commercially for design and evaluation purposes. These models can be configured to represent individual males and females and in different percentile dimensions for different populations. Many of the models have built-in human motion, posture simulations, and biomechanical strength as well as percentile force exertion prediction capabilities. Crash test dummies resembling human biomechanical characteristics are also used to evaluate the crashworthiness of vehicles in accident situations (Seiffert and Wech, 2003).

Integrated digital workplace and digital manikins and visualization tools are available in several software applications. Computer-aided design (CAD) tools with manikin models (digital human models), such as Jack/Jill, SAFEWORK, RAMSIS, SAMMIE, and the UM 3DSSP, have been used by different designers to assist in the product development process (Chaffin 2001, 2007; Reed et al., 1999, 2003; Badler et al., 2005; Human Solutions, 2010). Many of these tools are being updated to incorporate additional capabilities.

Before using any of the models in the design process, the ergonomics engineer should conduct validation studies to determine whether the population of the particular users of the product being designed can be accurately represented in terms of their dimensions, postures, motions, strength, and comfort. The postures assumed by the selected digital human model and their outputs should match closely with the postures and dimensions of real users under different actual usage situations.

Human Factors Checklists and Score Cards

Checklists and score cards are commonly used by human factors experts to evaluate products. The checklists aid in evaluating applications of human factors guidelines, design considerations, and requirements, whereas the scorecards help in summarizing findings of the human factors evaluations and enable tracking quantitative and comparative assessment of the product over time as the design progresses.

Human factors checklists typically include a series of questions related to meeting human factors guidelines. The product is usually evaluated by one or more human factors experts or the users of the product. Each evaluator uses the product and then answers each question. The answers to the questions (e.g., "yes" or "no") or a rating on how well the human factors guideline was met (e.g., using a 10-point scale, where 10=met very well, 1=did not meet the guideline) can be used to summarize the responses (e.g., percentage of human factors guidelines met, or percentage of guidelines met with ratings of 8 or above) in key areas such as locations of controls and displays, visibility of displays, legibility of displays, comprehension or interpretability of controls and displays, operation of controls, and feedback from controls. Examples of ergonomics checklists for evaluation of automotive controls and displays are presented in (Bhise, 2012).

Human factors score cards are created to provide feedback to the design teams on the ergonomics characteristics of the product being designed. Ergonomics experts systematically develop scoring criteria and evaluation procedures. The product design is analyzed by conducting evaluations based on objective analyses (e.g., measurements of task completion times, CAD analyses of hand reach to controls, visibility analyses of the instrument cluster seen through the steering wheel, and legibility predictions of letters and numerals in displays [Bhise, 2012]) as well as subjective analyses from ratings of one or more ergonomics experts for each ergonomic consideration. A score card is prepared by summarizing the results of the evaluations. The score cards are presented and discussed with the design, engineering, and program management teams during program review meetings.

The data gathered from completion of the checklists can be categorized for comparisons by type of users (e.g., based on familiarity/unfamiliarity with the product, male/female users, and young/mature users), and scores can be developed by type of users (e.g., older user's score card, women user's score card, and unfamiliar user's score card).

An example of an ergonomic score card for controls and displays in the interior of an automotive product is presented in Figure 21.1. The score card represents an ergonomics summary chart (called the *smiley faces chart*). The chart lists each control and display in different interior regions of the vehicle on the left-hand side of the table. The evaluation criteria are grouped into nine columns located in the middle of the table. The nine criteria groups are labeled as

1. Visibility, obscurations, and reflections
2. Forward vision down angle
3. Grouping, association, and expected locations
4. Identification labeling
5. Graphics legibility and illumination
6. Understandability/interpretability
7. Maximum and minimum reach distance
8. Control area, clearance, and grasping
9. Control movements, efforts, and operability

A five-point rating scale (with 5=highest score and 1=lowest score) is used to evaluate the vehicle on ergonomic guidelines in each of these nine groups. The ratings

Region/Location	No.	Interior items: Controls, displays and handles	Control and display evaluation criteria									Comments: Specific problems and suggestions
			Visibility, obscurations and reflections	Forward vision down angle	Grouping, Association and Expected Loc.	Identification labeling	Graphics legibility and illumination	Understandability/interpretability	Max. And min. Comfortable reach distance	Control area, clearance and grasping	Control movements, efforts and operability	
Door mounted controls	1	Inside door handle	◉	⊘	☺	☺	☺	☺	☺	☺	☺	Door handle partially obscured by the steering wheel–requires head movement.
	2	Door pull handle	☺	⊘	☺	☺	☺	☺	◉	◉	⊘	Requires some chicken-winging–should be moved forward 25–50 mm. Cannot use full power grasp.
	3	Door lock	☺	☺	☺	☺	●	☺	☺	◉	☺	Lock symbol difficult to interpret. Difficult to push–touch area moves under bezel surface.
	4	Window controls	☺	☺	☺	☺	☺	☺	☺	☺	●	Pull-up and push down switches should be used to avoid accidental activation.
	5	Window lock	☺	☺	☺	☺	☺	●	☺	☺	☺	Window lock symbol is difficult to understand.
	6	Mirror control	☺	☺	☺	☺	●	☺	☺	☺	☺	Provide illuminated label.
Column/S/W	7	Turn signal stalk	☺	☺	☺	☺	☺	☺	☺	☺	☺	
	8	Wiper switch (right stalk)	☺	☺	☺	☺	☺	☺	☺	☺	☺	
	9	Ignition switch	◉	⊘	☺	☺	☺	☺	☺	☺	☺	Requires head movement to see the control during key insertion.
	10	Cruise controls	☺	☺	☺	☺	☺	☺	☺	☺	☺	
	11	Shifter	☺	⊘	☺	☺	☺	☺	☺	☺	☺	Located on console.
Left/P	12	Light switch (on left stalk)	☺	☺	☺	☺	☺	☺	☺	☺	☺	
	13	Panel dim	☺	☺	☺	☺	☺	☺	☺	☺	☺	On left side of I/P.
	14	Parking brake (on console)	☺	☺	☺	☺	☺	☺	☺	☺	☺	
	15	Hood release	◉	⊘	☺	●	☺	☺	☺	☺	☺	Difficult to see from the driver's seat. Not labeled.
Cluster	16	Tachometer	☺	☺	☺	☺	☺	☺	⊘	⊘	⊘	
	17	Speedometer	☺	☺	☺	☺	☺	☺	⊘	⊘	⊘	
	18	Temperature	☺	☺	☺	☺	☺	☺	⊘	⊘	⊘	
	19	Fuel gauge	☺	☺	☺	☺	☺	☺	⊘	⊘	⊘	
	20	Oil pressure							⊘	⊘	⊘	Not on this vehicle.
	21	PRNDL Display	☺	☺	☺	☺	☺	☺	☺	⊘	◉	Push-button controls are somewhat unexpected for this function.
Center of instrument panel/console	22	Radio	◉	◉	☺	☺	◉	☺	☺	☺	☺	1/2 – 2/3 of the radio controls below 35 deg. Silver buttons on silver background difficult to locate quickly.
	23	Climate control	☺	◉	☺	☺	◉	☺	☺	☺	☺	Mode selector symbols difficult to read on silver background. Low location and disassociated display.
	24	Clock (top in center stack)	☺	☺	☺	☺	☺	☺	☺	☺	☺	
	25	Backlight defrost	☺	☺	☺	☺	◉	☺	☺	☺	☺	Symbol difficult to read on silver background.
	26	Windshield defrost	☺	☺	☺	☺	◉	☺	☺	☺	☺	Symbol difficult to read on silver background.
	27	Traction control										Not on this vehicle.
	28	Parking brake	☺	☺	☺	☺	☺	☺	☺	☺	☺	On console.
	29	Hazard switch	☺	☺	☺	☺	☺	☺	☺	☺	☺	On center above the CD slot.
	30	Ash tray	☺	●	☺	☺	☺	☺	☺	☺	☺	Located too low.
	31	Cigarette lighter	☺	●	☺	☺	☺	☺	☺	◉	☺	Located too low. Not enough clearance for finger grasp.
	32	Cupholder	☺	●	☺	☺	☺	☺	●	☺	☺	Requires chicken-winging. Located too close and too low.
	33	Shifter	☺	☺	☺	☺	☺	☺	☺	☺	☺	
Other	34	Glove box latch	☺	⊘	●	☺	☺	☺	☺	☺	☺	Latch opening push button is not located on the glove box. The centerstack location is not expected by most drivers.
	35	Cruise control on/off	☺	☺	☺	☺	☺	☺	☺	☺	☺	On right spoke of S/W.
	36	Trunk release	◉	☺	☺	☺	☺	☺	☺	☺	☺	Difficult to see from the driver's eyepoint.
	37	Fuel fill door release	☺	☺	☺	☺	☺	☺	☺	☺	☺	On floor–left side.

			Key:	⊘	●	◉	☺			
				Not Appl.	Low	Mid	High			
			Rating:	–	1–2	3	4–5			

FIGURE 21.1   Ergonomics evaluation score card.

are usually obtained by trained ergonomists using inputs from (a) measurements obtained from a 3-D-CAD model of the occupant package (e.g., reach distances, down angles), (b) ergonomics review by sitting in an interior buck (if available), and (c) results from applicable design tools and models (e.g., Society of Automotive Engineers [SAE] design practices [SAE, 2009; Bhise, 2012]). The ratings are graphically displayed using a graphic scale of "smiley" faces for each of the nine groups, for each item listed in each row. The chart provides an easy-to-view format that can be used to provide the overall ergonomics status of a vehicle interior and was found by the author to be a useful tool in various design and management review meetings. The objective of the ergonomics engineer is to convince the design team during the design review meetings to remove as many "black dots (for ratings of 1 and 2) and black donuts (for rating of 3)" from the charts and increase the number of "smiley faces (rating of 4 and 5)" by making the necessary design changes.

Task Analysis

Task analysis is one of the basic tools used by ergonomists in investigating and designing tasks associated with operating a product or a process. It provides a formal comparison between the demands that each task places on the human operator and the capabilities that the human operator possesses to respond to the demands. Task analysis can be conducted with or without a real product or a process, but it is easier if the real product or equipment is available and the task can be performed by actual (representative) users under real usage situations to understand the details within the tasks.

The analysis involves breaking the task or operation into smaller units (called the *subtasks*) and analyzing subtask demands with respect to the user capabilities. The subtasks are the smallest units of behavior that need to be differentiated to solve the problem at hand. Some examples of subtasks are: grasp a handle, read a display, select a control setting, and adjust the control to a desired setting. The user capabilities that are considered here are generally sensing, use of memory, information processing, and response execution (body movements, reaches, accuracy, postures, forces, time constraints, and so forth).

Task analysis can be conducted using different formats. The tabular format for task analysis presented by Drury (1983) was found by the author to be useful during the automotive design process (Bhise, 2012). The left-hand columns of the task analysis table describe the subtasks involved in the task along with the purpose of each subtask. Thus, the description of each subtask makes the analyst think about the need for each subtask and, perhaps, even suggest a better way to do the task. The right-hand columns of the table force the analyst to consider various human functional capabilities, such as searching and scanning, retrieving information from the memory, interpolating (information processing), and manipulating (e.g., hand finger movements) required in performing each subtask. The last column requires the analyst to think about possible errors that can occur in each subtask. This last column is the most important output that can be used to improve the task and/or the product to reduce the errors, problems, and difficulties involved in performing the task.

If the task analysis is performed on a product that already exists (or if its mockup, prototype product, or simulation is available), then a number of user trials can be performed, and information can be gathered on how different users perform each subtask and the problems, difficulties, and errors experienced by the users. The information then can be used in creating the task analysis table. Even if a product is not available, the task analysis can be performed on an early product concept by predicting the possible sequence of subtasks needed to use the product in performing each task (or product usage). Bhise (2012) provides examples of task analysis conducted using this format.

Human Performance Evaluation Models

Human factors researchers have developed a number of models to predict the performance of users under different product usage or work situations. The models are generally based on statistical analysis (e.g., regression models) of data gathered from experimental research studies. An available human performance model, if found to apply to a given product usage situation, can be applied to predict the performance of different users under different product usage situations. The model predictions, along with the results of additional experimental research, can be used to narrow down a number of product concepts and designs during the early stages of the product development processes.

For example, one of the oldest operator performance models to predict the time requirements of factory jobs is the methods-time measurement (MTM) (Maynard et al., 1948). The model is based on breaking down a given task into a series of predefined micro-motions. The times required to perform different micro-motions (e.g., reach, move, grasp, release, turn, position, assemble, turn body, leg motions, eye motions, steps, and side-steps) are provided in tables. The applicable times from the tables can be added to predict the time to complete a given task. The times are for an experienced operator handling and assembling small parts. Other predetermined time prediction models are also used in the industry (Konz and Johnston, 2004).

A human information processing model for prediction of human operator time requirements in information processing tasks was developed by Card et al. (1980, 1983). The basic approach involves top-down, successive decomposition of a task. The analyst divides the task into logical steps. For each step, the analyst identifies the human and device task operators. This approach assumes error-free performance, well-learned tasks, and particular locations of controls.

Many other models are also available to predict human errors. Leiden et al. (2001) have provided a review of a number of cognitive models, such as ACT-R, Air MIDAS, Core MIDAS, APEX, COGNET, D-OMAR, EPIC, GLEAN, SAMPLE, and Soar, for prediction of human error. Many information processing and workload assessment models are also available in the human factors literature (see Bhise, 2012). Such models, once the limitations on their applicability are understood, can be used to evaluate user performance with a given product under a specified usage situation.

Bhise (2012) has described a number of models used for ergonomic evaluations in the automotive design process. The models include driver positioning and occupant packaging practices incorporated into the SAE practices (SAE, 2009). Bhise (2012) also described a number of human vision models based on visual contrast thresholds, and disability and discomfort glare prediction equations. The models have been used to evaluate automotive headlighting systems (the comprehensive headlamp environment systems simulation [CHESS] model; Bhise et al., 1977, 1988, 1989), legibility of displays (Bhise and Hammoudeh, 2004), and veiling glare effects from sunlight reflections into vehicle windshields (Bhise and Setumadhaven, 2007, 2008a,b; see Chapter 22).

Hankey et al. (2001) developed a model to predict the level of demand placed on the driver while using in-vehicle devices (e.g., an entertainment system or a navigation system). The model is identified as the IVIS-DEMAnD model. Jackson and Bhise (2002) applied the model to evaluate the relative sensitivity of vehicle parameters and other external factors, such as traffic demand and driver age, in evaluating figure of demand during various driving tasks.

Human factors models in general are based on modeling relationships between many independent variables and assumed characteristics of human operators. The relationships are usually developed based on analyses of data collected in experimental research. The models are thus approximations of human performance and should be used with caution. The model results should be validated by performing studies using real subjects and products under actual usage situations. This issue of testing with real subjects is covered in the next section.

Laboratory, Simulator, and Field Studies
Early product concepts, product prototypes, and final products (production versions) are generally evaluated using real subjects performing a number of tasks to verify and validate the product design. Product evaluation under actual usage situations involving field testing is generally preferred. However, due to a number of reasons, such as high costs of building a working model of the product, costs in recruiting subjects, and time required to perform field tests, other research approaches, such as laboratory tests, product simulations, use of early prototypes, or testing with simulators, are commonly considered.

Data collection methods employed in human factors studies include (a) observing subjects performing given tasks with the product, (b) communicating with the subjects to obtain ratings on selected product features or asking the subjects to describe problems and difficulties in using the product, (c) measuring the performance of the subjects and the product during product uses, (d) measuring the physiological state of the subjects while completing different usage tasks, and (e) obtaining subjective ratings to measure operator workload.

Human Performance Measurement Methods
Human performance in laboratory, field, and simulator studies can be measured using a combination of methods of observation, communication, and experimentation. These methods are described in Bhise (2012). Some examples of measurements in these evaluations include

1. *Observable Human Responses*: We can observe the human operator's responses, such as his/her visual information acquisition behavior, through measurements of eye movements, head movements, eye glances, and time spent in viewing different objects (e.g., displays), and control movements through measurements of movements of body parts, such as hand and foot,

while operating various controls. We can also measure the physiological state of the operator by measuring variables such as the operator's heart rate, sweat rate, and pupil diameter.

2. *Operator's Subjective Responses*: We can also develop a structured questionnaire and ask the operator a number of questions at different points in a test procedure (if an experimenter is present) or at the end of the procedure to understand the operator's problems, difficulties, confusions, frustrations, and situational awareness issues. We can also ask the operator to provide ratings on his/her workload, comfort, ease in using different controls and displays, and so forth.

3. *State of Product/Equipment*: We can also record the state of the product (or the equipment being used) by installing measuring instruments in the product to measure product outputs (as functions of time) such as speed, control positions and movements, distance traveled, vibrations, temperatures, and energy consumption.

OBJECTIVE MEASURES AND DATA ANALYSIS METHODS

Depending on the task used to evaluate a product, the task performance measurement capabilities, and the instrumentation available, the ergonomics engineer designs an experiment and procedure to measure dependent (or response) variables. Objective measures can be based on physical measures such as time (taken or elapsed), distance (position or movements in lateral, longitudinal, or vertical directions), velocities, accelerations, events (occurrences of predefined events), and measures of the user's physiological state (e.g., heart rate). The recorded data are reduced to obtain values of the dependent measures and their statistics, such as means, standard deviations, minimum, maximum, and percentages above and/or below certain preselected levels. The values of the dependent measures are then used for statistical analyses based on the experiment design selected for the study. Some examples of applications involving objective measures are provided in a later section of this chapter.

SUBJECTIVE METHODS AND DATA ANALYSIS

Subjective methods are used by engineers because in many situations, (a) the subjects are better able to perceive characteristics and issues with the product, and thus, they can be used as the measurement instruments, (b) suitable objective measures do not exist, and (c) subjective measures are easier to obtain.

Pew (1993) made several important points regarding subjective methods. Subjective data must come from the actual user rather than the designer; the user must have an opportunity to experience the conditions to be evaluated before providing opinions; care must be taken to collect the subjective data independently for each subject; and the final test and evaluation of a system should not be based solely on subjective data.

The two most commonly used subjective measurement methods during the vehicle development process are (a) rating on a scale and (b) paired comparison–based methods. These two methods are presented in the following subsections.

RATING ON A SCALE

In this method of rating, the subject is first given instructions on the procedure involved in evaluating a given product, including explanations of one or more of the product attributes and the rating scales to be used for scaling each attribute. Interval scales are used most commonly. Many different variations are possible in defining the rating scales. The interval scales can differ due to (a) how the end points of the scales are defined, (b) the number of intervals used (note: an odd number of intervals allows the use of a midpoint), and (c) how the scale points are specified (e.g., without descriptors vs. with word descriptors or numerals). Bhise (2012, 2014) presents a number of examples of interval scales.

Table 21.2 illustrates how the direction magnitude and the 10-point acceptance scales together can be used to evaluate a number of interior dimensions in a vehicle package. The distribution of responses on each direction magnitude scale provides feedback to the designer on how the dimension corresponding to the scale was perceived in terms of its magnitude, and the ratings on the acceptance scale provide the level of acceptability of the magnitude of the dimension. For example, if the ratings on item number 5 (gas pedal lateral location) in Table 21.2 show that 80% of the subjects rated the gas pedal location as "Too much to the left" on the direction magnitude scale, and the average rating on the 10-point acceptance scale was 4.0, the designer can conclude that the gas pedal needs to be moved to the right to improve its acceptability. The author found that such use of dual scales was very helpful in fine-tuning the vehicle dimensions in the early stages of the vehicle design process.

PAIRED COMPARISON–BASED METHODS

The method of paired comparison involves evaluating products presented in pairs. In this evaluation method, each subject is essentially asked to compare two products in each pair using a predefined procedure and is asked to simply identify the better product in the pair on the basis of a given attribute (e.g., comfort or usability). (If the respondent states that there is no difference between the two products, the instruction will be to randomly pick one of the products in the pair. The idea is that, if there truly is no difference in that pair among the respondents, the result will average out to 50:50.) The evaluation task of the subject is, thus, easier as compared with rating on a scale. However, if n products have to be evaluated, then the subject is required to go through each of the $n(n-1)/2$ possible numbers of pairs and identify the better product in each pair. Thus, if five products need to be evaluated, the total number of possible pairs will be $5(5-1)/2 = 10$. The major advantage of the paired comparison approach is that it makes the subject's task simple and more accurate, as the subject only has to compare the two products in each trial and identify the better product in the pair. The disadvantage of the paired comparison approach is that as the number of products (n) to be evaluated increases, the number of possible paired comparison judgments that each subject needs to make increases rapidly (proportional to the square of n), and the entire evaluation process becomes very time consuming.

We will review two commonly used methods based on the paired comparison approach: (a) Thurstone's method of paired comparisons and (b) the analytical

TABLE 21.2

Illustration of Vehicle Package Evaluation Using Direction Magnitude and Acceptance Rating Scales

Item No.	Driver Package Consideration	Rating Using Direction Magnitude Scale			Acceptance Rating: 1 =Very Unacceptable, 10 =Very Acceptable
1	Steering wheel longitudinal (fore/aft) location	Too close	About right	Too far	
2	Steering wheel vertical (up/down) location	Too low	About right	Too high	
3	Steering wheel diameter	Too small	About right	Too large	
4	Gas pedal fore/aft location	Too close	About right	Too far	
5	Gas pedal lateral location	Too much to left	About right	Too much to right	
6	Lateral distance between the gas pedal and the brake pedal	Too small	About right	Too large	
7	Gas pedal to brake pedal lift-off	Too small	About right	Too large	
8	Gearshift lateral location	Too much to left	About right	Too much to right	
9	Gearshift location longitudinal location	Too close	About right	Too far	
10	Height of the top portion of the instrument panel directly in front of the driver	Too low	About right	Too high	
11	Height of the armrest on driver's door	Too low	About right	Too high	
12	Belt height (lower edge of the driver's side window)	Too low	About right	Too high	
13	Space above the driver's head	Too little	About right	Too generous	
14	Space to the left of the driver's head	Too little	About right	Too generous	
15	Knee space (between instrument panel and right knee with foot on the gas pedal)	Too little	About right	Too generous	
16	Thigh space (between the bottom of the steering wheel and the closest lower surface of the driver's thighs).	Too little	About right	Too generous	
17	Outboard edge of the rocker panel during ease in entry and exit	Too far out	About right	Too close	
18	Driver seating location	Too close to door	About right	Too far from door	
19	Seat height from the ground	Too high	About right	Too low	
20	Visibility over the hood	Too little	About right	Too generous	

hierarchy method. Thurstone's method allows us to develop scale values for each of the n products on a z-scale (z is a normally distributed variable with mean equal to zero and standard deviation equal to one) of desirability (Thurstone, 1927), whereas the analytical hierarchy method allows us to obtain relative importance weights of each of the n products (Satty, 1980). Both the methods are simple and quick to administer and have the potential of providing more reliable evaluation results as compared with other subjective methods whereby a subject is asked to evaluate one product at a time.

THURSTONE'S METHOD OF PAIRED COMPARISONS

Let us assume that we have five products (or designs or issues) that need to be evaluated. The five products are named L, W, N, P, and R. The 10 possible pairs of the product are (1) L and W, (2) L and N, (3) L and P, (4) L and R, (5) W and N, (6) W and P, (7) W and R, (8) N and P, (9) N and R, and (10) P and R. The steps to be used in the procedure are presented in the following subsections.

Step 1: Select an Attribute for Evaluation of the Products

The purpose of the evaluation is to order the five products along an interval scale based on a selected attribute. Let us assume that the five products are five different layouts of center stack controls for the operation of navigation, audio entertainment, and climate controls in a vehicle. The attribute selected is "ease of operation of the center stack controls."

Step 2: Prepare the Products for Evaluation

It is further assumed that five test cars have been built for the evaluation of the center stack controls with identical vehicle models, features, and interiors. The center stack part of the instrument panel in each vehicle is fitted with the center stack controls in a different layout.

Step 3: Obtain Responses of Each Subject on All Pairs

It is also assumed that 80 subjects will be selected randomly from the population of the likely owners of the vehicle for the evaluation study.

Each subject will be brought into the test area separately by an experimenter. The experimenter will provide instructions to the subject and ask the subject to operate controls in the center stack by performing a series of preselected tasks related to use of the navigation system, audio system, and climate control system while driving each vehicle on a preselected route. Each subject will be asked to evaluate 10 pairs of vehicles selected in a random order. After completing the test drives for each pair of vehicles, the subject will be asked to select the center stack layout that is easier to use in each pair.

The responses of an individual subject are illustrated in Table 21.3. Each cell of the table contains Yes or No depending on whether the center stack layout shown in the column was better (easier to use) than the center stack layout shown in the row. It should be noted that only the 10 cells above the diagonal (marked by x) need to be evaluated.

TABLE 21.3
Responses of an Individual
Subject for the Ten Possible
Product Pairs

	L	W	N	P	R
L	x	No	No	No	No
W		x	No	No	Yes
N			x	No	Yes
P				x	Yes
R					x

Note: A "Yes" response indicates that the product shown in the column is better than the product in the row. A "No" response indicates that the product shown in the row is better than the product shown in the column.

Step 4: Summarize Responses of All Subjects in Terms of Proportion of Product in the Column Better Than the Product in the Row

After all the subjects have provided responses, the responses are summarized as shown in Table 21.4 by assigning a 1 to a Yes response and a 0 to a No response. Thus, the cell corresponding to W column and L row indicates that only 1 out of the 80 subjects judged center stack layout W to be better than center stack layout L.

The complements of the summarized ratings in Table 21.4 are entered in the cells below the diagonal, as shown in Table 21.5. For example, the complement of "1/80 responses of product W better than product L" is "79/80 responses of product L better than product W."

TABLE 21.4
Number of Subjects Preferring
Product in the Column over
Product in the Row Divided by
Number of Subjects

	L	W	N	P	R
L	x	1/80	3/80	2/80	4/80
W		x	3/80	30/80	50/80
N			x	30/80	50/80
P				x	60/80
R					x

TABLE 21.5

Response Ratio Matrix with Lower Half of the Matrix Filled with Complementary Ratios

	L	W	N	P	R
L	x	1/80	3/80	2/80	4/80
W	79/80	x	3/80	30/80	50/80
N	77/80	77/80	x	30/80	50/80
P	78/80	50/80	50/80	x	60/80
R	76/80	30/80	30/80	20/80	x

TABLE 21.6

Proportion of Preferred Responses (p_{ij})

		i=1	i=2	i=3	i=4	i=5
		L	W	N	P	R
j=1	**L**	x	0.013	0.038	0.025	0.050
j=2	**W**	0.988	x	0.038	0.375	0.625
j=3	**N**	0.963	0.963	x	0.375	0.625
j=4	**P**	0.975	0.625	0.625	x	0.750
j=5	**R**	0.950	0.375	0.375	0.250	x

The proportions in Table 21.5 are expressed in decimals in Table 21.6. Each cell in the matrix presented in Table 21.6 thus represents proportion p_{ij}, indicating the proportion of responses in which the product in the ith column was preferred over the product in the jth row.

Step 5: Adjusting p_{ij} Values

To avoid the problem of distorting the scale values (computed in Step 6) of the products when p_{ij} values are very small (close to 0.00) or very large (close to 1.00), the p_{ij} values in Table 21.6 above 0.977 are set to 0.977 and the p_{ij} values below 0.023 are set to 0.023, as shown in Table 21.7.

Step 6: Computation of Z-values and Scale Values for the Products

In this step, the values of the proportions (p_{ij}) in each cell are converted into Z-values using the table of standardized normal distribution found in any standard statistics textbook. For example, the value of $p_{21} = 0.023$ is obtained by integrating the area under the standardized normal distribution curve (with mean equal to 0 and standard deviation equal to 1.0) from minus infinity to -1.995. Thus, a Z-value of -1.995 provides a p-value of 0.023. The Z-values can also be obtained using a function called NORMINV by setting its parameters as (p_{ij},0,1) in Microsoft Excel. The Z-values

TABLE 21.7

Adjusted Table of p_{ij} (If $p_{ij} > 0.977$, then set $p_{ij} = 0.977$; and if $p_{ij} < 0.023$, then set $p_{ij} = 0.023$)

		$i=1$	$i=2$	$i=3$	$i=4$	$i=5$
		L	W	N	P	R
$j=1$	L	x	0.023	0.038	0.025	0.050
$j=2$	W	0.977	x	0.038	0.375	0.625
$j=3$	N	0.963	0.963	x	0.375	0.625
$j=4$	P	0.975	0.625	0.625	x	0.750
$j=5$	R	0.950	0.375	0.375	0.250	x

TABLE 21.8

Values of Z_{ij} Corresponding to Each p_{ij} and Computation of Scale Values (S_i)

		$i=1$	$i=2$	$i=3$	$i=4$	$i=5$
		L	W	N	P	R
$j=1$	L	x	-1.9953933102	-1.7804643417	-1.9599639845	-1.644853627
$j=2$	W	1.9953933102	x	-1.7804643417	-0.318639364	0.318639364
$j=3$	N	1.7804643417	1.7804643417	x	-0.318639364	0.318639364
$j=4$	P	1.9599639845	0.318639364	0.318639364	x	0.6744897502
$j=5$	R	1.644853627	-0.318639364	-0.318639364	-0.6744897502	x
$\sum Z_{ij} =$		7.3806752634	-0.2149289685	-3.5609286834	-3.2717324627	-0.3330851488
$S_i =$		2.0875702114	-0.0607910924	-1.0071827277	-0.9253856842	-0.094210707

Note: Z_{ij} = Value of NORMINV(p_{ij},0,1) function from Microsoft Excel.

(Z_{ij}) obtained by converting all the proportion (p_{ij}) values in Table 21.7 using this conversion procedure are shown in the matrix on the top part (Z matrix) of Table 15.9.

The Z-values obtained in each column are summed (i.e., summed over all js), and the scale values for each product (S_i) are obtained using the formula (see the last two rows of Table 21.8)

$$S_i = \left(\sqrt{2/n}\right) \sum Z_{ij}$$

where n = number of products used in paired comparisons

The bottom row of Table 21.8 presents the scale values (S_i) for each product (note: using $n=5$ in the formula). It should be noted that the sum of the scale values (i.e., summing over all is) computed from the formula is equal to 0.0 (i.e., $\sum S_i = 0.0$).

Figure 21.2 presents a bar chart of the scale values (S_i) of the five products shown in Table 21.8. Thus, using Thurstone's method of paired comparisons, scale values

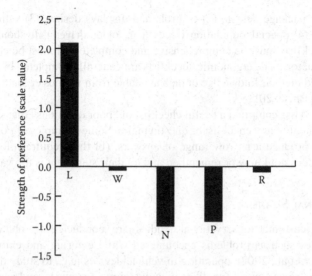

FIGURE 21.2 Scale values of the five products.

of the five products are obtained. The scale values indicate the strength of the relative preference of each of the products in the set of n products. The unit of the scale values is number of standard deviations, and the zero value on the scale corresponds to the point of indifference (i.e., the product with the zero scale value is neither liked [preferred] nor disliked [not preferred]). Thus, in this example, product L is the best (most preferred) among the five products, and product N is least preferred.

ANALYTICAL HIERARCHY METHOD

In the analytical hierarchy method, the products are also compared in pairs. However, the better product in each pair is also rated in terms of the strength of the attribute it possesses in relation to the strength of the same attribute in the other product in the pair. The strength of the attribute is expressed using a ratio scale. The scale (or the weight) value of 1 is used to denote equal strength of the attribute in both the products in the pair; the scale value of 9 is used to indicate extreme or absolute strength of the attribute in the better product; and the product with the weaker strength is assigned the inverse of the scale value of the better product. Two examples illustrating the application of the analytical hierarchy method are provided in Chapter 17.

SOME APPLICATIONS OF EVALUATION TECHNIQUES IN AUTOMOTIVE DESIGN

CHECKLISTS

A checklist is used to check that the product being designed meets each applicable guideline (or principle or requirement) in the area covered by the checklist. The checklist approach is commonly used during the design of many areas, (1) interior and exterior package design, (2) controls and displays design, (3) vehicle lighting design, and (4) special population issues (e.g., older drivers). It should be noted that the checklists must be comprehensive and complete and must be completed by trained evaluators. The ergonomic checklists are generally completed by ergonomics experts based on their knowledge or data available from various ergonomic analyses and studies (Bhise, 2012).

Pew (1993) has compiled a useful checklist of "poor questions" that should guide the development of any checklist or questionnaire. Some examples of poor questions are (a) they produce a narrow range of answers, (b) they require information the respondent does not know or remember, and (c) their statement is too vague.

OBSERVATIONAL STUDIES

Driver and customer observational studies are conducted to obtain information on issues such as problems encountered while entering and exiting vehicles (Bodenmiller et al., 2002), operating in-vehicle devices (e.g., to study driver under-standing of various control functions in audio, climate controls, and navigation systems), and performing vehicle service tasks (e.g., checking fluids, changing fuses and bulbs, refueling, changing a tire) (see Bhise, 2012).

VEHICLE USER INTERVIEWS

Drivers and other vehicle users are interviewed individually or in groups (e.g., focus group sessions) to understand their concerns, issues, and wants related to various vehicle features. For example, Bhise et al. (2005) asked drivers to develop layouts of center stack and console areas through a structured interview technique (a method of communication).

RATINGS ON INTERVAL SCALES

Rating methods using different interval scales are used for ergonomic evaluations of issues such as (a) interior and exterior package dimensions, (b) characteristics of controls and displays (e.g., acceptability of locations, sizes, and grasp areas; haptic feedback during movement or activation of controls), and (c) interior materials (e.g., visual and tactile characteristics of materials on instrument panels, door trim, seat areas, armrests, and steering wheels) (Bhise et al., 2006, 2008, 2009).

STUDIES USING PROGRAMMABLE VEHICLE BUCKS

Programmable vehicle bucks are used in early package evaluation studies to assess exterior and interior dimensions such as vehicle width, windshield rake angle, seating reference point location (e.g., longitudinal distance from the accelerator pedal, lateral location from the vehicle centerline, and height from the ground), driver eye location, visibility over the instrument panel, hood, and side windows, and height of armrest (Richards and Bhise, 2004).

Driving Simulator Studies

Driving simulators are now routinely used in many automotive companies to evaluate driver workload issues in operating various in-vehicle devices (Bertollini et al., 2010). All the three methods of observation, communication, and experimentation can be used during the simulator tests.

Field Studies and Drive Tests

Various studies under actual driving situations on test tracks and public roads under different road, traffic, lighting, and weather conditions are conducted for evaluation of issues in areas such as seat comfort, field of view, vehicle lighting, controls and displays usage, and driver workload (Jack et al., 1995; Owens et al., 2010; Tijerina el al., 1999).

System and Component Verification and Vehicle Validation Methods

Entities requiring objective tests typically require physical testing with test equipment and measurement instruments. Many physical tests are conducted for verification of functional requirements on vehicle entities related to mechanical, electrical, and electronic functions, which usually involve extensive laboratory and field tests. Tests are also conducted over a large number of operational cycles for durability and reliability evaluations. Evaluations of software applications are generally conducted using a variety of simulation tests followed by real-world field tests. Every vehicle manufacturer has detailed test procedures and test equipment to conduct verification tests at component, subsystem, system, and vehicle levels.

Vehicle validation methods, which generally involve using customers as the evaluators, are covered in Chapter 14.

CONCLUDING REMARKS

Product evaluation activities should be considered as an integral part of the product design. The design of any entity in a vehicle must be evaluated to ensure that it meets its stated requirements and is liked by its intended customers. Early evaluations can be conducted using computer-aided engineering (CAE) methods and design reviews by experts from disciplines related to the functions of the entities. As the physical components (prototype or early production) are available in the latter half of the product development process, physical testing and evaluation by experts in the real world are preferred. Physical tests are generally very costly in comparison with CAE tests. However, the testing of actual components under real field conditions provides opportunities to evaluate a number of effects of variables related to real road-, traffic-, and weather-related environments. It should also be realized that while objective evaluations with measurement equipment and instrumentation can provide more precise data, some vehicle attributes, such as ride, comfort, styling, and ergonomics, also need subjective evaluations by experts as well as customers.

REFERENCES

Badler, N., J. Allbeck, S.-J. Lee, R. Rabbitz, T. Broderick, and K. Mulkern. 2005. *New Behavioral Paradigms for Virtual Human Models*. Paper 2005-01-2689. Presented at the 2005 SAE Digital Human Modeling Conference, Iowa City, IA.

Bertollini, G., L. Brainer, J. Chestnut, S. Oja, and J. Szczerba. 2010. *General Motors Driving Simulator and Applications to Human Machine Interface (HMI) Development*. SAE Paper no. 2010-01-1037. Presented at the 2009 SAE World Congress, Detroit, Michigan.

Besterfield, D. H., C. Besterfield-Michna, G. H. Besterfield, and M. Besterfield-Scare. 2003. *Total Quality Management*, 3rd edn. Upper Saddle River, NJ: Prentice Hall. ISBN 0-13-099306-9.

Bhise, V., R. Boufelliga, T. Roney, J. Dowd, and M. Hayes. 2006. *Development of Innovative Design Concepts for Automotive Center Consoles*. SAE Paper no. 2006-01-1474. Presented at the SAE 2006 World Congress, Detroit, MI.

Bhise, V., R. Hammoudeh, J. Dowd, and M. Hayes. 2005. Understanding Customer Needs in Designing Automotive Center Consoles. In *Proceedings of the Annual Meeting of the Human Factors and Ergonomics Society*, Orlando, Florida.

Bhise, V., S. Onkar, M. Hayes, J. Dalpizzol, and J. Dowd. 2008. Touch feel and appearance characteristics of automotive door armrest material. *Journal of Passenger Cars— Mechanical Systems*, SAE 2007 Transactions.

Bhise, V., V. Sarma, and P. Mallick. 2009. *Determining perceptual characteristics of automotive interior materials*. SAE Paper no. 2009-01-0017. Presented at the 2009 SAE World Congress, Detroit, MI.

Bhise, V. and S. Sethumadhavan. 2008a. Effect of Windshield Glare on Driver Visibility. *Transportation Research Record (TRR), Journal of the Transportation Research Board*, No. 2056, Washington, DC.

Bhise, V. and S. Sethumadhavan. 2008b. Predicting effects of veiling glare caused by instrument panel reflections in the windshields. SAE Paper no. 2008-01-0666. *International Journal of Passenger Cars: Electronics Electrical Systems*, 1(1): 275–281. Society of Automotive Engineers, Inc., Warrendale, PA.

Bhise, V. D. 2007. Effects of veiling glare on automotive displays. In *Proceedings of the Society of Information Display Vehicle and Photons Symposium*, Dearborn, MI.

Bhise, V. D. 2012. *Ergonomics in the Automotive Design Process*. Boca Raton, FL: CRC Press. ISBN: 978-1-4398-4210-2.

Bhise, V. D. 2014. *Designing Complex Products with Systems Engineering Processes and Techniques*. Boca Raton, FL: CRC Press. ISBN: 978-1-4665-0703-6.

Bhise, V. D., E. I. Farber, C. S. Saunby, J. B. Walnus, and G. M. Troell. 1977. *Modeling Vision with Headlights in a Systems Context*. SAE Paper No. 770238. Paper presented at the 1977 SAE International Automotive Engineering Congress, Detroit, MI, 54 pp.

Bhise, V. D. and R. Hammoudeh. 2004. A PC based model for prediction of visibility and legibility for a human factors engineer's tool box. In *Proceedings of the Human Factors and Ergonomics Society 48th Annual Meeting*, New Orleans, Louisiana.

Bhise, V. D. and C. C. Matle. 1989. Effects of Headlamp Aim and Aiming Variability on Visual Performance in Night Driving. *Transportation Research Record*, 1247: 46–55, Transportation Research Board, Washington, DC.

Bhise, V. D., C. C. Matle, and E. I. Farber. 1988. *Predicting Effects of Driver Age on Visual Performance in Night Driving*. SAE Paper no. 881755 (also no. 890873). Paper presented at the 1988 SAE Passenger Car Meeting, Dearborn, MI.

Bodenmiller, F., J. Hart, and V. Bhise. 2002. *Effect of Vehicle Body Style on Vehicle Entry/Exit Performance and Preferences of Older and Younger Drivers*. SAE Paper no. 2002-01-00911. Paper presented at the SAE International Congress in Detroit, MI.

Bridger, R. S. 2008. *Introduction to Ergonomics*. Boca Raton, FL: CRC Press. ISBN: 978-0-8493-7306-0.

Card, S. K., T. P. Morgan, and A. Newell. 1980. The Keystroke-Level Model for User Performance Time with Interactive Systems. *Communications of the ACM*, 23(7): 396–410.

Card, S. K., T. P. Morgan, and A. Newell. 1983. *The Psychology of Human-Computer Interaction*. Hillsdale, NJ: Lawrence Erlbaum Associates.

Chaffin, D. B. 2001. *Digital Human Modeling for Vehicle and Workplace Design*. Warrendale, PA: SAE International. ISBN: 978-0-7680-0687-2.

Chaffin, D. B. 2007. Human motion simulation for vehicle and workplace design: Research articles. *Human Factors in Ergonomics & Manufacturing*, 17(5): 475–484.

Chapanis, A. 1959. *Research Techniques in Human Engineering*. Baltimore, MD: The Johns Hopkins Press.

Creveling, C. M., J. L. Slutsky, and D. Antis, Jr. 2003. *Design for Six Sigma: In Technology and Product Development*. Upper Saddle River, NJ: Prentice Hall PTR.

Drury, C. 1983. Task Analysis Methods in Industry. *Applied Ergonomics*, 14: 19–28.

Garrett, J. W. 1971. The Adult Human Hand: Some Anthropometric and Biomechanical Considerations. *Human Factors*, 13(2): 117–131.

Hankey, J. M., T. A. Dingus, R. J. Hanowski, W. W. Wierwille, and C. Andrews. 2001. *In-Vehicle Information Systems Behavioral Model and Design Support: Final Report*. Report No. FHWA-RD-00-135 sponsored by the Turner-Fairbank Highway Research Center of the Federal Highway Administration, Virginia Tech Transportation Institute, Blacksburg, Virginia.

Human Solutions. 2010. RAMSIS model applications. Website: www.human-solutions.com/automotive/index_en.php (Accessed: December 1, 2016).

Jack, D. D., S. M. O'Day, and V. D. Bhise. 1995. *Headlight Beam Pattern Evaluation: Customer to Engineer to Customer—A Continuation*. SAE Paper no. 950592. Presented at the 1995 SAE International Congress, Detroit, MI.

Jackson, D. and V. D. Bhise. 2002. *An Evaluation of the IVIS-DEMAnD Driver Attention Model*. SAE Paper no. 2002-01-0092. Paper presented at the SAE International Congress in Detroit, MI.

Jurgens, H., I. Aune, and U. Pieper. 1990. *International Data on Anthropometry*. Geneva, Switzerland: ILO.

Kolarik, W. J. 1995. *Creating Quality: Concepts, Systems, Strategies, and Tools*. New York: McGraw-Hill.

Konz, S. and S. Johnson. 2004. *Work Design-Industrial Ergonomics*, 6th edn. Scottsdale, AZ: Holcomb Hathaway.

Kroemer, K. H. E., H. B. Kroemer, and K. E. Kroemer-Elbert. 1994. *Ergonomics: How to Design for Ease and Efficiency*. Englewood Cliffs, NJ: Prentice Hall.

Lee, S. E., E. Llaneras, S. Klauer, and J. Sudweeks. 2007. *Analyses of rear-end crashes and near-crashes in the 100-car naturalistic driving study to support rear-signaling countermeasure development*. Project sponsored by the National Highway Traffic Safety Administration, Washington, DC. Report no. DOT HS 810 846.

Leiden, K., K. R. Laughery, J. Keller, J. French, W. Warwick, and S. D. Wood, 2001. *A Review of Human Performance Models for the Prediction of Human Error*. Boulder, CO: Micro Analysis & Design.

Maynard, H., G. Stegemerten, and J. Schwab. 1948. *Methods-Time Measurement*. New York: McGraw-Hill.

NHTSA. 2015. Federal Motor Vehicle Safety Standards. Website: www.nhtsa.gov/cars/rules/import/FMVSS/ (Accessed: July 5, 2015).

Owens, J. M., S. B. McLaughlin, and J. Sudweeks. 2010. *On-road Comparison of Driving Performance Measures When Using Handheld and Voice-control Interfaces for Cell Phones and MP3 Players*. SAE Paper no. 2010-01-1036. Presented at the 2010 SAE World Congress held in Detroit, MI.

Pew, R. W. 1993. *Experimental Design Methodology Assessment*. BBN Report No. 7917, Cambridge: Bolt Beranek & Newman.

Pheasant, S. and C. M. Haslegrave. 2006. *BODYSPACE: Anthropometry, Ergonomics and the Design of Work*, 3rd edn. London: CRC.

Reed, M. P., M. B. Parkinson, and D. B. Chaffin. 2003. A new approach to modeling driver reach. Technical Paper 2003-01-0587. *SAE Transactions: Journal of Passenger Cars—Mechanical Systems* (112): 709–718.

Reed, M. P., R. W. Roe, and L. W. Schneider. 1999. Design and development of the ASPECT manikin. Technical Paper 990963. SAE Technical Paper No. 1999-01-0963. Society of Automotive Engineers, Inc., Warrendale, PA. (Also published in SAE Transactions: Journal of Passenger Cars, Vol. 108).

Richards, A. and V. Bhise. 2004. *Evaluation of the PVM Methodology to Evaluate Vehicle Interior Packages*. SAE Paper no. 2004-01-0370. Also published in SAE Report SP-1877, SAE International, Inc., Warrendale, PA.

SAE. 2009. *SAE Handbook*. Warrendale, PA: Society of Automotive Engineers.

Sanders, M. S. 1983. *U.S. Truck Driver Anthropometric and Truck Work Space Data Survey*. Report no. CRG/TR-83/002. West Lake Village, CA: Canyon Research Group.

Sanders, M. S. and E. J. McCormick. 1993. *Human Factors in Engineering and Design*, 7th edn. New York: McGraw-Hill.

Satty, T. L. 1980. *The Analytic Hierarchy Process*. New York: McGraw Hill.

Seiffert, U. and L. Wech. 2003. *Automotive Safety Handbook*. Warrendale, PA: SAE International.

Thurstone, L. L. 1927. The method of paired comparisons for social values. *Journal of Abnormal and Social Psychology*, 21: 384–400.

Tijerina, L. E. Parmer, and M. J. Goodman. 1999. *Driver Workload Assessment of Route Guidance System Destination Entry while Driving: A Test Track Study*. East Liberty, OH: Transportation Research Center.

Van Cott and R. G. Kinkade, eds. 1972. *Human Engineering Guide to Equipment Design*. Washington: McGraw-Hill/U. S. Government Printing Press.

Wickens, C., S. E. Gordon, and Y. Liu. 1998. *An Introduction to Human Factors Engineering*. Upper Saddle River, NJ: Pearson Prentice Hall (Addison Wesley Longman, Inc.).

Woodson, W. E. 1992. *Human Engineering Design Handbook*, 2nd edn. New York: McGraw-Hill.

Yang, K. and B. El-Haik. 2003. *Design for Six Sigma: A Roadmap for Product Development*. New York: McGraw-Hill.

Zikmund, W. G. and B. J. Babin. 2009. *Exploring Market Research*, 9th edn. Boston, MA: Cengage Learning.

Section III

**Applications of Tools:
Examples and Illustrations**

22 Evaluation Studies

INTRODUCTION

During the development of a new vehicle, hundreds of studies are performed to understand issues, evaluate and select alternatives, and verify and validate the vehicle design. The purpose of this chapter is to provide the reader with additional examples of applications of various tools and methods used and types of studies that are conducted before, during, and after product development programs. The descriptions of the studies covered in this chapter are purposely short, with the intention of making the reader realize the breadth of issues and the types of evaluations that are conducted to obtain the required information for making decisions during the vehicle program.

The examples illustrated in this chapter include (1) benchmarking of low-cost vehicles, (2) photo-benchmarking of storage areas in sports utility vehicles (SUVs), (3) computer-aided design (CAD) outputs to illustrate various vehicle design and assembly considerations, (4) observational studies to design center consoles, (5) a legibility model to predict letter sizes in displays, (6) modeling of veiling glare caused by reflection of the sunlit instrument panel, (7) driving simulator and laboratory and field tests, (8) package evaluation surveys, and (9) market research clinics for vehicle concept selection.

BENCHMARKING OF LOW-COST VEHICLES

Benchmarking studies are conducted by all engineering departments and design teams to learn from the vehicle's best competitors and to improve their designs. The benchmarking study described here was conducted by a group of graduate students in an Automotive Systems Engineering program to determine the feasibility of marketing the Tata Nano in the U.S. market (Hussain and Randive, 2010). The Tata Nano was developed originally as a low-cost vehicle for the Indian market. Its price was about $2500 (INR 200,000). This study was conducted in 2010. The project began with a benchmarking of low-cost vehicles sold in 2009 in the United States.

Benchmarking was conducted as a comparison tool to analyze the competing vehicles in the U.S. market based on their cost, performance, features, and systems configurations. The cheapest vehicles sold in the U.S. market and selected for the benchmarking were the Hyundai Accent Blue ($9985), the Nissan Versa ($9990), and the Chevrolet Aveo ($11,965). The Honda Fit ($14,900) was included in the benchmarking study because of its size and popularity. The manufacturer's suggested retail prices (MSRP) included in the analysis were the starting prices of their base models, that is, base MSRP (excluding freight charges, tax, title, license, dealer fees, and optional equipment).

Table 22.1 presents the price, exterior and interior dimensions, overall weight, and body type of the vehicles. The Tata Nano, with 122 in overall length, was significantly shorter than the benchmarked vehicles, which ranged between 159 and 176 in. Table 22.2 presents powertrain performance and details. The Tata Nano has a very small two-cylinder 34 hp engine, as compared to the other vehicles, which had four-cylinder and over 100 hp engines. Other technical details, such as fuel economy and emissions standards, safety provisions, and features offered, are provided in the report prepared by Hussain and Randive (2010).

The data in the study showed that the Tata Nano was the shortest and lightest vehicle among the five vehicles. The Honda Fit had the highest price, because it had many extra features and safety systems (e.g., a security system, a vehicle stability with traction control system, a cruise controls system, air-conditioning, a tilt-telescopic steering column, an AM/FM/CD audio system, and power door locks and windows) as the standard equipment in the base model. The other vehicles did not provide some of these features in their base model, and this was compensated by their lower price.

The benchmarking exercise identified $10,000 as the base price of a low-cost vehicle in the U.S. market, and it also raised the question of what features must be included in the base model as the standard equipment. Although the relation between cost and features seems obvious, it is important to understand the answers to questions such as: What are the minimum features to be provided? How important are the provided features to the customers in this low-cost vehicle market segment? How much would the customer be willing to pay for the provided features? Such questions were not discussed in depth during the initial definition of the low-cost vehicle.

PHOTO-BENCHMARKING

Photo-benchmarking is a useful and effective tool for providing pictorial information to allow comparison between products and their features. A team of students in the author's automotive systems engineering class were asked to compare the storage spaces provided in the interiors of large SUVs. Figure 22.1 shows pictures obtained to compare storage spaces within the center console area, such as small item storage bins and cup holders, and the main console storage space in three large SUVs: the Ford Explorer, the GMC Acadia, and the Toyota Highlander. Similarly, Figure 22.2 shows pictures of storage areas in the driver's door and rear cargo area behind the third row seat and with the second and third row seats folded to provide the maximum cargo space. The side-by-side photo comparisons provide information on differences in possible configurations of these storage areas. The figures also provide approximate storage volumes for all the spaces.

QUALITY FUNCTION DEPLOYMENT

Table 22.3 presents a partial quality function deployment (QFD) chart (showing the relationship diagram and importance ratings) conducted to translate (i.e., relate) the customer needs of the interior storage spaces to functional specifications of the storage spaces to functional specifications of the storage spaces illustrated in Figures 22.1 and 22.2.

TABLE 22.1
Comparison of Exterior and Interior Dimensions of the Low-cost Vehicles

Specifications		Vehicle Model				
		Hyundai Accent 3-door	Nissan Versa 1.6 Base Sedan	Chevy Aveo LS—Sedan	Honda Fit	Tata Nano
Price ($)		9,970	9,990	11,965	14,900	2,500
Dimensions	Overall length (in)	159.3	176	169.7	161.6	122.01
	Overall width (in)	66.7	66.7	67.3	66.7	58.98
	Overall height (in)	57.9	60.4	59.3	60	63.5
	Wheelbase (in)	98.4	102.4	97.6	98.4	87.8
	Track width (in) front	57.9	58.3	57.1	58.7	52.2
	Track width (in) rear	57.5	58.5	56.3	58.1	51.8
	Headroom (in) Front/rear	39.6/37.8	40.6/37.9	39.3/37.4	40.4/39.0	93 cm
	legroom (in) Front/rear	42.8/34.3	41.4/38.0	41.3/35.4	41.3/34.5	Rear—80 cm max and 62 cm min
	shoulder room (in) Front/rear	53.5/53.1	53.5/50.7	53.6/52.8	52.7/51.3	N/A
	hip room (in) Front/rear	50.2/48.8	48.8/47.2	51.6/52.8	51.5/51.3	N/A
	passenger volume (cu. ft.)	92.2	94.3	N/A	90.8	N/A
	Cargo volume (cu. ft.)	15.9	13.8	12.4	20.6/57.3	N/A
Body	Type	3-door hatchback	4-door sedan	4-door sedan	5-door hatchback	5-door hatchback
	Construction	Uni-body	Uni-body	Uni-body	Uni-body	Uni-body
Weight	Curb (lb)	2,467	2,516	2,568	2,489	1,323

TABLE 22.2

Powertrain and Performance Comparisons of the Low-cost Vehicles

Specifications		Vehicle Model				
		Hyundai Accent 3-door	Nissan Versa 1.6 Base Sedan	Chevy Aveo LS—Sedan	Honda Fit	Tata Nano
Price ($)		9,970	9,990	11,965	14,900	2,500
Powertrain	Model	N/A	HR16DE	LS 1LS ECOTEC	N/A	N/A
	Type	Inline 4-cylinder	4-cylinder DOHC	Gas 4-cylinder	Inline 4-cylinder	2-cylinder
	Fuel	Gasoline	Gasoline	Gasoline	Gasoline	Gasoline
	horsepower @ rpm	110@6,000	107@6,000	108@6,400	117 @ 6,600	34.52 @ 5,250
	Torque (lb-ft)	106@4,500	111@4,600	105@4,000	106@4,800	26.11@3,000
	Power/weight ratio (hp per ton)	98.095	93.561	92.523	103.415	57.44
	Displacement (liters)	1.6	1.6	1.6	1.497	0.624
	Valve train	16-valve	16-valve (continuous variable valve timing)	Variable valve timing (VVT)	16-valve SOHC i-VTEC®	4-valve SOHC
	Fuel system	Multi-point programmed fuel injection	Nissan direct injection system	Sequential fuel injection	Multi-point fuel injection	Multi-point fuel injection

	Ignition system	Electronic	Iridium tipped spark plugs	N/A	Direct ignition system with immobilizer	N/A
	Emission class	ULEV	ULEV	ULEV	ULEV-2	EURO III
Drivetrain	Drive configuration	Front-wheel drive	Front engine/front-wheel drive	Front-wheel drive	Front-wheel drive	Rear-wheel drive
	Type	5-speed manual (4-speed automatic not available in this model as option)	5-speed manual (4-speed automatic and 6-speed manual not available)	5-speed manual transmission	5-Speed manual transmission	4 forward + 1 reverse
Performance	Max speed (mph)	112	113	110	115	65
	0–60 mph (s)	9.4	9.4	9.4	9.5	0–60 km/h (37 mph): 8 s

Center console storage area	Ford Explorer	GMC Acadia	Toyota Highlander
Small items storage bin	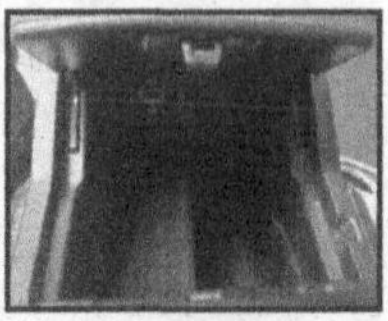 2 L	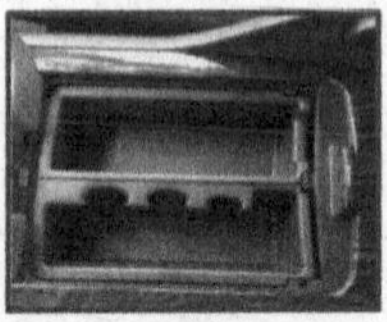 0.5 L	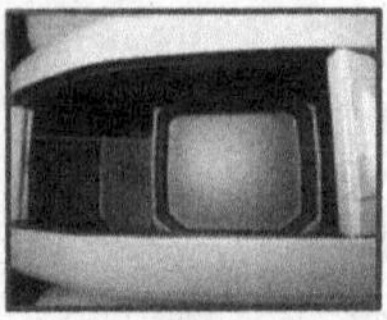 0.45 L
Cupholder storage	 87 mm depth	 73 mm depth	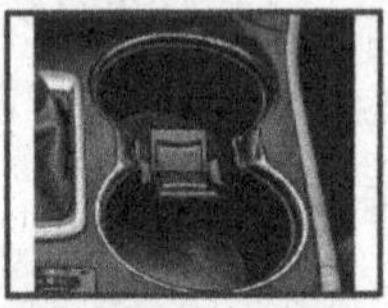 90 mm depth
Main storage bin	 7.68 L	 6.5 L	 23.75 L

FIGURE 22.1   Comparison of storage areas in center console.

The three most important functional specifications from the QFD analysis are (a) volume and dimensions of the storage spaces, (b) meeting the minimum and maximum reach zone requirements (i.e., the storage items should not be too close to the occupants or too far to reach), and (c) operability (i.e., the storage areas should be easy to operate and use).

CAD EVALUATIONS

CAD systems allow comparisons between designs of many different vehicles and their systems by (a) creating views for side-by-side comparisons or (b) superimposing views of the same systems in different vehicles obtained from the same viewing point. Sequential views can be created to illustrate the effect of changes in systems over time. Comparisons with addition and/or deletion of certain entities in the vehicles can also help in visualizing problems such as packaging space and the assembly process. These views can also be generated from various different viewing locations to visualize and get a better understanding of spaces, clearances, and interferences between systems.

SUPERIMPOSED DRAWINGS

Superimposing similar views of different vehicles by coinciding on a preselected common reference point (e.g., the accelerator heel point, the seating reference point [SgRP], driver eye points) is a powerful tool to understand differences and

Interior storage area	Ford Explorer	GMC Acadia	Toyota Highlander
Door bottle storage bin	 4.2 L	 1.1 L	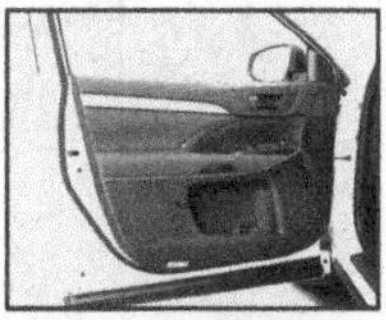 1.7 L
Storage space behind third-row seat	 595 L	 682 L	 391 L
Maximum cargo capacity	 2285 L	 3287 L	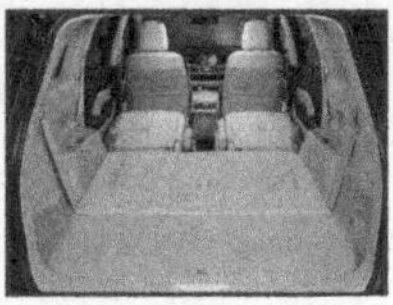 2371 L

FIGURE 22.2 Comparison of storage areas in driver's door and rear cargo compartment.

similarities between different characteristics, such as vehicle entry/exit space for feet, legs, and torso, or field of view available to the drivers, of the vehicles. For example, Figure 22.3 illustrates how a rear door opening can be modified to provide additional room during entry into a vehicle. The original and modified door openings were superimposed with respect to the SgRP of the rear occupant (as a common reference point).

COMPOSITE VIEWS OF LEFT SIDE AND RIGHT SIDES OF DIFFERENT VEHICLES

Creating two half models (left half and right half) and placing the halves together on a common vertical XZ plane passing through the vehicle centerline allows comparisons of dimensions and differences between the exteriors of two vehicles (see Figure 22.4).

SEQUENTIAL VIEWS OF ASSEMBLY

Views can be created at different points in time during the product development, production, and/or assembly processes to provide a better idea of the sequence of events and differences between different events over time. For example, Figure 22.5 presents 15 sequential pictures showing how a vehicle based on a space-frame design is assembled. The first picture shows the floor section of the space frame. The second picture shows the addition of the front part of the space frame. The third picture

TABLE 22.3
Partial QFD Chart for Interior Storage Areas of Large SUVs

Customer Needs (What)			Importance of Rating 1=Least 10=Most	Ergonomics		Craftsmanship		Packaging		Materials and Styling		Easy to Clean		Cost
Primary	Secondary	Tertiary		Min and Max Reach Criteria	Operability	No Visible Fasteners	No Gaps and Precise Fit	Volume and Dimensions	Compartmentalization	Leather or Vinyl	Flocking or Mat	Materials	Removable Mat	Manufacturing Process and Materials
Good interior storage	Features	Console cup holder size/adjustability	8	⊕	⊕	○	⊕	⊕		▲	▲	▲	⊕	○
		Console main bin storage	9	○	⊕	○	⊕	⊕	○		○	⊕	○	⊕
		Console bin coin storage	6	○	▲	○		▲	⊕		○	▲		▲
		Console small item storage	8	⊕	○	○	○	○	○			○		○
		Door bottle holder storage	4	▲	▲	▲		▲		⊕		⊕		▲
		Seat pockets	3	▲	▲	▲		○						
		Coat hooks	3	▲	▲	○								

Functional Specifications (How)

Phone storage	7	Θ	Θ	O		◄	O			O	◄		O
IP glovebox	6	◄	O	O	Θ	O			◄		O		
Sunglass storage	7	Θ	O	O	Θ	◄			◄	O	◄		Θ
Bin/cup holder for 3rd row passengers	5	O	◄	◄	Θ								
Capacity — Cargo trunk storage capacity	7		◄	◄		Θ				O	◄		
Cargo maximum capacity	6	◄	◄	◄	Θ	Θ				O	◄		
Safe and Secure — Locked storage	4	◄	◄	O	◄	◄			Θ		◄		Θ
Absolute importance ratings		363	317	199	304	394	105	35	93	206	124		213
Relative importance ratings		15	14	9	13	17	4	1	4	9	5		9

Key:

Θ = 9 (strong relationship)

O = 3 (Medium relationship)

◄ = 1 (Weak relationship)

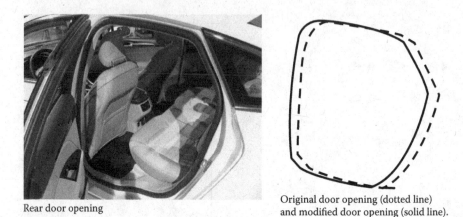

Rear door opening

Original door opening (dotted line)
and modified door opening (solid line).

FIGURE 22.3 Illustration of superimposed door opening outlines for improved rear door entry and exit.

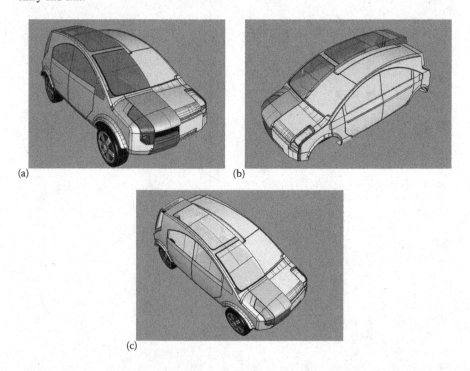

FIGURE 22.4 Composite views of SUV version (left) and hatchback sedan version (right).

shows the addition of the rear part of the space frame. The fourth picture shows the installation of the roof members. The fifth picture shows the addition of the engine block and the fuel tank. The following pictures progressively show the installation of engine components, exhaust pipe, suspensions, wheels, instrument panel, floor pan, seats, and other components. Finally, the 15th picture shows the state of the

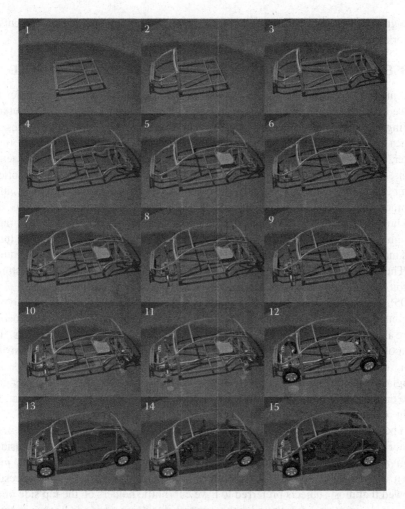

FIGURE 22.5   Sequential views of a vehicle during assembly.

completed vehicle just prior to installation of the exterior body panels, lamps, and trim components.

DYNAMIC ACTION SIMULATIONS/VIDEOS

Most CAD models (with computer-aided engineering [CAE] capabilities) can simulate the motion of various components (e.g., displacements and deformations of components, or a human operator's hand and body motions during completion of a task) and dynamically display changes in states or variables such as force or stress level, temperatures, or fluid flows during the motion by using different color codes as functions of time. Such video simulations are useful in visualizing and understanding the behavior and/or performance of various entities under different vehicle motion conditions. Additional examples of videos are (a) outputs of crash tests captured by use of high-speed video recordings (Autoblog, 2010) and human operator movements in a work station illustrated by use of Jack models (Siemens, 2015).

OBSERVATIONAL STUDIES IN DESIGNING A CENTER CONSOLE

A center console that extends rearward from the instrument panel and occupies the space between the two front passenger seats of automotive products started out as a styling and convenience feature in sporty cars in the 1950s. Early console designs typically incorporated floor-mounted gear shift levers, hand brakes, and ash trays. Later, in the 1980s, cup holders and storage spaces for items such as audio cassettes, coins, and so on were incorporated. Now, the latest automotive center consoles offer a number of controls for features such as heated seats, power windows and locks, infotainment (e.g., multi-function controls such as the BMW's iDrive), power points, and CD holders. Recent advances in technologies (e.g., digital, wireless, and data storage), design trends, and customer demands have created a number of additional features, and many of them could be incorporated in the center stack and center console units.

Thus, a project was undertaken to develop a methodology to understand customer needs so that future consoles can be designed to satisfy the customers (Bhise et al., 2005). Two studies were conducted in the project. The first study involved an observational survey of 150 vehicles in three parking lots to determine what items people store in their vehicles and the observed locations of the items in the vehicles. The data obtained from the survey provided a list of all the stored items, their distribution, and their locations inside the vehicle. Papers, bottles, cups, books, bags, and sunglasses were the most frequently observed items in the vehicles. The center console area was the most frequently used, followed by the front passenger's seat, the right rear passenger seat, and the floor area in front of the front passenger seat.

The second study was conducted to determine storage preferences of items in the center console. A foam-core center console with Velcro surfaces was built inside a minivan. Thirty-six drivers were asked to select items that they would carry most often in their vehicles and place them on the center console surfaces. The results showed that most subjects preferred to have cup/bottle holders on the top side of the console, large storage areas under the armrest (for items such as CD cases, coins, purses, phones, maps, and tissues), and an area for various power and data ports (i.e., 120 V, 12 V, USB, and headphones). Many also wanted trunk, fuel release, and seat controls on the console, paper holders on the sides of the console, a pen storage area on the top side of the console, and an entertainment screen on the rear part of the console for the rear passengers. Subjects differed considerably regarding the locations of cell phones, garage door openers, and sunglasses. The resulting layouts of stored items were summarized. The summary data were provided to four teams of industrial design and engineering students to create design concepts for future automotive center consoles, which are presented in Bhise et al. (2006).

MODELS FOR ERGONOMIC EVALUATIONS

Human factors models in general are based on modeling the relationships between many independent variables, the characteristics of the human operators, and their responses (performance or behaviors) in performing certain tasks. The relationships are usually developed from the data collected in experimental research. The models are thus approximations of human performance and should be used with caution. The model results should be validated by performing studies using real subjects and products under actual usage situations.

Bhise (2012) has described a number of models used for ergonomic evaluations in the automotive design process. The models include driver positioning and occupant packaging practices incorporated in the Society of Automotive Engineers (SAE) practices (SAE, 2009). Bhise (2012) also described a number of human vision models based on visual contrast thresholds, and disability and discomfort glare prediction equations. The models have been used to evaluate automotive headlighting systems (Comprehensive Headlamp Environment Systems Simulation [CHESS] model; Bhise et al., 1977a,b, 1985, 1988, 1989), legibility of displays (Bhise and Hammoudeh, 2004) and veiling glare effects from sunlight reflections into vehicle windshields (Bhise, 2007; Setumadhaven, 2008a,b).

A legibility model and a veiling glare prediction model are briefly described in the following subsections.

LEGIBILITY PREDICTION MODEL

The visibility model developed by Bhise (2012) allows a user to evaluate the visibility of targets illuminated by a light source by inputting its intensity and other variables (target size, distance of the target from the source, distance of the observer from the target, reflectance of the target and the background, ambient illumination, observer age, percent confidence, and ease in detecting). For legibility computations, the model provides options to evaluate externally illuminated or backlighted displays. The target (letters or numerals) to be read is specified by its size (i.e., letter height and stroke width [see Figure 22.6]), and other input variables are viewing distance, observer's age, level of confidence and ease in reading, and target illumination

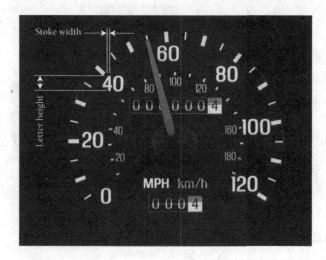

FIGURE 22.6 Speedometer display showing letter height and strike width measurements.

characteristics. The user is provided with three options for entering illumination data: inputting (a) light source intensity, distance from source to target in English or metric units, (b) illumination directed at the target (letter or numeral to be read) and its background and the reflectance characteristics of the target and its background, or (c) the luminance of the target and its background. Depending on the method of data entry selected, the program shows open data entry boxes and grayed-out boxes (when not needed) for required reflectance or transmittance values. If Option (c) is selected, then the luminance of the target and its background need to be measured using a photometer (see Figure 22.7).

Two versions of the model are available: one Excel based and the other an executable file from a Visual Basic program. The Excel-based program is useful for students and researchers who want to understand the basic computational procedure. They can also easily modify this version for their own custom applications. The Visual Basic version provides easy-to-input screens and options to exercise the model when inputs are provided under different conditions (e.g., directly inputting target luminance as compared with computing from source candlepower, distance, and target reflectance). Detailed information on the model can be found in Bhise and Hammoudeh (2004).

Figure 22.8 presents a graph developed from repeatedly running the model to illustrate the letter height required as a function of the observer's age for easy reading of a typical automotive speedometer-type display. The easy reading condition was simulated to achieve target-to-background contrast at least five times that of the threshold contrast level for the observer with the given age. The figure shows three curves for the following conditions: (a) letter to background luminance contrast (C) = 5, illumination falling on the face of the speedometer (E) = 300 lux, (b) C = 20 and

FIGURE 22.7   Luminance measurements using a Spectra radiometer.

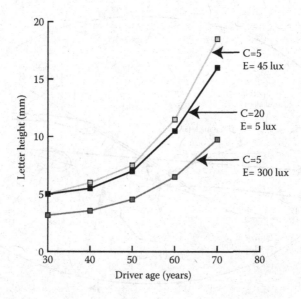

FIGURE 22.8 Letter height required for easy reading of speedometer numerals as a function of driver age.

E = 5 lux, and (c) C = 5 and E = 45 lux. (Note: most automotive speedometers have miles per hour printed with white numerals about 6 mm high on a black background).

WINDSHIELD VEILING GLARE PREDICTION MODEL

When a vehicle is driven toward the sun (i.e., when the sunlight passing through the windshield strikes the top of the instrument panel, and the illuminated top of the instrument panel reflects into the windshield), the driver looking through the reflection of the instrument panel top in the windshield will have the veiling luminance of the reflection superimposed on his visual scene. The veil decreases the visibility of the objects in the driver's visual field. The upper part of Figure 22.9 shows the light path from the incident sun rays to the reflected light from the top of the instrument panel seen by the driver (driver's eyes located by use of the SAE eyellipses) and the driver's sightline through the windshield to view an outside target. When the outside target is dimly lit, such as inside a tunnel (or a parking structure), the veil created by the incident sunlight can reduce the visibility of the target (see the lower part of Figure 22.9).

This veiling glare effect has been measured and was modeled by Bhise (2007) and Setumadhaven (2008a,b). The model can be used to study the effects of relevant parameters such as level of illumination falling on the windshield, sun angle, instrument panel angle, windshield angle, gloss level of the top of the instrument panel material, reflection characteristics of the windshield, and driver vision characteristics (age and visual contrast thresholds). The model can be used to predict trade-offs between the windshield angle, the instrument panel angle, and the gloss level of the

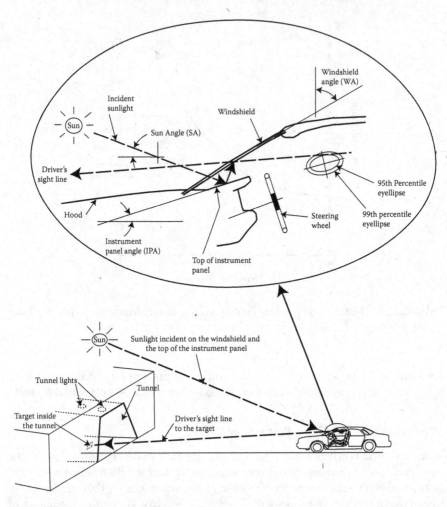

FIGURE 22.9   Veiling glare situation of a driver approaching a target in a darker tunnel while the sunlight falls on the windshield.

material to ensure that the driver is able to see all important targets in the types of situation shown in Figure 22.9.

SIMULATOR, LABORATORY, AND FIELD STUDIES

Early product concepts, product prototypes, and final products (production versions) are generally evaluated by using real subjects performing a number of tasks to verify and validate the product design. Product evaluation under actual usage situations involving field testing is generally preferred. However, due to safety and other reasons, such as the high cost of building a working model of the product and the costs and time required to perform field tests, other research approaches, such as laboratory tests, product simulations, the use of early prototypes, or testing with simulators, are commonly considered. The data collection methods employed in the human factors studies conducted in laboratory and driving simulators include (a) observing subjects performing given tasks with the product (e.g., operating a new radio), (b) communicating with the subjects to obtain ratings on selected product features, or asking the subjects to describe problems and difficulties in using the product, (c) measuring the performance of the subjects (e.g., time taken to complete a task, or number and type of errors committed in performing the task) and the product during product uses, (d) measuring the physiological state of the subjects while completing different usage tasks, and (e) obtaining subjective ratings to measure operator workload.

DRIVING SIMULATORS

Driving simulators are increasingly used during the design and evaluation of many vehicle systems. In a driving simulator, a large number of driving maneuvers under different driving and traffic conditions can be generated, and the responses of many drivers and vehicle systems can be observed, measured, and recorded for further analyses. The simulators are especially useful to evaluate complex in-vehicle devices (e.g., display screens presenting outputs of navigation, radio, and climate control systems) that could increase driver workload.

Figure 22.10 shows a driving simulator used to evaluate controls and displays of various designs of audio products (Bhise et al., 2003). While driving the simulator and performing a number of tasks (e.g., turn on the radio and find a FM station with a specified frequency) with different radio designs, driver eye glances, lateral deviations in the lane, and vehicle speed variations were measured. The combination of the driving simulator, the use of working prototypes of the radios, a data acquisition system, and data analysis capabilities was found by the author to be a very powerful approach to the development process of driver information systems.

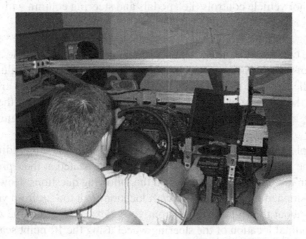

FIGURE 22.10 Subject operating a radio while driving in a vehicle simulator.

LABORATORY AND FIELD TESTS

Many laboratory tests are conducted to evaluate the performance of vehicle systems. Some examples of such tests are measurements of vehicle body characteristics, such as stiffness, deflections, and vibrations, in a laboratory; engine dynamometer tests in a laboratory to measure engine torque versus speed, fuel consumption, and emissions; and aerodynamic tests to measure air flows, drag, and wind noise in a wind tunnel. The laboratory tests help verify whether each of the tested systems meets its respective system design and performance requirements before the systems are installed in vehicles for whole-vehicle tests in later phases of the product development process.

Many field tests are also performed using early prototype vehicles to evaluate whole-vehicle performance in actual driving situations on test tracks and public roads. Some examples of evaluations are vehicle handling tests (in maneuvers such as lane changes, serpentine paths, braking in a straight line and in turns, and on pavements with different friction coefficients), ride and seat comfort, and ergonomics tests on ease of operability of in-vehicle system. In addition, several early prototype vehicles are used for crash tests to validate safety performance in meeting front, side, and rear collisions, and rollover situations. Such tests are described in Chapter 14.

PACKAGE EVALUATION SURVEYS

The vehicle package is generally developed by package engineers assigned to the vehicle program along with the inputs from design (styling), engineering, and marketing professionals. However, it is generally desirable to get an independent confirmation on the vehicle package by conducting a market research clinic in which a representative group of prospective owners are asked to evaluate the overall vehicle package. A full-size interior buck of the vehicle with the passenger compartment and trunk/cargo area is created. The buck includes all the interior surfaces (i.e., instrument panel, door trim and roof-liner surfaces) with storage areas and major vehicle controls (i.e., pedals and steering column with the steering wheel) and seats.

The package variables that need to be evaluated are selected by the vehicle design team with the concurrence of the chief vehicle program manager. Table 22.4 presents a partial list of vehicle package variables along with details on the evaluation ratings data to be collected. The ratings using the direction magnitude scales are used to get insights into the perception of the magnitudes of the variables, and the acceptance ratings using a 10-point scale provide the acceptability of the magnitudes of the variables.

For example, the first variable shown in Table 22.4 is the longitudinal (fore/aft) location of the steering wheel. The participants, once seated at their preferred driving position in the interior buck, are asked the following questions: How is the steering wheel positioned in terms of its fore/aft location? Is it too far from you, too close to you, or at about the right distance from you? Next, please rate the acceptability of the longitudinal location of the steering wheel using the 10-point scale, where 10 equals very acceptable and 1 equals very unacceptable.

TABLE 22.4

Package Evaluation Considerations and Ratings

Item No.	Vehicle Package Variable To Be Evaluated	Rating Using Direction Magnitude Scale			Acceptance Rating: 1=Very Unacceptable, 10=Very Acceptable
1	Steering wheel longitudinal (fore/aft) location	Too close	About right	Too far	
2	Steering wheel vertical (up/down) location	Too low	About right	Too high	
3	Steering wheel lateral (left/right) location	Too far to left	About right	Too far to right	
4	Steering wheel diameter	Too small	About right	Too large	
5	Gas pedal fore/aft location	Too close	About right	Too far	
6	Gas pedal lateral location	Too far to left	About right	Too far to right	
7	Lateral distance between the gas pedal and the brake pedal	Too small	About right	Too large	
8	Gas pedal to brake pedal lift-off	Too small	About right	Too large	
9	Gearshift lateral location	Too far to left	About right	Too far to right	
10	Gearshift location longitudinal location	Too close	About right	Too far	
11	Height of the top portion of the instrument panel directly in front of the driver	Too low	About right	Too high	
12	Height of the armrest on driver's door	Too low	About right	Too high	
13	Height of the armrest on the center console	Too low	About right	Too high	
14	Belt height (lower edge of the driver's side window)	Too low	About right	Too high	
15	Space above the driver's head	Too little	About right	Too generous	
16	Space to the left of the driver's head	Too little	About right	Too generous	
17	Longitudinal location of the header	Too close	About right	Too far	
18	Knee space (between instrument panel and right knee with foot on the gas pedal)	Too little	About right	Too generous	
19	Thigh space (between the bottom of the steering wheel and the closest lower surface of the driver's thighs)	Too little	About right	Too generous	

(Continued)

TABLE 22.4 (CONTINUED)

Package Evaluation Considerations and Ratings

Item No.	Vehicle Package Variable To Be Evaluated	Rating Using Direction Magnitude Scale			Acceptance Rating: 1 = Very Unacceptable, 10 = Very Acceptable
20	Height of the top of the instrument panel in front of the driver	Too low	About right	Too high	
21	Height of the rocker panel during entry and exit	Too low	About right	Too high	
22	Foot movement space during entry and exit	Too little	About right	Too generous	
23	Obscuration caused by the left A-pillar	Too small	About right	Too large	
24	Obscuration caused by the right A-pillar	Too small	About right	Too large	
25	Obscuration caused by the left B-pillar	Too small	About right	Too large	
26	Obscuration caused by the right B-pillar	Too small	About right	Too large	
27	Longitudinal location of the B-pillar	Too far forward	About right	Too much rearward	
28	Obscuration caused by the right C-pillar	Too small	About right	Too large	
29	Obscuration caused by the left C-pillar	Too small	About right	Too large	
30	Storage space of the glove compartment	Too little	About right	Too generous	
31	Storage spaces in the instrument panel	Too little	About right	Too generous	
32	Storage spaces in the center console	Too little	About right	Too generous	
33	Luggage/cargo space behind the second row or trunk compartment	Too little	About right	Too generous	
34	Number of cup holders accessible from front seats	Too few	About right	Too many	

Usually, about 100–300 participants are invited to evaluate the vehicle buck. An interviewer in the market research clinic asks each participant a series of questions on all the variables covered in the survey. The responses of each participant are recorded and summarized over all the participants. Table 22.5 presents an example of the summarized outputs showing percentages of responses to each of the three levels of the direction magnitude scales and average rating values on the 10-point acceptability scale for each of the variables included in Table 22.4. The items with low average acceptance ratings scores (five or below) revealed the following problems experienced by many drivers: (a) the gear shifter was located too close to the driver, (b) the armrest on the center console was too low, (c) the space above the driver's head was too low, (d) the obscuration from the left A-pillar in the driver's forward field of view was perceived to be too large, and (e) the obscuration from the right C-pillar in the driver's rear field of view was perceived to be too large.

CONCEPT SELECTION MARKET RESEARCH

This section covers a case study involving a market research clinic conducted to evaluate three vehicle concepts created to replace an existing vehicle. The existing vehicle, called the *reference vehicle*, was evaluated along with two other two leading competitor vehicles, Competitor #1 and #2. Let us assume that 150 current owners and principal drivers of the reference vehicle were invited to participate in the market research clinic.

Each participant was asked to rate each of the subattributes of the attributes of each vehicle using the 10-point scale (where 10 = excellent—liked very much and 1 = very poor—disliked very much) presented in Table 22.6. The averages of ratings provided by participants are also provided in the table.

The overall results show that Concepts W and P were liked better than the reference vehicle. However, these two leading concepts had the same overall rating average (7.3) as Competitor #1, which had the best overall package rating (7.3). The ratings on the subattributes of each of the three attributes provided in Table 22.6 present further insights into possible improvements that should be incorporated into the two top concepts. Concepts W and P both received higher ratings for exterior styling and appearance; however, they received lower ratings for rear legroom. The design team need to discuss and understand the results and decide on which concept to select. The selected concept should be improved further by considering the ratings data on various subattributes.

CONCLUDING REMARKS

This chapter provided several examples of design and evaluation problems addressed by the application of a number of product evaluation tools and methods at different stages of the vehicle development process. Such applications must be included in the system design and evaluation procedures manuals and communicated to the design team members to ensure that the right tools are used at the right time during the vehicle development process. It is very important to create a thorough plan for vehicle development

TABLE 22.5

Results of the Vehicle Package Evaluation

Item No.	Vehicle Package Variable To Be Evaluated	Percentage of Participants Providing the Rating			Acceptance Rating: 1=Very Unacceptable, 10=Very Acceptable
1	Steering wheel longitudinal (fore/aft) location	Too close 5%	About right 90%	Too far 5%	9
2	Steering wheel vertical (up/down) location	Too low 8%	About right 85%	Too high 7%	8
3	Steering wheel lateral (left/right) location	Too far to left 3%	About right 95%	Too far to right 2%	10
4	Steering wheel diameter	Too small 8%	About right 90%	Too large 2%	8
5	Gas pedal fore/aft location	Too close 7%	About right 85%	Too far 8%	8
6	Gas pedal lateral location	Too far to left 12%	About right 81%	Too far to right 7%	7
7	Lateral distance between the gas pedal and the brake pedal	Too small 8%	About right 85%	Too large 7%	8
8	Gas pedal to brake pedal lift-off	Too small 2%	About right 73%	Too large 25%	6
9	Gearshift lateral location	Too far to left 5%	About right 71%	Too far to right 24%	6
10	Gearshift location longitudinal location	Too close 35%	About right 40%	Too far 25%	5
11	Height of the top portion of the instrument panel directly in front of the driver	Too low 7%	About right 86%	Too high 7%	9
12	Height of the armrest on driver's door	Too low 16%	About right 65%	Too high 19%	8
13	Height of the armrest on the center console	Too low 35%	About right 65%	Too high 0%	5
14	Belt height (lower edge of the driver's side window)	Too low 15%	About right 75%	Too high 10%	7
15	Space above the driver's head	Too little 40%	About right 60%	Too generous 0%	4
16	Space to the left of the driver's head	Too little 20%	About right 70%	Too generous 10%	6
17	Longitudinal location of the header	Too close 30%	About right 60%	Too far 10%	7
18	Knee space (between instrument panel and right knee with foot on the gas pedal)	Too little 20%	About right 80%	Too generous 0%	7

19	Thigh space (between the bottom of the steering wheel and the closest lower surface of the driver's thighs).	Too little 22%	About right 73%	Too generous 5%	6
20	Height of the top of the instrument panel in front of the driver	Too low 7%	About right 83%	Too high 10%	8
21	Height of the rocker panel during entry and exit	Too low 7%	About right 85%	Too high 8%	8
22	Foot movement space during entry and exit	Too little 25%	About right 75%	Too generous 0%	6
23	Obscuration caused by the left A-pillar	Too small 0%	About right 65%	Too large 35%	5
24	Obscuration caused by the right A-pillar	Too small 0%	About right 75%	Too large 25%	6
25	Obscuration caused by the left B-pillar	Too small 3%	About right 80%	Too large 17%	7
26	Obscuration caused by the right B-pillar	Too small 0%	About right 78%	Too large 22%	6
27	Longitudinal location of the B-pillar	Too far forward 20%	About right 80%	Too far rearward 0%	7
28	Obscuration caused by the right C-pillar	Too small 0%	About right 65%	Too large 35%	5
29	Obscuration caused by the left C-pillar	Too small 0%	About right 70%	Too large 30%	6
30	Storage space of the glove compartment	Too little 10%	About right 80%	Too generous 10%	8
31	Storage spaces in the instrument panel	Too little 30%	About right 65%	Too generous 5%	6
32	Storage spaces in the center console	Too little 15%	About right 72%	Too generous 13%	7
33	Luggage/cargo space behind the second row or trunk compartment	Too little 4%	About right 90%	Too generous 6%	9
34	Number of cup holders accessible from front seats	Too few 0%	About right 100%	Too many 0%	10

TABLE 22.6
Summary of Ratings Data Collected in Market Research Clinic

Vehicle Attribute	Subattribute	Concept W	Concept P	Concept J	Reference Vehicle	Competitor #1	Competitor #2
Exterior styling and appearance	Front view	9	6	10	7	9	6
	Side view	10	8	7	5	8	5
	Rear view	7	10	8	8	7	7
	Front quarter view	8	7	6	5	8	8
	Rear quarter view	6	9	8	7	7	7
	Average score	8.0	8.0	7.8	6.4	7.8	6.6
Interior styling and appearance	Instrument panel	7	9	8	6	9	7
	Driver's door	8	8	5	6	7	6
	Center console	8	7	6	7	8	7
	Rear door	6	8	6	6	5	9
	Driver's seat	6	7	6	5	6	7
	Gear-shifter knob	8	9	6	7	8	6
	Steering wheel	8	8	8	5	5	7
	Gas and brake pedal	6	4	7	6	5	6
	Average score	7.1	7.5	6.5	6.0	6.6	6.9
Vehicle package	Interior spaciousness	8	7	8	4	8	9
	Front legroom	7	5	6	5	8	7
	Front headroom	6	7	7	6	7	6
	Shoulder room	8	7	6	5	9	7
	Front door armrest	7	6	7	7	8	6
	Center armrest	8	7	8	4	8	7
	Console storage space	6	8	7	6	7	8
	Rear legroom	5	4	6	7	8	7
	Rear headroom	6	6	5	6	5	7
	Rear shoulder room	6	7	6	8	7	7
	Trunk space	7	8	7	6	8	8
	Average score	6.7	6.5	6.6	5.8	7.5	7.2
All above	Average score	7.3	7.3	7.0	6.1	7.3	6.9

with the help of attribute engineering managers and systems engineers to ensure that the vehicle being designed meets all vehicle attribute requirements. The systems engineering management plan should include details of the content and timings of all important steps to be followed by the design team. The tools and methods to be used during the design, verification, and validation processes must be also documented.

REFERENCES

AutoBlog. 2010. Videos: Ford's 2011 Explorer and 2012 Focus go off the wall. Website: www.autoblog.com/2010/12/28/videos-fords-2011-explorer-and-2012-focus-go-off-the-wall/#continued (Accessed: June 27, 2016)

Bhise, V. and S. Sethumadhavan. 2008a. Effect of windshield glare on driver visibility. *Transportation Research Record (TRR), Journal of the Transportation Research Board*, No. 2056, Washington, DC.

Bhise, V. and S. Sethumadhavan. 2008b. Predicting effects of veiling glare caused by instrument panel reflections in the windshields. SAE paper no. 2008-01-0666. *International Journal of Passenger Cars: Electronics Electrical Systems*, 1(1): 275–281. Society of Automotive Engineers, Warrendale, PA.

Bhise, V. D. 2007. Effects of veiling glare on automotive displays. *Proceedings of the Society of Information Display Vehicle and Photons Symposium*. Dearborn, MI.

Bhise, V. D. 2012. *Ergonomics in the Automotive Design Process*. Boca Raton, FL: CRC Press.

Bhise, V. D., R. Boufelliga, T. Roney, J. Dowd, and M. Hayes. (2006). *Development of Innovative Design Concepts for Automotive Center Consoles*. SAE Paper no. 2006-01-1474. Paper presented at the 2006 Annual Meeting of the Society of Automotive Engineers, Detroit, MI.

Bhise, V. D., E. I. Farber, and P. B. McMahan. 1977a. Predicting target detection distance with headlights. *Transportation Research Record*, 611: 1–16. Transportation Research Board, Washington, DC.

Bhise, V. D., E. I. Farber, C. S. Saunby, J. B. Walnus, and G. M. Troell. 1977b. *Modeling Vision with Headlights in a Systems Context*. SAE Paper No. 770238, 54 pp. Presented at the 1977 SAE International Automotive Engineering Congress, Detroit, MI.

Bhise, V. D. and R. Hammoudeh. 2004. A PC based model for prediction of visibility and legibility for a human factors engineer's tool box. *Proceedings of the Human Factors and Ergonomics Society 48th Annual Meeting*, New Orleans, Louisiana.

Bhise, V. D., R. Hammoudeh, J. Dowd, and M. Hayes. 2005. Understanding customer needs in designing automotive center consoles. *Proceedings of the Annual Meeting of the Human Factors and Ergonomics Society* held in Orlando, Florida, September 2005.

Bhise, V. D. and C. C. Matle. 1985. Review of driver discomfort glare models in evaluating automotive lighting. Presented at the 1985 SAE International Congress, Detroit, MI.

Bhise, V. D. and C. C. Matle. 1989. Effects of headlamp aim and aiming variability on visual performance in night driving. *Transportation Research Record*, 1247, Transportation Research Board, Washington, DC.

Bhise, V. D., C. C. Matle, and E. I. Farber. 1988. *Predicting Effects of Driver Age on Visual Performance in Night Driving*. SAE paper no. 881755 (also no. 890873). Presented at the 1988 SAE Passenger Car Meeting, Dearborn, MI.

Bhise, V. D., E. Smid, and J. D. Dowd. 2003. *ACE Driving Simulator and Its Applications to Evaluate Driver Interfaces*. SAE Paper 2003-01-0124. Paper presented at the 2003 Annual Meeting of the Society of Automotive Engineers, Detroit, MI.

Hussain, T. and S. Randive. 2010. *Defining a Low Cost Vehicle for the U.S. Market*. Published by the Institute for Advanced Vehicle Systems, College of Engineering and Computer Science, the University of Michigan-Dearborn, Dearborn, MI. Website: www.engin.umd.umich.edu/IAVS/books/A_Low_Cost_Vehicle_Concept_for_the_U.S._Market.pdf (Accessed: September 13, 2015).

Siemens. 2015. NX human modeling and posture prediction. Website: http://m.plm.automation.siemens.com/en_us/Images/7172_tcm1224-4287.pdf (Accessed: July 17, 2015).

23 Developing a Passenger Car
A Case Study

INTRODUCTION

The case study presented in this chapter is based on a number of projects completed by graduate students on developing proposals for planning a target car for introduction into the U.S. market as a 2021 (MY) vehicle. The reference car for the project was a late MY (2016 or 2017 MY) vehicle sold in the United States, and the target car was developed to replace the reference car. Two competitors' vehicles were also selected for benchmarking. The competitors' vehicles were also late models of leading competitors in the same market segment as the reference vehicle.

The objective of the chapter is to illustrate outputs of analyses conducted by the students as class projects in AE 500, a one-semester course in Automotive Systems Engineering application in vehicle development. The course is entitled "Automobile: An Integrated System." Appendices 1 through 5 provide descriptions of the class projects. Selected portions of the project reports prepared by Thodupunuri et al. (2016) were modified and edited for inclusion in this chapter. The vehicles selected for the project were (a) 2021 Ford Focus as the target car, (b) 2016 Ford Focus as the reference car, and (c) 2017 Hyundai Elantra and 2016 Toyota Corolla as the two competitors' vehicles. Figure 23.1 presents pictures of the 2016 Ford Focus and its two current competitors.

The outputs covered in this chapter include (a) customer characteristics, (b) customer needs, (c) market segment, (d) benchmarking of the vehicles, (e) specification of the target vehicle, (f) Pugh diagrams comparing the target vehicle characteristics with those of the reference and competitors' vehicles, (g) a technological plan, including changes, design challenges, and key open issues in the proposed vehicle development plan, (h) program timings and gateways, and (i) sales forecasts and financial analysis.

CUSTOMER CHARACTERISTICS, NEEDS, MARKET SEGMENT, BENCHMARKING, AND VEHICLE SPECIFICATION

Technical, market, and customer information about the reference and benchmarked vehicles was collected from several sources, which included vehicle brochures, manufacturers' websites, car magazines, visits to local dealers, and customer interviews.

2016 Ford Focus

2017 Hyundai Elantra

2016 Toyota Corolla

FIGURE 23.1 Three 2016 benchmarked vehicles.

CUSTOMER CHARACTERISTICS

The characteristics of the customers who currently purchase a compact sedan were as follows:

1. Customers are of all ages and belong mostly to the middle class.
2. Affordability is one of the main criteria for a buyer of this car. The price range (under about $23,000) and the mileage per gallon of gas (about 30–32 mpg combined city and highway driving) are important considerations for the customers.
3. These customers come from many social groups and are influenced by recommendations of friends, colleagues, and family members.
4. It is also a popular vehicle among students and commuters.

CUSTOMER NEEDS

Customers in this market segment are looking for a compact affordable vehicle with the following characteristics:

1. Compact size
2. Better styling and appearance
3. Good fuel economy
4. Reliability; for example, the car should be able to start even at very low temperatures
5. Improved driver comfort
6. Improved quality of interior features
7. Better ergonomics and seats
8. Simplified and less complicated infotainment system
9. More entertainment features
10. Increased safety features
11. Comfortable rear passenger compartment
12. Good vehicle handling
13. Responsive engine
14. Low noise and vibrations
15. Increased availability of optional features, for example, heated seats, Wi-Fi hot spot, voice recognition in navigation. and keyless access control
16. Better climate control and internal air circulation system
17. Rear passenger window defogger
18. Increased security systems

MARKET SEGMENT

The market segment for the project was the compact economy car. *Compact car* is a largely North American term denoting an automobile smaller than a mid-size car, but larger than a subcompact car. Compact cars usually have wheelbases between 100 (2540 mm) and 109 in (2769 mm). The U.S. Environmental Protection Agency (EPA) defines a compact car as measuring between 100 (2.8 m³) and 109 cu. ft (3.1 m³) of combined passenger and cargo volume capacity. Currently, this compact market segment contains about a 16% share of the U.S. light vehicle market. The Ford Focus falls in the compact-size car segment. Examples of other cars that fall in this segment are the Honda Civic, the Toyota Corolla, the Chevrolet Cruze, and the Hyundai Elantra.

BENCHMARKING

The benchmarking comparison of our reference model (2016 Ford Focus) and the two competitors (2017 Hyundai Elantra and 2016 Toyota Corolla) is presented in Table 23.1. The specifications of the target vehicle (developed by the graduate students) are also included in this table to facilitate comparisons. The primary changes

TABLE 23.1
Comparison of Reference Vehicle with Two Benchmarked Competitors and the Target Vehicle

Parameter/Feature	2016 Ford Focus Titanium (Reference Vehicle)	2017 Hyundai Elantra	2016 Toyota Corolla	2021 Ford Focus Titanium (Target Vehicle)
MSRP (US$)	23,225	22,350	23,055	27,000
Curb weight (lb)	3,055	2,976	2,865	2,800
Fuel economy (mpg)	26 city/38 highway/30 combined	28 city/37 highway/32 combined	29 city/37 highway/32 combined	30 city/40 highway/35 combined
CO_2 Emission (est.) (g/mile)	386.66	366.66	353.33	220
Exterior Dimensions				
Exterior body length (mm)	4,539	4,569	4,651	4,539
Exterior body width (mm)	1,824	1,801	1,775	1,824
Exterior body height (mm)	1,468	1,435	1,455	1,468
Wheelbase (mm)	2,649	2,700	2,700	2,649
Front track (mm)	1,554	1,549	1,519	1,554
Rear track (mm)	1,534	1,557	1,521	1,534
Ground clearance (mm)		135	142	142
Interior Dimensions				
Front legroom (mm)	1,095	1,072	1,074	1,095
Rear legroom (mm)	843	907	1,052	907
Front headroom (mm)	973	986	965	986
Rear headroom (mm)	965	947	942	965
Front hip room (mm)	1,369	1,356	1,346	1,369
Rear hip room (mm)	1,341	1,318	1,115	1,341
Front shoulder room (mm)	1,412	1,427	1,392	1,427
Rear shoulder room (mm)	1,336	1,405	1,392	1,380

Volumes				
Cargo/trunk volume (cu. ft)	13.2	14.4	13	13.2
Passenger volume (cu. ft)	90	95.8	97.5	90
Fuel tank (gal)	12.4	14	13.2	12.4
Powertrain				
Engine volume (L)	2.0	2.0	1.8	1.5
Cylinder configuration	I-4	I-4	I-4	I-4
Fuel injection system	Direct injection, flex fuel	Multi-point fuel injection	Multi-point fuel injection	Direct injection
Valvetrain	DOHC with variable valve timing	DOHC with variable valve timing	DOHC with variable valve timing	DOHC with variable valve timing
Drive type	Front-wheel drive	Front-wheel drive	Front-wheel drive	Front-wheel drive
Transmission	6-speed powershift automatic (DCT)	6-speed automatic (AT)	CVT with shift modes	6-speed powershift automatic (DCT)
Horsepower (hp)	160 @ 6,500	147 @ 6,200	132 @ 6,000	155 @ 6,000
Torque (lb-ft)	146 @ 4,500	132 @ 4,500	128 @ 4,400	146 @ 4,000
Turbocharger	N/A	N/A	N/A	Yes
Chassis/Body				
Front suspension	Independent MacPherson strut with 23.5 mm stabilizer bar	Independent MacPherson strut	Independent MacPherson strut front suspension with stabilizer bar	Independent MacPherson strut with 23.5 mm stabilizer bar
Rear suspension	Control Blade™ multilink independent with 19.0 mm stabilizer bar	Coupled torsion beam axle	Torsion beam rear suspension with rear stabilizer bar	Control Blade™ multilink independent with 19.0 mm stabilizer bar
Steering activation	Rack-and-pinion, electric power-assisted	Electric motor–driven power steering	Electric power steering—rack-and-pinion	Rack-and-pinion, electric power-assisted
Turning radius (ft)	36	34.78	35.6	35

(Continued)

TABLE 23.1 (CONTINUED)
Comparison of Reference Vehicle with Two Benchmarked Competitors and the Target Vehicle

Parameter/Feature	2016 Ford Focus Titanium (Reference Vehicle)	2017 Hyundai Elantra	2016 Toyota Corolla	2021 Ford Focus Titanium (Target Vehicle)
Brakes	Power-assisted front discs, rear drums	Power-assisted all discs	Power-assisted all discs	Power-assisted all discs
Tires	215/50R17 BSW	225/45R17	P215/45R17	205/50R17 BSW
Wheel material	Aluminum alloy	Alloy wheels	Alloy wheels	Aluminum alloy
Wheel diameter (in.)	17	17	17	17
Wheel width (mm)	215	225	215	215
Body material	Galvanized steel	Galvanized steel	Galvanized steel	Galvanized high-strength steel and aluminum doors
Features				
Instrument display	Analog	Analog	Analog	Analog + digital
Driver's information panel	Yes	Yes—color LCD	Yes—basic monochrome	Yes—high-resolution display
Engine temperature gauge	Yes	Yes	Yes	NA
Headlight beam	Halogen MFR	Halogen projectors	LED low beam with Halogen high beam	LED low beam with halogen high beam
Rain-detecting wipers	N/A	N/A	N/A	YES
Rear seat folding	60/40	60/40	60/40	60/40
Cup holders	2 + 2	2 + 2	2 + 2	2 + 2
Climate control system	Dual zone climate control	Dual automatic temperature control with clean air ionizer and auto defogging system	Automatic climate control with dust and pollen filter	Dual automatic temperature control with clean air ionizer and auto defogging system
Seats	Electric driver seat, power lumbar and heated front seats	Electric driver seat, power lumbar and heated front seats, with optional heated rear seats	Electric heated front seats	Electric heated front seats Turnable front seats

Sunroof	N/A	Available	Available	Available
Reverse sensing system	Proximity sensors	Reverse camera	Reverse camera	Reverse camera with proximity sensors
ABS	Yes	Yes	Yes	Yes
Number of air bags	8	7	8	8
Traction control	Yes	Yes	Yes	Yes
Brake assist	Yes	Yes	Yes	Yes
Security system	Keyless entry with push-button start	Keyless entry with push-button start	Keyless entry with push-button start	Keyless entry with push-button start
Daytime running lights	Yes	Yes	Yes	Yes
Exterior light control	Manual	Manual	Manual	Automatic
EBD	Yes	Yes	Yes	Yes
Cold-cranking amps	590	550	390	390
Radio	Yes	Optional premium infinity audio with eight channels including subwoofer with touch screen	Optional premium sound with 6.1 in. touch screen	Optional premium sound with touch screen
MP3	Yes	Yes	Yes	Yes
CD/DVD	Yes	Yes	Yes	No
Steering wheel audio controls	Yes	Yes	Yes	Yes
Bluetooth connectivity	Yes	Yes	Yes	Yes
Aux	Yes	Yes	Yes	Yes
USB port	Yes	Yes	Yes	Yes

Note: ABS, antilock braking system; AT, automatic transmission; CVT, continuous variable transmission; DCT, dual-clutch transmission; DOHC, double overhead camshaft; EBD, electronic brakeforce distribution; LCD, liquid crystal display; LED, light-emitting diode.

in the target vehicle are in the areas of fuel efficiency (to meet EPA regulations for 2021 MY passenger cars), fuel capacity, rear passenger legroom, horsepower, weight, and electrical and safety features.

DESCRIPTION OF THE TARGET VEHICLE

The 2021 Ford Focus compact sedan car will be launched as a global vehicle in many countries. The 2021 Focus will have a powerful but fuel-efficient engine and a tuned transmission with enhanced mileage, reduced emissions, better aerodynamics, alluring interior and exterior styling, and improved brakes. It will be competitively priced to project high value by considering its price and operating expenses. The key features of the target vehicle are presented in Table 23.2.

The unique characteristics of the target vehicle are

1. *Engine Downsizing*: The engine has been downsized from 2.0 to 1.5 L and has been optimized to give best-in-class emissions without much compromise on performance. Other technologies, such as turbo charging and direct injection, have been employed for improvement of fuel efficiency and performance with a downsized engine.
2. *Weight Reduction*: Weight reduction measures of material change from steel to aluminum body panels have been employed. The doors, hood, and tailgate have been made out of aluminum to improve the overall weight, saving by about 260 lb.

TABLE 23.2

Basic Characteristics of the Target Vehicle

Engine	1.5 L I-4 turbo-charged direct injection gasoline with flex-fuel
Performance	155 hp @ 6000 RPM, 146 lb-ft @ 4000 RPM
Emissions	220 g/mile
Fuel economy	30 city/40 highway/35 combined
Drivetrain	Six-speed power-shift automatic (DCT), front-wheel drive
Brakes	Four-wheel power-assisted ventilated disc brakes with ABS, EBD, and traction control
Suspension/steering	Front: independent MacPherson strut with 23.5 mm stabilizer bar. Rear: Control Blade™ multilink independent with 19.0 mm stabilizer bar. Rack and pinion electric-assisted power steering with 35 ft curb–curb turn radius.
Wheels and tires	17 in. aluminum alloy wheels with 205/50R17
Curb weight	2800 lb
NHTSA crash rating	5 stars
Ground clearance	5.4 in.
Interior volume	90 cu. ft
Fuel tank capacity	12.4 gal

3. *Safety*: The body and chassis design has been improved to promote overall crash safety. New knee and door air bags have been employed to improve the crashworthiness and occupant safety of the vehicle. The target for National Highway Traffic Safety Administration (NHTSA) safety is set at 5 stars as compared with the 4.5 stars for the reference car.

4. *Electrical and Electronics*: The cold-cranking current has been reduced to 390 A by the downsized engine, and external light-emitting diode (LED) headlights are used to reduce the power generation capacity of the alternator.

CHANGES IN THE TARGET VEHICLE

Table 23.3 shows the technology plan for the target vehicle. It presents major changes in different vehicle systems proposed for the target vehicle. Technological challenges and open issues that will be faced during the implementation of the changes are described briefly in the third and fourth columns of this table.

ASSESSMENT OF TARGET VEHICLE

This section presents three separate Pugh diagrams to compare the proposed 2021 MY vehicle and its current two competitors by using the reference vehicle as the "datum." The three Pugh diagrams presented in Tables 23.4 through 23.6 compare the vehicles based on customer needs, vehicle attributes, and vehicle systems, respectively. Comparison of the total scores presented in the bottom lines of each of the tables shows that the 2021 MY vehicle would be substantially improved over the existing vehicles.

CUSTOMER NEEDS PUGH DIAGRAM

The data in Table 23.4 show that the 2021 vehicle has a total score of 14, which means that the 2021 model will be perceived by the customers as satisfying many of their needs at higher levels as compared with the current 2016 reference car.

VEHICLE ATTRIBUTES PUGH DIAGRAM

The total score presented in the last row shows that the 2017 Hyundai Elantra's score of 9 is very close to the 10 score of the target car. This finding should sound a wake-up alarm to the design team, as it shows that their design would not be perceived to be very advanced. Additional improvements in many attributes are needed.

VEHICLE SYSTEMS PUGH DIAGRAM

The data in Table 23.6 show that the 2021 vehicle has a total score of 14, which means that the 2021 model will be perceived by the customers to have improved vehicle systems as compared with the current 2016 reference car.

TABLE 23.3
Technology Plan

Vehicle System	Changes Planned	Technological Challenges	Comments on Key Open Issues
Powertrain system (engine, transmission, drivetrain)	Reduced size to 1.5 L turbo charged direct injection engine	NHTSA emission norms, increased temperature of exhaust gases, durability and reliability of the engine	Engine size for packaging, and costs. Exhaust and air induction system have to be re-engineered as intake and exhaust parameters change
	Higher specific power output per liter of engine capacity	Increased peak firing pressure, increased bearing loads, increased piston loads	Material research on the block material, improved sintered bearings for better lubrication and life, improved oil additive specification should be provided
	Direct fuel injection	High exhaust gas temperatures, high corrosion-resistant materials	Increased cost for implementation of new technology will have effect on other parts (e.g., fuel pump), and maintaining good quality will be supplier's responsibility. Trade off cost versus performance. Supplier to be considered: Bosch
	Electric waste-gate turbo charger	Complete calibration to be done on vehicle level, and control of boost pressure is to be accurately done	Higher cost of electric waste-gate, reliability and durability are unknown.
	Reduced emissions and increased fuel economy	Improved compression ratio along with turbo charging demands for materials with higher strength and thermal capacity	Using tubular stainless steel for the exhaust manifold instead of cast parts. Emission technology has to be updated to satisfy 2021 EPA regulations as particulate matter emission level increases with turbo charged direct injection engines. Supplier for exhaust systems: Faurecia
	Increased warranty	Quality of components delivered	Maintaining in-house and supplier quality. Trade off customer satisfaction versus cost
Body/chassis system	Weight reduction of the body and chassis components	Use of lightweight aluminum alloy for body panels; use high-strength steels for frame sections (pillars, roof rails, rocker sections, cross-car beams, and so forth)	High cost of development of aluminum body parts. Vendor development is difficult, but can be achieved through study of other vehicles with lightweight materials and optimized for multi-material structures

	Introducing ventilated disc brakes on all four wheels	Meet FMVSS braking standards and achieve best-in-class 60–0 mph stopping distances	Supplier to be considered is Bosch
	Tire-pressure monitoring system	Accurate measurement of the tire pressure	Failure of the pressure sensors can lead to reliability issues. Supplier quality will be a point of concern
	All new aluminum alloy wheels	Strength, styling, cost	Bought out item, supplier quality issues. Trade off styling versus cost
Seat system	Introducing lumbar support in front two seats	Quality of the material used, ergonomic positioning	Satisfaction of the customers. Trade off customer satisfaction versus cost. Supplier to be considered: Johnson controls
	Reduction in overall weight	Optimized contouring and low thickness of rear padding to be provided to give good comfort, but low weight	Vendor development is to be done for the new contouring and slim seats. Trade off: cost versus weight reduction
Headlighting system	Adaptive headlights	Sensors needed to track vehicle's side-to-side movement and motors attached to each headlight	Installation of all the sensors and motors in exact locations. Supplier to be considered: Denso
Fuel system	Material change of fuel tank to help reduce weight of overall vehicle	Changing the material of the tank requires extensive durability and reliability cycles	Trade-off to be considered: cost of material to vehicle weight
Climate control system	Introducing dual zone climate control for customer satisfaction	Individual temperature sensors for each zone, lots of extra hidden ducting to carry the air where it's needed, and extra vents—lots and lots of extra vents	Constructing the lightweight parts for HVAC to reduce overall weight. Supplier to be considered: Delphi, Visteon
	Rear AC vents for effective cooling/heating	Installing rear ducts and opting for high-power brushless motor for blower, also reducing noise out of blower	Packaging of rear AC vents. Power consumption of the blower can be high
Safety and security systems	Brake assist	Radar sensors, active or passive infrared sensors, laser sensors, reliability issues	Current location of the sensors will be crucial, as it affects the braking performance. Supplier to be considered: Delphi systems

(Continued)

TABLE 23.3 (CONTINUED)
Technology Plan

Vehicle System	Changes Planned	Technological Challenges	Comments on Key Open Issues
	Blind spot sensor	Installing radar sensors behind the bumper cover, reliability issues	Anything blocking the radar will impair its performance—ice, mud, snow, and especially a fresh coat of paint. Supplier to be considered: Robert Bosch Corp.
	Forward collision alert	This system uses radar sensors, active or passive infrared sensors, or laser sensors along with a mounted camera to control the speed of the vehicle in case of emergency; a small miscalculation in the operations can end up in fatal accidents. Reliability might be an issue	Extreme weather conditions need to be taken into consideration while designing the sensors. So supplier quality and product quality are to be verified. Supplier to be considered: Robert Bosch Corp.
	Lane keep assist	Video sensors, laser sensors, infrared sensors, reliability issues	Further research is required as the sensors can't decipher missing or incorrect lane markings. Markings covered in snow or old lane markings left visible can hinder the ability of the system. Sensor should not give any alert while vehicle is in parking condition. Supplier to be considered: Delphi systems.
	Air bag system	Additional air bag is to be installed in doors to protect from side impact and in seatbelt to protect the passenger in all modes of impact	Packaging air bag in door might be an issue. Current suppliers need to come up with advanced technology to design and tune the seat belt air bag system. Supplier to be considered: Autoliv Corp.
Driver and infotainment system, telematics, and convenience features	Multi-touch screen for audio, climate, and navigation controls	Ergonomic challenges in placing the controls, touch sensitivity, and size of the screen	Research and development cost, quality of the supplier, and reliability. Supplier to be considered: Johnson Controls

Introducing CarPlay and Android Auto for better connectivity with Android and IOS mobiles	Recruiting new or training existing resources to install this new software	Usage and acceptance of the software by the customers
Aesthetic instrument clusters	Fit and finish, texture of the material	Vendor development, supplier quality will be a concern. Supplier to be considered: Johnson controls
Improved navigation system	Accuracy of the system, compatible with Bluetooth devices	Bought-out part. Supplier quality must be monitored for better customer satisfaction. Supplier to be considered: Panasonic
Voice-activated controls	Sensor should be capable of receiving commands in many commonly spoken languages and accents	A new version of voice recognition software with automatic updating so that the software keeps on improving over time
Onboard data storage for audio system	Memory device placement and occupied place, compatibility with small memory devices	As a consequence of the high data access speed of HDD, the basic navigation abilities have been improved. The time required to search the route, the speed of map scrolling, and the response of voice recognition was improved. One of the most striking characteristics of this navigation system is the music server function using a rewritable storage device. Supplier to be considered: Harman
Sunroof	Adding sunroof with reduced intrusion in head clearance space and weight	Trade-off: convenience feature versus cost. Supplier to be considered: Arvin Meritor Inc.
Steering wheel–mounted controls	Ergonomic positioning of buttons, controls. Ability to use with gloves	Research and development cost involved, packaging issues with buttons

Note: AC, air conditioning; HDD, hard disk drive; HVAC, heating, ventilation, and air conditioning.

TABLE 23.4
Pugh Diagram for Customer Needs

Customer Needs	2016 Ford Focus Titanium (Datum)	2017 Hyundai Elantra	2016 Toyota Corolla	2021 Ford Focus Titanium (Target Car)
Styling and appearance		+	S	+
Packaging and interior quality		+	+	+
Safety		+	S	+
Infotainment system and center console		+	+	+
Keyless access control		+	+	+
Climate control system		+	−	+
Fuel economy		+	+	+
Vehicle handling		−	−	+
Engine performance		−	−	S
Noise and vibrations		+	−	S
Driver comfort and ease of use		+	S	+
Rear passenger comfort		+	+	S
Battery and cold cranking		−	−	+
Cost		+	+	−
Voice recognition		S	S	S
Wi-fi		S	S	+
Rear door window defogger		S	S	+
Memory rear view mirror		S	S	+
Heated steering		S	S	+
Security		S	S	+
Sum of (+)		11	6	15
Sum of (−)		3	5	1
Sum of (S)		6	9	4
Total score		8	1	14

PROGRAM TIMINGS, SALES, AND FINANCIAL PROJECTIONS

PROGRAM TIMINGS

The vehicle program to develop the proposed 2021 Ford Focus sedan was estimated to begin with the formation of the core product development team at approximately 40 months before Job#1. Job# 1 is scheduled on September 15, 2020.

PROJECTED SALES

It is estimated that the company will sell about 120,000 Focus sedan vehicles per year. The manufacturer's suggested retail price (MSRP) of the 2021 Focus is estimated to be about $28,000.

TABLE 23.5
Pugh Diagram for Vehicle Attributes

Vehicle Attribute	2016 Ford Focus Titanium (Datum)	2017 Hyundai Elantra	2016 Toyota Corolla	2021 Ford Focus Titanium (Target Car)
Styling and appearance		+	+	+
Vehicle package		+	S	+
Safety		+	S	+
Driver information and interface		+	S	+
Climate control		+	–	+
Performance		S	–	+
Vehicle handling		–	S	+
Noise, vibrations and harshness		+	–	S
Ergonomics		+	–	+
Emissions		+	+	+
Cost		+	S	–
Weight		+	+	+
Security		S	S	+
Sum of (+)		10	3	11
Sum of (–)		1	4	1
Sum of (S)		2	6	1
Total score		9	–1	10

FINANCIAL PROJECTIONS

Over the estimated 5 year sales period (MY 2021–2026), the total sales of the Focus are estimated at 484,500 vehicles with a total revenue of about $14.2 billion and net earnings of about $7.8 billion. The maximum cumulative cost incurred in the total Focus program (all Focus models) is estimated at about $0.5 billion just prior to Job#1. The cost breakeven point is estimated to occur at about 6 months after Job#1.

CONCLUDING REMARKS

The goal of the outputs presented in this chapter was to provide the students with an understanding of the early part of the automotive product development by performing the data-gathering and decision-making tasks. The projects required the students to develop specifications for the 2021 Ford Focus by undertaking benchmarking of the latest available models of the Ford Focus, Hyundai Elantra, and Toyota Corolla, studying applicable government regulations, and researching future trends in design and technologies.

It should be realized that the data provided in the tables included in this chapter were gathered by the students by searching through various sources, such as

TABLE 23.6

Pugh Diagram for Vehicle Systems

Vehicle System	2016 Ford Focus Titanium (Datum)	2017 Hyundai Elantra	2016 Toyota Corolla	2021 Ford Focus Titanium (Target Car)
Body-in-white		S	S	S
Closures (doors, hood, and trunk)		S	S	+
Seat system		+	S	+
Instrument panel system		+	+	+
Exterior lamp system		S	+	+
Rear vision system		S	S	+
Underbody frame		S	S	S
Suspension system		−	−	S
Steering system		S	S	+
Braking system		S	S	S
Wheels and tires		+	−	S
Engine		S	−	−
Transmission		−	−	+
Fuel system		+	+	S
Electrical system		−	−	+
Climate control system		S	−	+
Seat belts and airbag system		S	S	+
Wiping and defrosting system		S	S	+
Security lighting and locking system		S	S	+
Driver assistance system		S	S	+
Audio system		+	+	+
Navigation system		S	S	+
Sum of (+)		5	4	15
Sum of (−)		3	6	1
Sum of (S)		14	12	6
Total score		2	−2	14

the vehicle brochures, the websites of the vehicle manufacturers, and automotive magazines, in a very limited time (all the projects presented in Appendices 1 through 5 were completed within one semester by a team of no more than four graduate students). The Pugh analyses were conducted using data gathered under conditions of severely restricted time and scarcity of benchmarked vehicles. Similarly, the financial analyses were conducted using a number of assumptions on items such as time required to perform various design tasks, manpower needs, pay rates and costs associated with different evaluations, and so forth. Therefore, the data and the information provided in this chapter are expected to be very approximate and crude. However, from the viewpoint of educational value, it was felt that readers of the book and students should understand the need for and use of such data during the development of a new automotive product.

The important observations from the projects are briefly described in the following list:

1. The use of benchmarking competitive products provides an understanding of similarities and dissimilarities between vehicle dimensions (exterior and interior), various vehicle systems and features, their configurations, and issues related to packaging and interfacing. The similarity in vehicle dimensions and major vehicle systems among the benchmarked vehicles makes one realize that all vehicle manufacturers must be closely studying and learning from their competitors' products and using similar design considerations.

2. The need to meet the NHTSA/EPA corporate average fuel economy (CAFE) and emissions requirements described in Chapters 3 and 6 will have a major impact on design of the powertrain and the vehicle body structure. The engines need to be smaller but must generate more power from turbo-boosting or by supplemental electric motors (hybrid powertrains). In addition, the use of lightweight materials, power-saving features (e.g., stop/start, low-friction tires and bearings), or regeneration of power during deceleration must also be considered.

3. The creation of a technology plan is a "must," as it helps the design team understand all the design changes, challenges, and risks involved in developing a new vehicle.

4. The program timing chart and milestones provide an understanding of time constraints, scheduling of various activities, and the need for simultaneous engineering.

5. The financial plan allows the program team and the company management to understand the overall resource needs to cover the cumulative costs, breakeven point, and revenue potential.

6. The need for and role of product evaluations during vehicle development are quickly realized during the preparation of Pugh diagrams. Pugh diagrams based on customer needs, vehicle attributes, and vehicle systems allow the design team to get a better idea about the positioning of the target vehicle in comparison with the reference vehicle and other competitor vehicles.

7. The projects quickly make the students realize the value of teamwork, involvement of professionals from different disciplines, and coordination of systems engineering tasks in automotive product development.

REFERENCE

Thodupunuri, S., K. Mehta, and S. Yetachina. 2016. Development of 2021 Ford Focus. Projects P-1 to P-4 conducted for "AE 500: Automobile an Integrated System" course at the University of Michigan-Dearborn in winter term 2016.

24 Developing a Pickup Truck

A Case Study

INTRODUCTION

The purpose of this chapter is to provide a case study of the product development of a future model of a pickup truck. The target vehicle selected for this exercise was the 2021 Ford F-150 Pickup truck. Thus, the current model (2016 model year [MY]) of the Ford F-150 was used as the reference vehicle, and the 2016 Chevrolet Silverado and the 2016 Dodge Ram 1500 were used as the competitive vehicles for benchmarking.

The analysis presented here was originally prepared as a series of class projects by Ludwick et al. (2016) for the author's course entitled "Automobile: An Integrated System (AE 500)" taught in the 2016 winter term. Appendices 1 through 5 provide descriptions of the class projects. Selected portions of the project reports were modified and edited for inclusion in this chapter.

The outputs covered in this chapter include (a) customer characteristics, (b) customer needs, (c) market segment, (d) benchmarking of the vehicles, (e) specification of the target vehicle, (f) Pugh diagrams comparing the target vehicle characteristics with those of its reference and competitors' vehicles, (g) a technological plan including changes, design challenges, and key open issues in the proposed vehicle development plan, (h) program timings and gateways, and (i) sales forecasts and financial analysis.

The technical, market, and customer information about the reference and benchmarked vehicles was collected from several sources, which included vehicle brochures, manufacturers' websites, automotive magazines, visits to local dealers, and customer interviews. Figure 24.1 provides pictures of the three 2016 MY pickup trucks.

CUSTOMER CHARACTERISTICS AND NEEDS, MARKET SEGMENT, BENCHMARKING, AND VEHICLE SPECIFICATION

CUSTOMER CHARACTERISTICS

The F-150 has great diversity in its customer base. Its customer base includes businesses (small to large), farmers, construction crews, government agencies, people who own campers or boats, people who need a vehicle to haul material around, and people who just like driving trucks because of the higher view of the road. To keep the same number of anticipated customers as in previous years (if not to expand it),

2016 Ford F-150

2016 Cheverolet Silvarado 1500

2016 Dodge Ram 1500

FIGURE 24.1 Three pickup trucks.

the 2021 F-150 will continue to offer a large range of trims and options to appeal to all customers.

Most customers and drivers (about 75%–85%) are expected to be males, and this should be kept in mind during the entire product development and marketing. However, it should still be possible for a small (e.g., 5th percentile) female to drive the vehicle. The ages of the principal drivers of these vehicles predominantly range from about 36 to 55. The median annual household income of pickup truck owners is around $66,000. About 45% of truck owners spend at least $1000 on customization, and 17% spend at least $3000. About 40% of owners have given nicknames to their trucks (they are very passionate about their trucks—just like pets), and about 64% consider their truck as an extension of their personality. The majority of them work full time, are married, and live in single-family homes. Many owners use their vehicles for work as well as family purposes. These vehicles are popularly used in building construction and hauling materials. Many of the vehicles are also used occasionally for towing trailers and plowing snow, especially in northern states.

CUSTOMER NEEDS

Below is a list of the most important customer needs for the 2021 F-150:

1. A vehicle that can haul at least 1600 lb of payload
2. A vehicle that can tow at least 7600 lb
3. A vehicle that has features to help keep the occupants safe
4. A vehicle that is available at an affordable price
5. A vehicle that is comfortable and enjoyable to drive
6. A vehicle that provides good fuel economy compared with other trucks
7. A vehicle that is able to handle tough terrain (4 × 4, ground clearance, angles of approach, departure and ramp break-over, and tire quality)
8. A vehicle that is maneuverable (min. turn radius)
9. A vehicle that has convenience feature options (power lock, power windows)
10. A vehicle that can seat from two to six people
11. A vehicle that has an attractive appearance and looks tough

The specific needs of the vehicle will depend on its usage, as follows:

1. *Use as Work Vehicle*
   a. Payload and towing capabilities
      i. Payload: min: 1500 lb; class leading: 3270 lb
      ii. Towing: min: 5000 lb; class leading: 12,200 lb
   b. Low-end torque
   c. Trailer brake
   d. Trailer backup assist
   e. Durability:
      i. Might overload accidentally at some point.
      ii. Owners expect to be able to load up the vehicle every day if necessary.
      iii. Owners expect to be able to use vehicle in adverse/harsh conditions without an issue.
   f. Bed-liner option
   g. Step into truck bed
      i. Corner bumper step
      ii. Tailgate ladder
   h. Selectable gear shifting for automatic transmission
2. *Use as Commuter/Everyday Vehicle*
   a. Front/rear sensors and rear camera because of size of vehicle.

  b. Good view of the road:
    i. High seating position.
    ii. Good view out of windshield and side windows.
  c. Luxury options:
    i. Adaptive cruise control.
    ii. Heated/cooled seats.
    iii. Power seats.
    iv. Heated steering wheel.
    v. Navigation system.
    vi. Leather interior.
    vii. Premium stereo.
   viii. Bluetooth.
    ix. iPod/mp3 player connectivity.
  d. Competitive fuel economy:
    i. Not too worried about it, but can't be way off from the competitors.
  e. Optional power folding side view mirrors.
  f. Space in cab for some cargo that can't be transported in the bed (e.g., groceries).
  g. Precise steering to be able to place vehicle where the customer wants on the road, but not so precise or fast that the vehicle feels twitchy (like a sports car). This steering characteristic is particularly important to avoid while pulling a trailer or when the vehicle is weighed down.
  h. Handling characteristics need to make vehicle feel planted but not too aggressive or nervous on the road.

3. *Off-Road/Inclement Weather Use*
  a. Competitive approach, departure, and break-over angles
  b. Tires that allow vehicle to operate in snow/mud/sand up to reasonable point
  c. Four-wheel drive option
  d. Locking rear differential
  e. Ability to turn traction control on/off

4. *Customizability for Specific Customer Needs*
  a. Multiple bed length options (short, medium, long in line with the competition)
  b. Multiple cab style options
    i. Regular cab
    ii. Super cab
    iii. Crew cab
  c. Multiple rear axle ratio options
  d. Multiple powertrain options
  e. Multiple max payload and towing options

5. *Safety Needs under All Uses*
  a. Airbags: front, rear, and side
  b. Night vision camera
  c. Lane change assist

MARKET SEGMENT

The pickup trucks listed in the Introduction and sold in the North American market fall within the light-duty (Class 2) pickup truck market segment. These vehicles have a gross vehicle weight rating (GVWR) range of 6,001–10,000 lb (2,722–4,536 kg). (Note: Class 2 is subdivided into Class 2a and Class 2b, with Class 2a being 6,001–8,500 lb [2,722–3,856 kg] and Class 2b being 8,501–10,000 lb [3,856–4,536 kg]. Class 2a is commonly referred to as a light-duty truck, with class 2b being the lowest heavy-duty class, also called the light heavy-duty class.) These pickup trucks are sold in the North American market with a maximum seating capacity of six occupants and with three bed lengths of about 5.5, 6.5, and 8.0 ft. They are available in 4 × 2 and 4 × 4 (wheels × drive wheels) configurations, and the combined fuel consumption of 2016 MY pickups ranged between about 15 and 24 mpg. The price of these vehicles is approximately $26,000–$60,000.

BENCHMARKING AND VEHICLE SPECIFICATION

Table 24.1 presents benchmarking data comparing 2016 MY vehicles with the 2021 target vehicle. The table presents data on various vehicle characteristics to allow comparisons between the 2016 MY vehicles and the 2021 MY target vehicle (shown in the last column). The exterior dimensions of the 2021 F-150 are unchanged from the current 2015 F-150. The driver's headroom and hip room are increased for the 2021 F-150. The vehicle body, chassis, and powertrain are improved to maintain the overall vehicle weight at 5577 lb in spite of the added weight due to the 3.5 L hybrid engine and larger battery. The payload and towing capacity of 2021 F-150 is substantially increased over the three 2016 benchmarked vehicles.

DESCRIPTION OF TARGET VEHICLE

The 2021 Ford F-150 target vehicle will be designed to give tough competition in the light-duty pickup truck segment. To continue forward as a market leader in sales within the segment, cutting-edge features, new technologies, and numerous innovations will be added to the 2021 Ford F-150 to maintain the best-in-class position. The major features of the target vehicle developed in the projects are briefly described in this section:

1. *Carbon-Fiber Body with Aluminum Frame*: Fully boxed aluminum frame with carbon-fiber body for reduced weight, durability, and better efficiency (25 mpg) to satisfy corporate average fuel economy (CAFE) standards. The new vehicle body exterior will also be attractively styled.
2. *Safety*: Designed for a 5-star National Highway Traffic Safety Administration (NHTSA) safety rating with features such as lane-keeping assistance, hill climb assist, collision avoidance system, air bag system (front, rear, and side impact), and seat belts with pre-tensioners.

TABLE 24.1

Benchmarking Comparison of 2016 Pickup Trucks with the Proposed 2021 Ford Pickup

Vehicle Characteristic		2016 Ford F-150 (Reference Vehicle)	2016 Chevrolet Silverado 1500	2016 Ram 1500	2021 Ford F-150 Hybrid (Target Vehicle)
Price	Base MSRP	$26,373	$26,895	$26,145	$32,000
Fuel consumption	Fuel economy (mpg): City–highway	16–22	17–20	15–24	25
Exterior dimensions	Length (in.)	231.9	230	229	231.9
	Width (in.)	79.9	80	79.4	79.9
	Height (in.)	76.9	74	77.6	69.2
	Wheelbase (in.)	145	143.5	140.5	145
	Front track (in.)	67.6	68.7	68.6	67.6
	Rear track (in.)	67.6	67.6	68	67.6
	Ground clearance (in.)	10.3	8.9	9.4	9.3
	Body panels	Aluminum alloy body with steel frame	Fully boxed high-strength steel frame	High-strength steel frame	Carbon-fiber body with aluminum frame
Interior dimensions	Headroom (front/rear) (in.)	40.8/40.3	42.8/38.7	41/39.7	42.8/40.3
	Shoulder room (Front/rear) (in.)	66.7/65.8	65.9/65.8	66/65.7	66.7/65.8
	Legroom (front/rear) (in.)	43.9/33.5	45.3/34.6	41/40.3	43.9/40.3
	Hip room (front/rear) (in.)	62./64.7	60.7/60.3	63.2/62.9	63.2/64.7
	Cargo volume (cu. ft.)	62.3	61	52	62.3
Body/chassis	Body panels	Aluminum alloy body with steel frame	Fully boxed high-strength steel frame	High-strength steel frame	Carbon-fiber body with aluminum frame
	Chassis type	Body-on-frame	Body-on-frame	Body-on-frame	Body-on-frame

Powertrain	Steering system	Electronic rack and pinion	Electronic rack and pinion	Electronic rack and pinion	Electronic rack and pinion
	Engine type/size	3.5 L V6 EcoBoost (standard engine twin-turbo charged DOHC 24-valve)	EcoTec3 4.3 L V6 (standard)	3.6 L Pentastar V6 24V variable valve Timing (VVT) (Standard) pushrod	3.5 L hybrid engine
	Engine power (hp @ RPM)	365 @ 5000	420 @ 5600	395 @ 5600	420 @ 5600
	Engine torque (lb-ft @ RPM)	420 @ 2500	460 @ 4100	410 @ 3950	460 @ 4100
	Transmission	Select shift 6-speed auto with range select	8-speed automatic transmission	8-speed automatic	Single-gear hybrid powertrain with different modes
Towing/ payload	Max. towing (lb)	11,500	11,800	10,160	12,200
Weight characteristics	Max. payload (lb)	2,020	2,050	1,550	3,270
	Curb (lb)	5,577	5,658	5,964	5,577
	%Front/%rear	56.1/43.9	58.3/41.7	54.6/45.4	54.6/41.7
	GVWR (lb)	7,000	7,200	6,900	7,000

Note: DOHC, double overhead camshaft.

TABLE 24.2
Technology Plan

Vehicle System	Major Changes	Major Technological Challenges	Key issues
Body system	Carbon-fiber body system	Manufacturing of carbon fiber	Research must be done to find new lighter and cheaper materials.
		Welding or joining the carbon-fiber body	Research to find out new efficient methods for joining carbon fiber
		Painting the carbon-fiber body	Research to find out a feasible way to paint the carbon-fiber body
		Production and assembly line for the new carbon-fiber body	Cost risk is involved since carbon fiber is expensive
		Training the workers to work with new material and technology	
		Technology costs involved.	
		Meeting CAFE standards	Cost versus performance trade-off
Chassis system	New tires with larger dimensions for better handling (265/70R17 all weather)	Increased tire dimensions affecting the economy (CAFE standards)	Research into new roll-resistant tire technology.
			Potential supplier: Goodyear
	Auto pilot steering mode	Developing auto pilot mode	Research and development in the field of autonomous vehicles to improve safety and control
	EBD brakes with electronic booster	Manufacturing electronic brakes with EBD	Search for suppliers to make new brakes. Potential suppliers: ZF and TRW
			Research with suppliers to use regenerative braking feature
			Cost versus performance trade-offs
	Steering system with low-resistance bearings	Changes in the existing steering wheel design for low resistance to improve the handling	Research to find newer methods to reduce the resistance.

	Double wishbone suspension system with gas-pressurized shocks	Changes to be made to existing design of the shocks	Cost risks involved. Potential supplier: Bosch Cost versus performance trade-off. Suspension must not add to weight. Research to make lighter suspension systems with glass fiber. Recommended to make changes to existing design to make new suspension. Potential supplier: Eibach
Powertrain system	3.5 L hybrid engine	The new engine should satisfy CAFE standards	Research to develop a hybrid engine which balances both performance and economy to meet CAFE standards. Cost versus performance trade-off.
		Hybrid mustn't compromise with performance and power	Cost risks. New technology such as start/stop. Preferred to develop engine within the organization
		Reduced emissions	Research into new filter technologies to reduce emissions, such as better filters and adsorbent materials. Filter supplier: Bosch
	Single-drive electric powertrain	Powertrain should have less resistance to reduce noise and have low weight	Research into new materials and lubrication technology for efficient power transmission
		Meet CAFE standards	Research and develop lightweight powertrain materials to improve efficiency
		Electric powertrain should have good performance	Research to implement different gear ratios for better performance and power in hybrid powertrain. Cost versus performance trade-off. Cost risks in research. Recommended to manufacture the powertrain within the organization

(Continued)

TABLE 24.2 (CONTINUED)
Technology Plan

Vehicle System	Major Changes	Major Technological Challenges	Key issues
Safety system	Automatic braking system (collision avoidance)	Performance in varying weather conditions. Variability in tire wear and brake rotor/pad wear. Determining threshold distance for braking	Possibility of early or unwanted activation. Research into how far from obstacle brakes should activate. Liability if brakes activate at wrong time. Amount of deceleration customers are willing to endure
	Inflatable seat belt	Dangers for small/low-weight passengers (such as children)	Increased cost. Research into how best to activate "air bag" in the event of a crash
	Adaptive headlamps	Performance in varying weather. Calibrating activation threshold.	Increased cost versus actual real-world benefit of system. Types of sensors and packaging space are also issues
	Blind-spot monitoring system	Performance in varying weather. Calibrating activation threshold	Packaging of sensors in side-view mirrors or body structure. Increased cost versus actual real-world benefit of system. Drivers may rely on system too much, causing liability issues
	Active head restraints	Dangers for passengers/drivers of varying heights	Packaging of extra components in seat/headrests and effects on passenger comfort. Increased cost versus actual real-world benefit of system
Electrical and vehicle lighting Systems	Fiber optic cables to replace copper wire	Compatibility with components. Special installation procedures.	Increased cost versus actual real-world benefit. Supplier availability and feasible production capacity
	Smart key with built-in vehicle controls	Vulnerability to cyber-attacks	Increased cost. Need to determine which functions customers might want to be able to control remotely. Limited physical or virtual space (memory) for vehicle functions
Driver interface systems	In-vehicle Wi-Fi system (hotspot)	Vulnerability to cyber-attacks. Strength of signal	Research into whether critical vehicle systems might be vulnerable should Wi-Fi be hacked and how best to protect systems from cyber-attacks if Wi-Fi is hacked. Cost to customer for data usage

Interior	Advanced windshield heads-up display (HUD)	Driver distraction. Information resolution and clarity. Performance through range of lighting conditions	Research into whether or not HUD might be too distracting for drivers. Increased cost versus actual real-world benefit
	Android Auto or Apple CarPlay Option	Added complexity. Compatibility with wide array of mobile devices	Cost to offer both options in vehicle. Compatibility with other electronics in vehicle. Risk if neither platform is accepted by users
	Custom configuration digital gauge cluster	Information clarity	Increased cost versus actual real-world benefit of system. Could be distracting to drivers if they are changing things while in motion. Need to develop screen or identify suitable supplier (e.g., Samsung)
	Heated steering wheel	Packaging of components in steering column and steering wheel. Increased power draw	Research into trade-off between packaging for heating components and packaging for certain steering wheel–mounted controls. Could heating components damage other components in steering wheel? Increased cost
	Heated and cooled driver and passenger seats	Packaging issues. Increased power draw	Increased cost. Need to determine whether components can be packaged along with motors for power-adjusting function
	Variable tint windows	Operation through large temperature range. Increased power draw	Need to determine reliability and "durability" of system to ensure that it will function through the life of the vehicle. Increased cost. Need to develop in house or determine suitable supplier
	Power-adjustable rear seats	Packaging of components. Increased power draw	Increased cost. Need to determine whether components can be packaged along with components for heating and cooling seats

3. *Powertrain*: All-new 3.5 L hybrid engine with a single-drive electric hybrid powertrain for improved performance and economy and reduced emissions (335 g/mile equivalent for CO_2).
4. *Electrically Boosted Brakes with Antilock Braking System (ABS) and Electronic Brake Force Distribution (EBD)*: Necessary improvements are made to the brakes to come up with ABS with EBD for better control and handling to satisfy Federal Motor Vehicle Safety Standards (FMVSS) requirements.
5. *Automatic Driver Controls*: Automatic driving controls (e.g., automatic braking for crash avoidance) are added to the vehicle to make the vehicle innovatively attractive.
6. *New Improved Infotainment System with Apple CarPlay*: A new touch-screen infotainment display with SYNC technology with Apple Car Play.
7. *Other Features*: Other features such as lane-keeping assist, parking assist, and automatic light-emitting diode (LED) headlamps.

CHANGES IN THE TARGET VEHICLE

Table 24.2 shows the technology plan for the target vehicle. It presents major changes in different vehicle systems proposed for the target vehicle. Technological challenges and open issues that will be faced during the implementation of the changes are described briefly in the fourth and fifth columns of the table.

ASSESSMENT OF THE TARGET VEHICLE

This section presents three separate Pugh diagrams to compare the proposed 2021 MY vehicle and its current two competitors by using the reference vehicle as the "datum". The three Pugh diagrams presented in Tables 24.3 through 24.5 compare the vehicles based on customer needs, vehicle attributes, and vehicle systems, respectively. Comparison of the total scores presented in the bottom lines of each of the tables shows that the 2021 MY vehicle would be substantially improved over the existing vehicles.

CUSTOMER NEEDS PUGH DIAGRAM

The total scores presented in the last row of Table 24.3 show that the proposed 2021 target vehicle with its total score of 14 will be perceived by its customers to be more satisfying than the three existing pickups.

VEHICLE ATTRIBUTES PUGH DIAGRAM

The total scores presented in the last row of Table 24.4 show that the proposed 2021 target vehicle with its total score of 7 will be perceived by its customers to be better than the three existing pickups. The proposed pickup will perform better than the existing pickup in over eight different attributes.

TABLE 24.3
Pugh Diagram of Customer Needs

Customer Needs	Ford F150 2016 (Datum)	Chevrolet Silverado 1500 2016	Ram 1500 2016	Ford F150 2021 (Target)
Payload capacity		+	–	+
Towing capacity		+	–	+
Engine torque		+	–	+
Fuel economy		–	–	+
Safety		–	–	S
Trailer brake		S	S	S
Trailer backup assist		S	S	S
Durability		–	–	+
Warranty		S	S	S
Bed liner		S	S	+
Step into truck		S	S	S
Selectable gear shift		–	–	+
Front camera		S	S	S
Rear camera		S	S	S
Seating position		S	S	S
Adaptive cruise control		S	S	S
Heated/cooled seats		S	S	S
Power seats		S	S	S
Heated steering wheel		S	S	S
Navigation		S	S	+
Leather interior		S	S	S
Premium stereo		+	+	+
Bluetooth		S	S	S
iPod/mp3 player connectivity		S	S	S
Power mirrors		S	S	+
Storage space in cab		S	S	S
Handling		S	S	S
All–weather tires		S	S	+
Four–wheel drive option		S	S	S
Rear differential		S	S	S
Traction control on/off		S	S	S
Customizability		–	–	+
Sum of +		4	1	14
Sum of –		5	8	0
Sum of S		23	23	18
Total score		–1	–7	14

TABLE 24.4
Pugh Diagram of Vehicle Attributes

Vehicle Attributes	Ford F150 2016 (Datum)	Chevrolet Silverado 1500 2016	Ram 1500 2016	Ford F150 2021 (Target)
Package		S	S	S
Ergonomics		S	S	+
Safety		–	–	S
Styling and appearance		S	–	+
Aerodynamics		S	S	+
Performance		+	–	+
Drivability		S	S	S
Dynamics		–	–	+
Noise, vibrations and harshness		S	S	+
Interior climate control		S	S	S
Weight		–	–	+
Security		S	S	S
Communication and entertainment		+	+	+
Costs		–	+	–
Emissions		–	–	+
Customer life cycle		S	S	S
Product life cycle		S	S	–
Sum of S		10	9	6
Sum of +		2	2	9
Sum of –		5	6	2
Total score		–3	–4	7

VEHICLE SYSTEMS PUGH DIAGRAM

The total scores presented in the last row of Table 24.5 show that the proposed 2021 target vehicle with its total score of 14 will be perceived by its customers to have improved vehicle systems compared with the three existing pickups. The proposed pickup will have at least 10 better vehicle systems than the existing pickups.

PROGRAM TIMINGS, SALES, AND FINANCIAL PROJECTIONS

PROGRAM TIMINGS

The vehicle program to develop the proposed 2021 Ford F-150 was estimated to begin with the formation of the core product development team at approximately 40 months before Job#1. Job# 1 is scheduled on September 15, 2020.

TABLE 24.5
Pugh Diagram of Vehicle Systems

Vehicle System	Ford F150 2016 (Datum)	Chevrolet Silverado 1500 2016	Ram 1500 2016	Ford F150 2021 (Target)
Body–in–white		–	–	+
Closures system		S	S	+
Seat system		S	S	+
Instrument panel		S	S	+
Exterior lamps		S	S	S
Glass system		S	S	S
Rear vision system		S	S	S
Door frame		–	–	+
Headlamps		S	S	S
Suspension system		S	S	S
Steering system		S	S	S
Braking system		–	–	+
Wheels and tires		–	–	S
Power side mirrors		S	S	+
Engine		+	+	+
Transmission		–	–	+
Fuel system		+	+	+
Battery		–	–	+
Alternator		S	+	S
Power controls		S	S	S
Climate control system		S	S	S
Safety system		–	–	+
Security system		S	S	S
Driver interface system		S	S	S
Audio system		+	+	+
Navigation system		S	S	S
CD player/iPod connectivity		S	S	S
ApplePlay		+	+	+
Sum of +		4	5	14
Sum of –		7	7	0
Sum of S		17	16	15
Total score		–3	–2	14

PROJECTED SALES

It is estimated that the company will sell about 700,000 F-150 vehicles per year. The manufacturer's suggested retail price (MSRP) of the 2021 F-150 hybrid is estimated to be about $32,000.

FINANCIAL PROJECTIONS

Over the estimated 3 year sale period (MY 2021–2023), the total sales of F-150 are estimated at 2.1 million vehicles with a total revenue of about $86.0 billion and net earnings of about $43.3 billion. The maximum cumulative cost incurred in the total F-150 program is estimated at about $ 2.1 billion just prior to Job#1. The cost break-even point is estimated to occur at about 3 months after Job#1.

CONCLUDING REMARKS

The goal of the outputs presented in this chapter was to provide the students with an understanding of the early part of automotive product development by performing the data-gathering and decision-making tasks. The projects required the students to develop specifications for the 2021 Ford F-150 by undertaking benchmarking of the latest available models of the Ford F-150, Chevrolet Silverado, and Ram 1500, studying applicable government regulations and researching future trends in design and technologies.

It should be realized that the data provided in the tables included in this chapter were gathered by the students by searching through various sources, such as the vehicle brochures, websites of the vehicle manufacturers, and automotive magazines in a very limited time (all the projects presented in Appendices 1 through 5 were completed within one semester by a team of no more than four graduate students). The Pugh analyses were conducted using data gathered under conditions of severely restricted time and scarcity of benchmarked vehicles. Similarly, the financial analyses were conducted using a number of assumptions on items such as time required to perform various design tasks, manpower needs, pay rates and costs associated with different evaluations, and so forth. Therefore, the data and the information provided in this chapter are expected to be very approximate and crude. However, from the viewpoint of educational value, it was felt that readers of the book and students should understand the need for and use of such data during the development of a new automotive product.

In addition to the observations included in the concluding remarks section of Chapter 23, the important observations from the projects are briefly described here:

1. The light pickup trucks covered in this chapter are an important and different type of automotive product. They are used for a greater variety of purposes (e.g., carrying cargo, towing trailers on construction sites, on farms, and in cities for commercial as well as personal use) than passenger cars. The crew cab version allows up to six passengers to be accommodated.
2. The number of models and powertrains offered on the pickup trucks is also greater than those offered in passenger cars and sports utility vehicles (SUVs).
3. A hybrid powertrain such as the one proposed in the analyses in this chapter may be necessary to meet government emissions and economy regulations, but success of the 2021 MY F-150 within the market is dependent on customers accepting this powertrain, which is difficult to predict because of constantly changing economic and oil price situations.

4. Radical weight savings as well as increased stiffness and safety are achieved by using carbon fiber and aluminum for the vehicle body and frame to keep the overall vehicle weight the same as for the current MY F-150 (due to the increase in powertrain weight from batteries and motors).
5. The use of carbon-fiber body components will be a challenge:
   a. A large supply of carbon fiber will be needed, along with a major production facility that can support the demand of a mass-market vehicle such as the F-150.
   b. Retooling of the vehicle production lines will be required to install the carbon-fiber components.
   c. Body shops and repair facilities will need to be trained to work with the carbon-fiber components to maintain consumer confidence that their vehicles can be repaired and maintained locally.

REFERENCE

Ludwick, T., M. Hubbard, and B. Manohar. 2016. Development of 2021 Ford F-150 Pickup. Projects P-1 to P-4 conducted for "AE 500: Automobile an Integrated System" course at the University of Michigan-Dearborn in winter term 2016.

25 Developing a Sports Utility Vehicle

A Case Study

INTRODUCTION

The purpose of this chapter is to provide a case study of the product development of a future model of a sports utility vehicle (SUV). The target vehicle selected for this exercise was the 2021 General Motors Truck Company (GMC) Acadia. Thus, the current model (2016 model year [MY]) of the Acadia was used as the reference vehicle, and the 2016 Ford Explorer and 2016 Honda Pilot were used as the competitive vehicles for benchmarking. Figure 25.1 presents exterior views of the reference and the two competitive vehicles.

The analyses presented here were originally prepared as a series of class projects by Subramanian et al. (2016) for the author's course entitled "Automobile: An Integrated System (AE 500)" taught in the 2016 winter term. Appendices 1 through 5 provide descriptions of the class projects. Selected portions of the project reports were modified and edited for inclusion in this chapter.

The outputs covered in this chapter include (a) customer characteristics, (b) customer needs, (c) market segment, (d) benchmarking of the vehicles, (e) specification of the target vehicle, (f) Pugh diagrams comparing the target vehicle characteristics with its reference and competitors' vehicles, (g) a technological plan, including changes, design challenges, and risks in the proposed vehicle development plan, (h) program timings and gateways, and (i) sales forecasts and financial analysis.

Technical, market, and customer information about the reference and benchmarked vehicles was collected from several sources, which included vehicle brochures, manufacturers' websites, car magazines, visits to local dealers, and customer interviews.

CUSTOMER CHARACTERISTICS AND NEEDS AND MARKET SEGMENT

CUSTOMER CHARACTERISTICS

The customers for SUVs can be classified primarily as millennials with families. They do not want a minivan. The customers are about evenly split between males and females. The rugged looks and sporty feel appeal to men, and females tend to like the ability to easily load and unload children and the large storage space in the rear. The drivers also like the high command seating position, which offers excellent visibility over the hood and the beltline. The majority of customers are college

2016 GMC Acadia SLT-2

2016 Ford Explorer Limited

2016 Honda Pilot Touring

FIGURE 25.1   Reference and two benchmarked vehicles used for developing 2021 GMC Acadia as the target vehicle.

educated and middle class, with an average annual income above about $55,000. They like to take long road trips with family/friends. They are typically well adapted to electronics, do not want to sacrifice style for utility, and are looking for good value for a family hauler, not for expensive extras. Because of their ruggedness, good handling, and power, some of these vehicles are also used for police work with special police package optional features.

CUSTOMER NEEDS

The customer needs of the 2021 model year SUVs are

1. Capacity to seat up to seven adults with adequate leg room in both second and third rows
2. Comfortable ride

3. Safety features for occupants in all rows
4. Good visibility from the driver's seat for driving and maneuvering in parking lots (i.e., command seating position)
5. Strong acceleration, handling, and braking capabilities (especially for the police package demands)
6. Cruise control for highway driving
7. Traction and stability control
8. Accident avoidance features (e.g., blind area sensors, forward collision automatic braking)
9. Mild off-road capabilities
10. Average fuel economy in mid- to upper 30s miles per gallon range
11. Climate control system capable of independent temperature and airflow controls in front and rear
12. Easy access to third row
13. Anchors for child car seats
14. Second and third row seats split to fold down independently
15. Large cargo area with second and third row seats folded down
16. Cargo tie down anchors throughout rear of vehicle
17. Storage bins and cup holders accessible from all seating positions
18. Roof rails for attaching and carrying cargo on the roof
19. Power and communication ports in all three rows of seats (12 V, USB, and 120 V)
20. Bluetooth connectivity for mobile/handheld devices (e.g., phones, tablets, music streaming devices)
21. Hands-free voice control for infotainment and climate control features
22. Ability to tow lighter loads (e.g., a small cargo trailer or jet ski)
23. Quiet interior
24. Stylish design
25. Quality product and very dependable
26. Good value (more features and quality for the price)

MARKET SEGMENT

The SUVs considered here belong within the affordable (nonluxury) large-size SUV market segment. At the high end of this segment are body-on-frame SUVs such as the Chevrolet Suburban, GMC Tahoe, and Ford Expedition. The basic market segment characteristics of these vehicles are (a) three-row seating (seven to eight occupants), (b) about 200 in. overall length, 75–80 in. overall width, and 70 in. overall height, (c) more than 80 ft^3 of cargo volume, and (d) greater than 30 mpg of combined fuel economy (in 2021 MY). These SUVs provide about 5000 lb of towing and about 150 ft^3 of interior passenger space.

Car-based SUVs (crossovers) provide better fuel economy and a more dynamic driving experience than truck-based (body-on-frame) vehicles. Higher-end trim levels provide additional luxury features for several thousand more dollars, but the base models offer the value side of this market segment. People shopping in this segment generally only want the utilitarian features without all of the expensive luxury add-ons. The majority of the vehicle sales are in North America, but these vehicles are also exported to Europe, China, and the Middle East. Explorers are also sold in North America to police departments (known as the Police Interceptor version).

DESCRIPTION OF THE TARGET VEHICLE

The important changes proposed for the development of the 2021 Acedia were as follows:

1. Engine downsizing to 2.5 L I4 with turbo-boost and slightly higher horsepower than the 2016 reference vehicle
2. Nine-speed automatic transmission
3. Reducing weight by about 800 lb
4. Meeting National Highway Traffic Safety Administration/ Environmental Protection Agency (NHTSA/EPA) fuel economy and emissions requirements for 2021 MY
5. Optional regenerative braking
6. Semiactive optional suspension
7. Advanced safety features (e.g., night vision system, lane-departure warning system)
8. Improved climate control system (three-zone temperature control)
9. Additional occupant comfort and convenience features

BENCHMARKING DATA

Table 25.1 provides data to compare the three 2016 MY SUVs with the 2021 target vehicle. The specifications for the 2021 vehicle were determined by the design team by considering (a) customer needs, (b) applicable government requirements, and (c) projected manufacturer's needs.

TECHNOLOGY PLAN

Table 25.2 presents the technology plan for the target vehicle. It presents changes in the major systems proposed for the target vehicle. Technological challenges and open issues that will be faced during the implementation of the changes are described briefly in the fourth and fifth columns of the table.

ASSESSMENT OF THE PROPOSED VEHICLE

This section presents three separate Pugh diagrams to compare the proposed 2021 MY vehicle and its current two competitors by using the reference vehicle as the "datum." The three Pugh diagrams presented in Tables 25.3 through 25.5 compare the vehicles based on customer needs, vehicle attributes, and vehicle systems, respectively. Comparison of the total scores presented in the bottom line of each of the tables shows that the 2021 MY vehicle would be substantially improved over the existing vehicles.

TABLE 25.1

Benchmarking Comparison of Characteristics of Three 2016 MY SUVs with 2021 Target Vehicle

Make, Model, Model Year, and Trim Level	Reference Vehicle GMC, Acedia, 2016, SLT-2 AWD	Competitor #1 Ford, Explorer, 2016, Limited 4WD	Competitor #2 Honda, Pilot, 2016, Touring AWD	Target Vehicle GMC, Acedia, 2021, SLT-2, AWD with Hybrid
Price (MSRP)	$44,295	$43,300	$42,970	$47,999
Curb weight (lb.)	4,850	4,629	4,303	4,055
Exterior Dimensions (in.)				
Wheelbase	118.9	112.8	111	111.4
Overall length	200.8	198.3	194.5	192.5
Overall width (including mirror)	78.9	90.2	81.8	75.4
Overall height	70.4	70	69.8	68.7
Track width	Front 67.3/rear 67.1	Front 67.0/rear 67.0	Front 66.3/rear 66.3	Front 64.5/rear 64.5
Turning center diameter (ft)	40.4	38.9	39.4	38.7
Ground clearance	7.6	8.3	7.3	7.6
Wheels	19" × 7.5" cast aluminum	20" premium painted aluminum	20" alloy	19" × 7.5" cast aluminum
Aerodynamics				
Drag coefficient (Cd)	0.36	0.35	0.3	0.32
Interior Dimensions (in.)				
Headroom (front)	40.3	41.4	39.5	40
Headroom (second row)	39.6	40.6	39.9	39.6
Headroom (third row)	38.4	37.8	38.9	37.2
Legroom (front)	41.3	42.9	40.9	40.9
Legroom (second row)	36.8	39.5	38.4	38.7
Legroom (third row)	33.2	32	31.9	31.1
Hip room (front)	58	57.3	59.1	55.7

(Continued)

TABLE 25.1 (CONTINUED)
Benchmarking Comparison of Characteristics of Three 2016 MY SUVs with 2021 Target Vehicle

Make, Model, Model Year, and Trim Level	Reference Vehicle GMC, Acedia, 2016, SLT-2 AWD	Competitor #1 Ford, Explorer, 2016, Limited 4WD	Competitor #2 Honda, Pilot, 2016, Touring AWD	Target Vehicle GMC, Acedia, 2021, SLT-2, AWD with Hybrid
Hip room (second row)	57.8	56.8	57.3	53.3
Hip room (third row)	48.3	40.7	44.6	42.9
Shoulder room (front)	61.6	61.5	62	59.4
Shoulder room (second row)	61	61	62	58.7
Shoulder room (third row)	57.8	50.8	57.6	54.3
Capacities				
Seating capacity	8	7	8	7
Interior passenger volume (cu. ft)	151.8	151.7	151.7	148.2
Fuel tank (gal.)	22	18.6	19.5	18.8
Maximum towing capacity (lb.)	5,200	5000	5000	5,000
Engine	3.6 L, V6, 24-valve, DOHC	3.5 L (TiVCT) V6	3.5 L, 24-Valve, SOHC i-VTEC® V-6	2.5 L inline four-cylinder turbo charged hybrid, four valves per cylinder
	Aluminum engine block variable valve timing	Cast aluminum engine block	Aluminum-alloy engine block	Aluminum-alloy engine block
	Spark-ignition direction-injection	Naturally aspirated	Direct injection	Variable valve timing
	Aluminum-alloy engine block	Sequential multi-port electronic fuel injection	Drive-by-wire throttle system	Spark-ignition direction-injection
			Eco assist™ system	Drive-by-wire throttle system
				Start/stop button technology
Horsepower	281 hp @ 6,300 rpm	290 hp @ 6,500 rpm	280 hp @ 6,000 rpm	285 hp @ 5,600 rpm
Torque	266 lb-ft @ 3,400 rpm	255 lb-ft @ 4,000 rpm	262 lb-ft @ 4,700 rpm	270 lb-ft @ 3,400 rpm

Emissions, CO_2 content	476 g/mile	475 g/mile	ULEV-2—405 g/mile	ULEV-2—230 g/mile
Fuel	Regular unleaded	Regular unleaded petrol/flex fuel	Regular unleaded	Regular unleaded
Fuel economy	16 city/23 highway/19 combined	16 city/23 highway/19 combined	19 city/26 highway/22 combined	36 city/44 highway/40 combined
Drivetrain	Hydra-matic 6-speed automatic with manual mode (6T75)	6-speed selectShift® automatic	9-speed automatic transmission	9-speed automatic transmission with paddle shifters
	AWD	Intelligent 4WD	AWD	AWD
Exterior features	Luggage rack, side rails, roof-mounted	Black exterior door handles	OneTouch turn indicators	Distinctive grill designs
	Rear spoiler	Black lower front and rear bumpers, wheel-lip moldings, and body-side cladding	Rear privacy glass	IntelliBeam automatic headlamp high-beam control
	Moldings, body-color body-side	Body-color liftgate spoiler	LED brake lights	Panoramic roof—optional
	Headlamps, dual halogen projector lamp	Grill, dark foundry gray with chrome bars	Remote entry system	Wraparound rear glass
	Headlamp control, automatic on and off	Headlamps, bifunctional projector-beam halogen	Body colored, heated, power side mirrors, including integrated LED turn indicators and expanded view driver's mirror	All-terrain model that features a unique sporty and stealthy appearance
	Fog lamps, front, halogen projector beam	LED tail lamps	Security system	Available projector-beam headlamps with LED daytime running lamps

(Continued)

TABLE 25.1 (CONTINUED)
Benchmarking Comparison of Characteristics of Three 2016 MY SUVs with 2021 Target Vehicle

Make, Model, Model Year, and Trim Level	Reference Vehicle GMC, Acedia, 2016, SLT-2 AWD	Competitor #1 Ford, Explorer, 2016, Limited 4WD	Competitor #2 Honda, Pilot, 2016, Touring AWD	Target Vehicle GMC, Acedia, 2021, SLT-2, AWD with Hybrid
	Mirrors, outside heated, power-adjustable, power-folding and driver-side auto-dimming, body-color with integrated turn signal indicators and driver-side memory	Automatic LED low-beam headlamps	Chrome door handles	C-shaped signature GMC lighting
	Glass, solar-ray deep-tinted (all windows except light-tinted glass on windshield and driver and front passenger side glass)	Liftgate appliqué—bright	Smart entry	Halogen fog lamps with projector technology
	Wipers, front intermittent with washers	Mirrors—black (MIC), power electric remote, manual folding with integrated blind spot mirrors (integrated blind spot mirrors not included when equipped with BLIS)	Fog lights	LED tail lamps
	Wiper, rear intermittent with washer	Roof side rails, black with black end caps	OneTouch power moonroof with tilt feature	Roof rails, body-colored cladding, and chrome door handles and exhaust tips
	Door handles, chrome (bright beltline molding)	Acoustic-laminate windshield glass	Power tailgate	Hands-free power liftgate

Interior features			
Body, power rear liftgate	Privacy glass, second and third rows	Acoustic windshield and front door glass	Rain sensing front/rear wipers
	Solar tinted front door and windshield glass	Roof rails	Rear spoiler
			UV-reducing solar glass
		Body-colored parking sensors (front/rear)	Body-colored, heated, power side mirrors, including integrated LED turn indicators and expanded view driver's mirror
			Acoustic windshield and front door glass
Console, front center with two cup holders and storage	Air filtration system	Push-button start	Heated leather-wrapped steering wheel with real aluminum trim
Cup holders: 10 with (ABB) 7-passenger (2-2-3 seating configuration) and 12 with (ABC) 8-passenger (2-3-3 seating configuration)	Manual single-zone climate control	Power windows with auto-up/down driver's and front passenger's window	Real wood accents
Floor mats, carpeted front, second and third row, removable	Rear auxiliary climate control	Power door and tailgate locks	Tri-zone automatic climate control system with humidity control and air filtration
Insulation, acoustic package	SYNC® with MyFord®	Cruise control	Power windows with auto-up/down driver's and front

(Continued)

TABLE 25.1 (CONTINUED)
Benchmarking Comparison of Characteristics of Three 2016 MY SUVs with 2021 Target Vehicle

Make, Model, Model Year, and Trim Level	Reference Vehicle GMC, Acedia, 2016, SLT-2 AWD	Competitor #1 Ford, Explorer, 2016, Limited 4WD	Competitor #2 Honda, Pilot, 2016, Touring AWD	Target Vehicle GMC, Acedia, 2021, SLT-2, AWD with Hybrid
	Steering wheel, heated, leather wrapped with (UK3) steering wheel–mounted audio controls	AM/FM single-disc CD player, MP3 compatible with six speakers	Tilt and telescopic steering column	Power windows with driver's one-touch auto-up/down feature
	Steering column, tilt and telescopic with brake/transmission shift interlock	Media hub with one smart charging USB port	Multifunctional center console storage	Cruise control with steering wheel–mounted switches
	Instrumentation, five-gauge with enhanced driver information center, outside temperature indicator, and digital compass display	12 V power points, four total (two first row, one second row, and one cargo area)	Lockable glove compartment	Steering column, tilt and telescopic with brake/transmission shift interlock
	Power windows	Cargo hooks	Sliding sun visors	Heated leather-wrapped steering wheel with cruise control, five-way controls, and secondary audio controls
				Passenger-assist grab handles
	Door locks, power programmable with lockout protection	Color-keyed, carpeted floor mats, front and rear	Remote fuel filler door release	Beverage holders—front row (five), second row (four), third row (four)
	Remote vehicle start	Cruise control	Rear-window defroster with timer	Multifunctional center console storage
	Cruise control	Daytime running lights	Rear-seat heater ducts	

Universal home remote	Dome/map lights in all three rows	Cargo area light	Lockable glove compartment
Air conditioning, tri-zone automatic climate control with individual climate settings for driver and right-front passenger and second/third row controls	Easy fuel® capless fuel filler	Hidden storage well	Integrated sunshades (second row)
Rear-window electric defogger	First-row center floor console with wrapped armrest and storage bin	Illuminated steering wheel–mounted controls for cruise/audio/phone	Air filtration system
Rearview auto-dimming mirror	Front and rear scuff plates embossed with word "EXPLORER"	Beverage holders—front row (five), second row (six), third row (four)	Driver's and front passenger's illuminated vanity mirrors
Visors, driver and front passenger, padded with cloth trim, color-keyed and illuminated vanity mirrors	Integrated key-fob	Capless fuel filler	Capless fuel filler
Lighting, interior with theater dimming, interior ambient light pipe in instrument panel, cargo compartment, reading lights for front seats, second-row reading lamps integrated into dome light, door- and tailgate-activated switches, and illuminated entry and exit feature	Leather-wrapped shift knob	Map lights (all rows)	Cargo area light

(Continued)

TABLE 25.1 (CONTINUED)
Benchmarking Comparison of Characteristics of Three 2016 MY SUVs with 2021 Target Vehicle

Make, Model, Model Year, and Trim Level	Reference Vehicle GMC, Acedia, 2016, SLT-2 AWD	Competitor #1 Ford, Explorer, 2016, Limited 4WD	Competitor #2 Honda, Pilot, 2016, Touring AWD	Target Vehicle GMC, Acedia, 2021, SLT-2, AWD with Hybrid
	Trim, interior, aluminum on instrument panel and steering wheel	Manual day/night rearview mirror	Tri-zone automatic climate control system with humidity control and air filtration	Trim, interior, aluminum on instrument panel and steering wheel
	Storage system, rear cargo area under floor	Manual tilt/telescoping steering column	HomeLink® remote system	Sliding sun visors
	OnStar guidance plan for 6 months, OnStar with 4G LTE and built-in Wi-Fi hotspot	MyKey®	Driver's and front passenger's illuminated vanity mirrors	Rearview auto-dimming mirror
		Overhead console with dome/map lights and sunglasses holder	Conversation mirror with sunglasses holder	Manual day/night rearview mirror
		Passenger-assist grab handles (one at front passenger seat, two in second row)	115 V power outlet	Universal home remote
		Power door locks	Integrated sunshades (second row)	Cargo area bag hooks
		Power windows with one-touch-up/-down driver's window	Automatic-dimming rearview mirror	OnStar with 4G LTE and built-in Wi-Fi hotspot

Seating			
Seat trim, leather-appointed seating on first and second rows	Rear view camera with washer	Leather-wrapped steering wheel	Divide-N-hide® cargo system (one divider with family package)
Seating, seven-passenger (2-2-3 seating configuration) (may be substituted with [ABC] eight-passenger (2-3-3 seating configuration), heated driver and front passenger seats	Sun visors, color-keyed, single blade with covered vanity mirrors	Courtesy door lights (Front-row)	10-way power driver and front passenger seats
	Steering wheel with cruise control, five-way controls, and secondary audio controls	Blue ambient LED lighting	One-touch second-row seats
Seat, eight-way power driver with power recline and lumbar control and memory	Multi-level heating driver and passenger seat	60/40 split, flat-folding third-row bench seat	Standard second-row heated seats
	Eight-way power driver seat and passenger seat	Adjustable front seat-belt anchors	
	Second-row 60/40 split-fold-flat bench seat	Head restraints at all seating positions	One-touch second-row seats
Seat adjuster, two-position memory for driver's seat and outside rearview mirrors, passenger eight-way power	Third-row 50/50 split-fold bench seat, manual	Driver's seat with 10-way power adjustment, including power lumbar support with two-position memory	
	Height-adjustable driver seat		

(Continued)

TABLE 25.1 (CONTINUED)
Benchmarking Comparison of Characteristics of Three 2016 MY SUVs with 2021 Target Vehicle

	Reference Vehicle	Competitor #1	Competitor #2	Target Vehicle
Make, Model, Model Year, and Trim Level	GMC, Acedia, 2016, SLT-2 AWD	Ford, Explorer, 2016, Limited 4WD	Honda, Pilot, 2016, Touring AWD	GMC, Acedia, 2021, SLT-2, AWD with Hybrid
	Reclining front buckets	Bucket front seats	One-touch second-row seats	Heated and ventilated driver and front passenger seats
		Height-adjustable passenger seat	Front passenger's seat with four-way power adjustment	Height-adjustable passenger seat
		Passenger seat with power adjustable lumbar support	Leather-trimmed interior	Second-row 60/40 split-power folding-flat bench seat
				Third-row 50/50 split-power folding-flat bench seat
		Driver seat with power-adjustable lumbar support	Heated front seats	Optional rear seat alert
		Split-folding rear seat back		
		Power folding split-bench third-row seats		
Audio/Entertainment	Color Touch radio with IntelliLink AM/FM/SiriusXM radio with CD player	Sony premium brand stereo system	Bluetooth® handsFreeLink®	Apple CarPlay and Android Auto compatibility4 8" diagonal color touch radio
	6.5" diagonal touch-screen display	12 total speakers	Bluetooth® streaming audio	Bose® audio system with AM/FM/CD
	USB port and auxiliary input jack	390 W stereo output	MP3/auxiliary input jack	Ten speakers, including two woofers

Audio system feature, Bose premium 10-speaker system with subwoofer	Satellite radio	MP3/WMA playback capability	Streaming audio via Bluetooth® wireless technology
USB port, dual charge only, located rear of center console	One subwoofer	RDS	Audio memory system
Audio system controls, rear with headphone jacks	6 months satellite radio service provided	SVC	SD-card slot
SiriusXM satellite radio	USB connection	HondaLink®	Radio data system
	Sony premium brand speakers	SMS text message function	Video monitor
	Video monitor	Pandora® compatibility	USB ports with 2.5 A charging capacity in all three rows
	AM/FM stereo	SBV	
	Surround audio (discrete)	SiriusXM® radio	Electrostatic touch screen and customizable feature settings
	Memory card slot	Second-row HDMI™ interface	Satellite navigation system
	Auxiliary audio input and USB with external media control	Audio interface with charging (one USB port in center console)	SMS text message function
		Audio interface with charging (two USB ports in front)	SVC
		Ports with high-speed charging (two USB ports in second row)	

(Continued)

TABLE 25.1 (CONTINUED)

Benchmarking Comparison of Characteristics of Three 2016 MY SUVs with 2021 Target Vehicle

Make, Model, Model Year, and Trim Level	Reference Vehicle GMC, Acedia, 2016, SLT-2 AWD	Competitor #1 Ford, Explorer, 2016, Limited 4WD	Competitor #2 Honda, Pilot, 2016, Touring AWD	Target Vehicle GMC, Acedia, 2021, SLT-2, AWD with Hybrid
			540 W audio system with 10 speakers, including subwoofer	
			8 in. display audio with high-resolution WVGA (800 × 480) with electrostatic touch screen and customizable feature settings	
			Honda satellite-linked navigation system™ with voice recognition and Honda HD digital traffic	
Security	Theft-deterrent system, vehicle, PASS-Key III, engine immobilizer	Battery-saver feature	Smart entry and programmable remote entry	Keyless open and start
	Remote keyless entry, programmable with two transmitters, panic button, and extended range	Perimeter alarm		Tow vision camera system
	Universal remote transmitter (for garage door, security system)	Remote keyless/illuminated entry		Face recognition keyless entry system

SecuriCode™ keyless entry keypad

Theft-deterrent system, vehicle, PASS-Key III, engine immobilizer

Universal remote transmitter (for garage door, security system)

SecuriLock® passive anti-theft system

Note: 4WD, four-wheel drive; AWD, all-wheel drive; BLIS, blind area information system; DOHC, double overhead camshaft; HDMI, high-definition multimedia interface; LED, light-emitting diode; LTE, long-term evolution; MIC, media company; RDS, radio data system; SOHC, single overhead camshaft; SVC, speed-sensitive volume control; TiVCT, twin independent variable camshaft timing; WMA, Windows Media® Audio; WVGA, wide video graphics array; ULEV, ultra low emissions vehicle.

TABLE 25.2

Technology Plan

Sl No.	Vehicle System	Major Changes Planned	Major Technological Challenges	Key Open Issues/Recommendations
1	**Powertrain System**			
	Engine	2.5 L four-cylinder combined with a hybrid motor to maintain emission levels as per CAFE and NHTSA standards for MY 2021	The hybrid motor should be well tested on the SUV, as buyers like a rugged vehicle in this segment. The new technology should be not just trouble free, but also rugged for excessive off-road driving.	Newer engine should be developed with future emission regulation considerations. Hybrid motor supplier must design battery package under the specified weight as per plan. BYD will make the battery package.
		Turbo charging to increase responsiveness and quicker acceleration	Packaging of the new turbo charging introduces new plumbing and attachment problems.	Turbo charger suppliers must have a suitable hardware for the specified engine capacity, and the product must integrate with the engine layout. Honeywell and Garett suppliers are being consulted.
		Adaptable engine shut-off system	The ECU, coupled with a new hybrid control system that switches between a gasoline engine and an electric motor, requires a long period of development and could cause some issues even in the later stages of development.	Vehicles in the GMC's product lineup do not have an adaptable engine shut off system; hence, new suppliers have to be sourced.OMRON would be approached.
		Stop/start engine shut-off feature to improve fuel efficiency	New feature requires larger battery, more durable starter motor, and additional wiring.	Packaging of additional new wiring and a different starter motor. A team from Denso's electrical integration will be responsible.
		Variable valve timing for varying the engine intake and exhaust timings for a balance between power and efficiency	The variable valve timing unit, though a very good feature, increases complexity of the intake and exhaust systems and is complex to repair.	The variable valve timing system is already developed by GMC for 2017 range of vehicles (thus, no major issues in feature incorporation).

	Low-friction bearings for reduced friction and increased efficiency	The current manufacturing plant does not have a low-friction manufacturing unit. In-house manufacturing may not be possible.	Manufacturing of low-friction bearings requires advanced machinery. Trade-off between additional costs, such as labor and cost efficiency, should be analyzed.
Transmission	Nine-speed automatic transmission with paddle shifters	Nine-speed transmission will be more complex than the six-speed due to increased moving components, and frequent upshifts and downshifts can cause more wear in the future, though drivability increases.	Packaging is a challenge. Will have to be studied for feasibility. The cost will be higher, as the new gearbox will also include the development costs. Consider a joint venture with another manufacturer to share the costs.
2 Chassis System			
Suspension system	Multilink independent rear with semiactive air suspension and continuous damping system	Air suspension systems are very complex and require frequent maintenance.	The new electronic suspension may be imported from a manufacturer from different country and could affect parts availability.
Steering system	Electro-hydraulic variable-effort power steering with active return	Newer steering system could be a tight fit, and packaging could be an issue, as the cavity for steering assembly needs to house additional components. Serviceability could also be harder.	The supplier who will supply the steering system hasn't filed a patent yet, which might affect vehicle testing, as the final report containing the details might vary.
Brakes system	Regenerative braking with active engine braking system	New regenerative braking system will require the braking system to be equipped with additional electrical components.	Suppliers will be finalized between Bosch and Brembo.
Front and rear axle	Four-wheel drive system with auto terrain adapt	The four-wheel drive system will have an electronic terrain adapt system. It has to be packaged in such a way that water or slush should not affect it during off-road applications.	Supplier yet to be decided.

(Continued)

TABLE 25.2 (CONTINUED)
Technology Plan

Sl No.	Vehicle System	Major Changes Planned	Major Technological Challenges	Key Open Issues/Recommendations
3	Wheels and tires	Wheel size will be increased to 19"×7.5" cast aluminum wheels	Newer intricate wheel design could have many joints, and manufacturing it is a challenge.	Newer rim size could mean costlier rims, so must be made in an attractive design and must be diamond cut for people to pay the extra price for the rims. Supplier such as Fuchs will be chosen as OEM alloy maker.
		Low-rolling resistance tires	The low-resistance tire's off-road capability is still under development and might not be as capable as off-road tires.	The current supplier must have a low–rolling resistant tire in the size required for MY 2021 GMC Acadia. The tire supplier can be Pirelli.
4	Fuel System	Introduction of a capless fuel filter	This new system, though implemented, is still under development and requires a slightly advanced manufacturing process.	Supplier needs to be decided, as it is a new design.
	Exhaust System	System carryover with extra insulation	The exhaust pipe heat has to have additional packaging to prevent heat from the exhaust gas affecting the hybrid system.	The same system can be modified to be used on the 2021 ACADIA.
5	Body Systems BIW	Overall vehicle weight reduction, as the body will be redesigned.	Lightweight and a stiffer body, requires a change in material used for the body, which in turn increases manufacturing costs and material costs.	Lighter body with lightweight materials with increased strength will increase the costs. GMC plant will be updated with newer chassis analysis equipment.
	Closures System	Kick-activated liftgate system	New kick-activated system in the tailgate requires additional wiring, sensing system—increases overall complexity.	As it is a newly developed system used only by Ford, suppliers who can make the system need to be identified.

	System	Feature	Challenge	Comment
	Seat system	One-touch second-row adjustable seats with memory	Additional wiring in the floor and seat frames will make the seat detachment difficult.	Suppliers currently quote very high prices for the unit. Possibility of development within GMC to be considered.
6	**Electrical System**	Addition of hybrid motor with regenerative braking system circuit	The whole wiring cavity will have to be redesigned to accommodate newer electronic integration.	Possibility of eliminating or simplifying some other system could make integration of this system easier.
7	**Lighting System**	IntelliBeam automatic headlamps and LED tail lamps	Automatic headlamp units may require a different current flow regulation when compared with conventional 12 V system. Redesign of the electrical lines will be expensive and time consuming.	Supplier finalized as North American Lighting.
8	**Electronic Control System**	Additional intelligent ECU for hybrid motor cutoff switch	Lack of space in current instrument panel may require modification in the layout. Need space to package additional ECU.	Advanced vehicle electronic system could make the reliability levels take a dip if not engineered properly.
9	**Driver Interface** Infotainment system	a) Apple CarPlay and Android Auto compatibility, 8" Touch interface	The center stack in the dashboard should be designed to accommodate a large 8 in. unit. Packaging is also an issue.	Google and Apple are approached to handle software development.
		b) Satellite navigation system		
		c) Pedestrian detection and night vision	The pedestrian detection system and its control unit require access space for servicing. It may violate the minimum clearance space required between sensors and wiring.	Will increase the overall cost, as the system requires infrared and cameras to be installed. Cost versus convenience trade-off to be examined.

(Continued)

TABLE 25.2 (CONTINUED)
Technology Plan

Sl No.	Vehicle System	Major Changes Planned	Major Technological Challenges	Key Open Issues/Recommendations
10	**Climate Control System**			
		Tri-zone automatic climate control system with humidity control and air filtration		Overall interior packaging levels must be higher to maintain temperature zones in the three rows of seating.	Implementation of a tri-zone climate control will be a trade-off between cost and comfort.
11	**Safety Systems**			
	Air bag system	Air bags—frontal and side—for driver and front passenger and extra curtain airbags		Due to the decrease in overall length, the space between one air bag's deployment and the other gets changed, which requires rework in the airbag inflation space.	Newer air bag system will be made by Takata. Trade-off between cost and safety to be examined. As it is a new system, suppliers need to be identified.
	Seat belt system	Three-point seat belts at all seating positions		Incorporating seat belt air bag in the new vehicle requires analysis and compliance with safety standards, which requires revising of the measurements.	More testing should be done so as to prevent injuries to occupants from seat belt air bag.
	Wiper and defroster	Rain-sensing front / rear wipers		The rain-sensing module will change the regular windshield assembly and sealing and makes the windshield manufacturing, assembly, and replacement harder.	Sensing units as per the windshield design to be bought from the supplier Saint Gobain.
12	**Security System**			
		Theft-deterrent system with face recognition and keyless entry system		The new security systems will alter the current load on the electrical circuit and require redesign. Also installation and programming of the face recognition system costs a lot of capital to develop, install, and register.	Could be considered very complex by senior citizens, and the system needs to be tested for face recognition during rainfall and snow conditions.

Note: CAFE, corporate average fuel economy; ECU, electronic control unit; OEM, original equipment manufacturer.

TABLE 25.3
Pugh Diagram of Customer Needs

Customer Needs	GMC Acadia 2016 (Reference Vehicle)	Ford Explorer 2016	Honda Pilot 2016	GMC Acadia 2021 (Target Vehicle)
Comfort		S	S	+
Convenience		+	S	+
Low cost of maintenance		+	+	S
Appealing aesthetics		+	S	+
Commanding seating position		–	+	S
Noise isolation		S	S	S
Sufficient interior storage		S	+	+
Less emission		S	+	+
Good fuel mileage		S	+	+
Good reliability		+	+	+
Must be ergonomically designed		S	S	+
Powerful engine		+	+	+
Comfortable ride		S	S	S
Safety		S	–	+
Large cargo space		–	–	–
Security		+	–	+
Total (+)		6	7	11
Total (–)		3	3	1
Total S		7	6	5
Total score		3	4	10

Table 25.3 shows that the total score of the target vehicle is 10 based on the customer needs, and it is substantially higher than the total scores of the three 2016 MY SUVs. The target vehicle can be further improved for customer needs in the following areas: noise isolation, comfortable ride, and low maintenance costs.

Table 25.4 presents a weighted Pugh diagram for comparison of the proposed vehicle with the three benchmarked vehicles on important vehicle attributes. The sum of the weighted scores shown in the last row of the table is highest for the target vehicle. However, the target vehicle can be further improved in the following vehicle attributes: package, costs, and product and process compatibility.

From the total scores data provided in Table 25.5 for vehicle systems, the target vehicle has a very high score of 18 in comparison with the three benchmarked vehicles. However, it should be noticed that both the 2016 MY competitors, the Ford Explorer and the Honda Pilot, have a higher total score of 7 over the datum, and they will continue to be tough competitors when the 2021 MY vehicles are introduced into the market. Thus, every effort should be made to improve many vehicle systems of the target vehicle, especially those where it has received a rating of S (same as the datum) (e.g., climate control system).

TABLE 25.4

Pugh Diagram of Vehicle Attributes

Vehicle Attributes	Importance Rating	Importance Weightage	Datum—GMC Acadia 2016 (Reference Vehicle)	Ford Explorer 2016	Honda Pilot 2016	GMC Acadia 2021 (Target Vehicle)
Package	5	0.0435		+1	+2	+1
Ergonomics	7	0.061		+1	+2	+2
Safety	10	0.0870		+3	+1	+4
Styling and appearance	9	0.0783		+2	+1	+2
Thermal and aerodynamics	5	0.0435		-2	0	+2
Performance and drivability	10	0.0870		+2	+3	+3
Vehicle dynamics	8	0.0696		0	+1	+2
Noise, vibrations, and harshness (NVH)	8	0.0696		-1	-1	+1
Interior climate comfort	7	0.0609		0	0	+1
Weight	5	0.0435		+1	+2	+3
Security	8	0.0696		+2	-2	+3
Emissions	3	0.0261		0	+1	+4
Communication and entertainment	8	0.0696		+1	+2	+3
Costs	10	0.0870		+1	+2	-3
Customer life cycle	6	0.0522		+2	+1	+2
Product and process complexity	6	0.0522		+1	+2	-2
Sum or weighted sum	115	1		1.0349	1.0783	1.6782

TABLE 25.5
Pugh Diagram of Vehicle Systems

Vehicle System	Subsystem	Datum—GMC Acadia 2016 (Reference Vehicle)	Ford Explorer 2016	Honda Pilot 2016	GMC Acadia 2021 (Target Vehicle)
Body system	Body-in-white		S	S	S
	Closures system		S	S	S
	Seat system		S	S	S
	Instrument panel		S	S	S
	Exterior lamps		S	S	+
	Glass system		S	+	+
	Rear vision system		S	S	S
Chassis system	Underbody framework		S	S	S
	Suspension system		S	S	+
	Steering system		S	S	S
	Braking system		S	+	+
	Wheels and tires		S	S	S
Powertrain system	Engine		+	+	+
	Transmission		S	+	+
	Shafts and joints		S	S	S
	Final drive and axles		S	+	+
Fuel system	Fuel tank		–	–	–
	Fuel lines		S	S	S
Electrical system	Battery		S	S	+
	Alternator		+	-	+
	Wiring harnesses		S	S	S
	Power controls		S	S	S
Climate control system	Heater		S	S	S
	Air conditioner		S	S	S
	Climate controls		S	S	S
Safety and security system	Air bag system		+	S	+
	Seat belt system		+	S	+
	Wiping and defroster systems		S	S	+
	Security lighting and locking system		+	S	+
	Driver assistance systems		+	+	+
Driver interface and infotainment system	Primary and secondary vehicle controls and displays		S	+	+
	Audio system		S	+	+
	Navigation system		+	+	+
	CD/DVD Player		S	–	–
Total (+)			10	11	21
Total (–)			3	4	3
Total (S)			29	27	18
Total score			7	7	18

TABLE 25.6
2021 Acadia Program Milestones

Month-Year	Code	Definition of Gateway	Description of Gateway
Apr-16	PD	Program definition	GMC Acadia MY2021 program defined
Sep-16	PKO	Program kick-off	Program proposal accepted by senior management. Program and team leaders selected
Oct-16	TF	Team formation	Teams formed for various vehicle systems and chunks. Benchmarking of selected competitive vehicles initiated
Dec-16	TS	Targets set	Functional specifications targets are set. Several concept designs created
Feb-17	CR	Concepts reviewed	Vehicle concepts selected by senior management reviewed with benchmarks
Apr-17	CS	Concept selection	Market research data and program team recommendations reviewed by senior management and concept selected for further development
Jun-17	EL	Engineering launch	System-level design initiated considering functional and engineering feasibility
Dec-17	SD	System design approval	Managers of various engineering offices review and approve the design
Apr-18	ES1	Engineering sign-off	Team leaders sign off current design for further detailed engineering work
Aug-18	PA	Program approval	Vehicle program reviewed and approved by senior management. Funds released
Oct-18	VT1	Verification tests 1	Initial verification tests conducted. System- and subsystem-level tests conducted and results incorporated in the design. Tests also conducted on benchmark vehicle components for comparison
Apr-19	PTV	Prototype test vehicles	Prototype vehicles developed for vehicle-level verification tests
May-19	VT2	Verification tests 2	Further component testing done and results incorporated in the design
Dec-19	PF	Final prototypes	Final production prototype vehicles developed and reviewed by experts
Feb-20	VT3	Verification tests 3	Final testing done to ensure durability for desired service life
Mar-20	MFS	Marketing and field support plan launch	Marketing field support personnel are provided with required information and tools
May-20	PR	Production readiness	Manufacturing plants retooled and tested for vehicle build
Nov-20	MS	Final manufacturing sign-off	Manufacturing plant managers sign off on the functionality and build quality

(Continued)

TABLE 25.6 (CONTINUED)
2021 Acadia Program Milestones

Month-Year	Code	Definition of Gateway	Description of Gateway
Nov-20	ES2	Engineering sign-off 2	All team leaders sign off on the functioning, reliability, and durability of the vehicles
Nov-20	JB1	Job 1	Management approves release of the production vehicles for sale
Jan-21	PSR	Program status reviews	Customer feedback, sales, warranty, and costs reviewed periodically
Sep-25	STOP	Termination of production	Production of vehicle terminated. Cars sold from inventory
Jan-26	END	Termination of project	Project terminated. Serviceable parts maintained in inventory to support in-use vehicles, plant retooled for production of other vehicles

PROGRAM TIMINGS, SALES, AND FINANCIAL PROJECTIONS

PROGRAM TIMINGS

The vehicle program to develop the proposed 2021 GMC Acadia was estimated to begin with the formation of the core product development team at approximately 50 months before Job #1. Job #1 is scheduled on November 20, 2020. The program timings and gateways are shown in Table 25.6.

PROJECTED SALES

It is estimated that the company will sell about 200,000 vehicles per year. The manufacturer's suggested retail price (MSRP) of the 2021 Acadia is estimated to be about $48,000.

FINANCIAL PROJECTIONS

Over the estimated 4 year sales period (MY 2021 to 2025), the total sales of the GMC Acadia are estimated at 726,000 vehicles with a total revenue of about $31.0 billion and net earnings of about $18 billion. The maximum cumulative cost incurred in the total Acadia program is estimated at about $2.0 billion just prior to Job #1. The cost breakeven point is estimated to occur about 3 months after Job #1.

CONCLUDING REMARKS

The goal of the outputs presented in this chapter was to provide the students with an understanding of the early part of automotive product development by performing the data-gathering and decision-making tasks. The projects required the students to develop specifications for the 2021 GMC Acadia by undertaking benchmarking of the latest available models of the GMC Acadia, Ford Explorer, and Honda Pilot, studying applicable government regulations, and researching future trends in design and technologies.

It should be realized that the data provided in the tables included in this chapter were gathered by the students by searching through various sources, such as vehicle brochures, websites of the vehicle manufacturers, and automotive magazines, in a very limited time (all projects presented in Appendices 1 through 5 were completed within one semester by a team of no more than four graduate students). The Pugh analyses were conducted using data gathered under conditions of severely restricted time and scarcity of benchmarked vehicles. Similarly, the financial analyses were conducted using a number of assumptions regarding items such as time required to perform various design tasks, manpower needs, and pay rates and costs associated with different evaluations. Therefore, the data and the information provided in this chapter are expected to be approximate and crude. However, from the viewpoint of educational value, it was felt that readers of the book and students should understand the need for and use of such data during the development of a new automotive product.

In addition to the observations included in the concluding remarks section of Chapter 23, the important observations from the projects are briefly described here:

1. The SUV segment covered in this chapter is an important and different type of automotive product. SUVs carry more passengers, have greater load-carrying capacity, and provide more flexibility in seating as compared with passenger cars.
2. Three major changes proposed to increase the fuel efficiency of the SUV were (a) engine downsizing, (b) implementation of a hybrid powertrain, and (c) reducing the overall weight of the vehicle.
3. The market share of SUVs has been steadily growing, and thus, this represents a substantially lucrative opportunity to increase future profits.

REFERENCE

Subramanian, A., V. Manoharan, U. Sundar, and S. Srinivasan. 2016. Development of 2021 GMC Acadia. Projects P-1 to P-4 conducted for "AE 500: Automobile an Integrated System" course at the University of Michigan-Dearborn in winter term 2016.

Appendix I: Benchmarking and Preliminary Design Specifications

OBJECTIVES

1. To conduct a benchmarking study of competitor vehicles with the reference vehicle selected for the AE 500 project work
2. To develop preliminary specifications of the target vehicle
3. To develop Pugh diagrams to determine how the target vehicle and benchmarked vehicles compare with the reference vehicle

PROCEDURE

a. Select a reference vehicle (for your 2021 model year [MY] target vehicle to be studied/developed for all projects). The reference vehicle must be a recent model year (2015 or 2016 MY) vehicle sold in the U.S. market.

b. Select at least two other recent vehicles that are leading competitors of your reference vehicle in the *same market segment* for benchmarking.

c. Conduct a benchmarking exercise (i.e., comparing the two competitor vehicles with the reference vehicle by considering all important dimensions and vehicle features and determining improvements needed to define the target vehicle). Collect data on vehicle dimensions and features from Internet searches (e.g., vehicle manufacturer's brochures), articles in automotive magazines and journals, visiting dealerships, the Detroit auto show, and your own measurements, observations, and photographs of the vehicles and their chunks and systems, for side-by-side comparisons (i.e., photo-benchmarking).

   Prepare a benchmarking table comparing the reference vehicle with its two benchmarked comparators based on exterior and interior dimensions, features, and characteristics of their corresponding systems (see Table 1.1 for listing of vehicle systems) and available standard and optional equipment.

d. *Describe the market segment* based on the characteristics of the target vehicle, its customers, and its market location.

   The market segment can be defined using categories such as vehicle size (subcompact, compact, intermediate, large), body-style (sports car, sedan, coupe, station wagon, sports utility vehicle [SUV], van, pickup truck), economy/entry-luxury/luxury/ultra-luxury, and countries where the vehicle will be sold and used. Other characteristics to consider are the seating capacity, weight, and cargo/trunk volume of the target vehicle, and its customers (i.e., owners and principal users). Also see https://en.wikipedia.org/wiki/Car_classification, https://en.wikipedia.org/wiki/Truck_classification, and https://en.wikipedia.org/wiki/Commercial_vehicle

The customer characteristics to consider are male-to-female ratio, ages, education, professions, life-stage (e.g., single/married, student, nonstudent/working, retired), lifestyle (e.g., activities for which they would like to use the vehicle: commuting to work, vacation trips, trips with kids), income, make and models of other vehicles in the household, and so forth.

e. *Prepare a list of customer needs*, that is, the needs of the anticipated customers of the target vehicle.

To prepare the list, observe how the reference vehicle and the benchmarked vehicles are used (e.g., for commuting, long trips with family, or hauling work equipment/materials) in the real world (e.g., observe in traffic, parking lots, gas stations, rest areas, airports, etc.). Also, talk to a few users/owners of the vehicles about their needs and the problems and features they like, dislike, and want in these vehicles. Based on the information gathered, develop a list of customer needs for your target vehicle and group the customer needs into categories of vehicle uses.

The customer needs list must cover all major customer wants. (Note: This list should be comprehensive and as complete as possible to define all the characteristics and features of the vehicle. Note that you will later use the customer needs to develop attribute and subattribute level requirements for the vehicle.)

f. *Target vehicle specifications*: Add a column (for 2021 MY target vehicle) to the benchmarking table prepared in part (d), and add rows to include a brief description and characteristics of each *major system* of the vehicle (see Table 1.1) along with the preliminary *specifications* of your target vehicle, the current reference vehicle, and the benchmarked competitors.

The specifications should cover the entire vehicle and all its major vehicle systems. The information provided in this column should be sufficient to communicate basic information on each vehicle system (such as system type/configuration [e.g., type of engine: 2.3 L gasoline with turbo-boost, McPherson suspension in the front and trailing arm suspension in the rear] and its capabilities [e.g., 240 hp, 0–60 mph acceleration in 8.0 s, 60–0 mph deceleration within 125 feet braking distance]) to the design engineers of each vehicle system.

The specifications should also include (a) exterior dimensions (e.g., overall length, width and height, wheelbase, front and rear track width, and ground clearance), (b) interior dimensions (e.g., legroom, headroom, shoulder room, and hip room for each seat row; luggage space volume), (c) capacities (e.g., engine size, horse power, torque), (d) characteristics (e.g., curb weight, front and rear suspension type, brake type and size), (e) capabilities (e.g., 0–60 mph acceleration time, 60–0 mph braking distance, miles per gallon city/highway/combined, and CO_2 emissions in grams per mile), and (f) features (e.g., navigation system, smart headlamps, lane-departure warning system). Thus, this last added column should include complete specifications of your target vehicle.

Refer to the Environmental Protection Agency/National Highway Traffic Safety Administration (EPA/NHTSA) final rule on fuel economy and emissions requirements (see Figures 3.1 through 3.4) and determine the minimum requirements (fuel consumption and emissions) for your target vehicle. The target vehicle *must meet the requirements based on its footprint*. Include the target vehicle footprint data and these requirements in your specification table.

g. *Develop three separate Pugh diagrams* to compare your target vehicle and the benchmarked vehicles with your reference vehicle (i.e., your reference vehicle is the "datum") based on (a) customer needs, (b) vehicle attributes, and (c) vehicle systems. (See Tables 2.2 and 1.1 for the list of vehicle attributes and vehicle systems, respectively.)

The left-hand column of each Pugh diagram should include *all* items (one in each row) in the category (i.e., customer needs, vehicle attributes, or vehicle systems) for which the Pugh diagram is to be created. Using the reference vehicle characteristics as the "datum" for each row, evaluate each vehicle (benchmarked and target vehicle) by comparing the item (customer need, vehicle attribute, or vehicle system assigned to the row) and assign +, −, or S symbols in the columns for the benchmarked vehicles and the target vehicle.

h. Describe important product *design/development issues and challenges* in creating the target vehicle, including footprint-based fuel economy (in miles per gallon) and emissions (in grams per mile of equivalent CO_2) targets and safety-related changes.

Appendix II: Quality Function Deployment, Requirements Cascade, and Interface Analysis for a Selected Vehicle System

OBJECTIVES

1. To understand development of functional specifications through the application of quality function deployment (QFD) to a selected vehicle system
2. To understand automotive systems, subsystems, and interfaces between systems and their requirements
3. To cascade vehicle attributes' and subattributes' requirements to a vehicle system and its subsystem requirements
4. To understand coordination in system design tasks between different design and engineering activities and issues associated with trade-offs

PROCEDURE

1. *Select one of the following vehicle systems* of your reference vehicle for this project:
   a. Body closures system (doors, hood, trunk-lid or liftgate, hinges, latches, glass, and so forth)
   b. Rear suspension system (control arms, linkages, axles, wheel hub, wheels and tires, shock absorbers, cradle/frame, and so forth)
   c. Electrical system (alternator, battery, wiring harnesses, switches, relays, fuses, and so forth)
   d. Instrument panel system (cross-car beam, brackets, air registers, displays and controls, passenger air bag, glove box, and so forth)
   e. Fuel system (fuel tank, fuel module with fuel pump, fuel filter, fuel level sensor, pressure release valve, gas cap, fuel pipe, fuel lines, and so forth)
2. *Study the selected system* in the selected reference vehicle and list all subsystems of the system. *Prepare a decomposition tree for the selected system.* Your decomposition tree should include all subsystems, sub-subsystems, and major components of the selected system. Also, list all other vehicle systems (e.g., body system) that interface with the selected system.

3. *Prepare a QFD chart for a subsystem of the selected vehicle system* with the following details:
   a.   Determine the selected subsystem's customer needs (interview at least six customers to understand what they would like and dislike in the selected system and its selected subsystem). Provide descriptions of each of the customer needs.
   b.   Determine its functional specifications/requirements through discussions in your team (members from different functional areas will provide more complete information on design issues). Provide a description of each of the functional specifications.
   c.   Develop relationship and correlation matrices (using the QFD symbols to convey strengths).
   d.   Determine (estimate) importance ratings of the customer needs.
   e.   Evaluate your selected system (of the reference vehicle) along with the same systems in two other competitor products and plot their ratings for customer needs and functional specifications.
   f.   Provide targets for each of the functional specifications.
   g.   Determine relative importance rating scores (last row) of functional specifications. Determine the top few most important specifications of the chunk.
4. *Prepare a table listing of all attributes and major subattributes of the reference vehicle, and provide at least one major vehicle-level requirement for each of the major subattributes.* (Note: The first column should list all vehicle attributes. The second column should list all major subattributes of each attribute. The third column should list at least one vehicle-level requirement for each of the subattributes. Thus, each row of the table presents an attribute, a subattribute, and a requirement for each subattribute.)
5. *Cascade the subattribute requirements (provided in Column 3 of the table created in Step 4), provide at least one engineering design requirement for each applicable subattribute for the selected system*, and include them as the fourth column in the table prepared in Step 4.
6. *Provide at least one requirement for each of three major subsystems of the selected system for each applicable subattribute*, and add them to the table prepared in Step 4 as Columns 5, 6, and 7. (Note: Include only three major subsystems to limit the size of the table.)
7. *Develop an interface diagram for the selected vehicle system including all major subsystems of the system and other vehicle systems.* All interfaces between the subsystems of the selected system and other vehicle systems in the vehicle should be shown by arrows. Specify interface code(s) for each link in the interface diagram by placing the applicable letter code next to each arrow. Use the letter codes functional (F), physical (P), spatial-packaging space (S), energy exchange (E), material flow (M), and information flow (I).
8. *Develop an interface matrix including all of the subsystems (of the selected system) and other vehicle systems.* Specify characteristics (e.g., type of interface: functional [F], physical [P], spatial-packaging space [S], energy

exchange [E], material flow [M], information flow [I], or none [0]) of each of the interfaces corresponding to each cell in the matrix.

9. Provide (describe) *two interface requirements for each subsystem* of the selected system.
10. *Describe at least two major trade-offs* that you need to consider in designing the selected system to fit and work with other systems in the vehicle.
11. *Prepare a report* including all the items from Steps 2 through 10, and summarize *your observations and insights* gained from this project.
12. Your *written report* (file in MS Word or pdf format) should include
    a. A decomposition tree of the selected system
    b. Your QFD analysis, including your *completed QFD analysis chart*, pictures/sketches of the selected system (showing parts/features), *description of customer needs* and *definitions of functional specifications*, descriptions of findings/observations, discussion of *three to five most important functional specifications and overall findings*, and *conclusions*.
    c. A requirements cascade table (Excel file) showing vehicle attributes, their major subattributes, vehicle-level subattribute requirements, and cascaded requirements on applicable subattributes for the selected system and its three major subsystems.
    d. An interface diagram showing all major subsystems of the selected system and all other interfacing vehicle systems.
    e. An interface matrix (Excel file) showing all major subsystems of the selected system and all other interfacing vehicle systems (all as rows and columns) and types of interface for each cell of the matrix.
    f. Two interface requirements for each subsystem of the selected system.
    g. Descriptions of two major trade-offs observed in the design of the selected vehicle system.

Appendix III: Business Plan Development

OBJECTIVES

To create a business plan for development of your target vehicle
Contents of the report:

1. The business plan should include
   a. A description of the proposed (target) vehicle including vehicle features, options, and unique characteristics of its vehicle systems (1 page).
   b. Competitors (makes and models) of the proposed vehicle and comparisons of key dimensions and features (1 page).
   c. A description of its market segment (0.5 pages).
   d. Characteristics of anticipated customers (0.5 pages).
   e. Selling price and sales projections (1 page).
   f. Timing plan and gateways (1–2 pages) (see Figure 2.4, illustrating a timing chart, and Table 2.1 for definitions of gateways. Time should be indicated in months with respect to Job #1).
   g. Costs and revenue summary table and plots of curves of life-cycle costs and revenues (with assumptions related to hourly rates, interest, and inflation) for the vehicle program (2–3 pages).
   h. Benchmarking, changes, and risks: a benchmarking table comparing the specifications of the target vehicle with the reference and benchmarked vehicles. The table should include two additional columns: a "comments" column and a "risks" column. The comments column should briefly describe important changes to be included in the target vehicle, and the "risks" column should briefly describe any major risks to the vehicle program in implementing the changes (2–3 pages).
2. Conclusions and discussions
   a. Summarize major accomplishments and findings. Describe why your vehicle with its proposed characteristics will sell well (1 page).
   b. Discuss what worked well and what failed or did not get done to your satisfaction, and describe lessons learned and recommendations for future work (1 page).

Page limits on your P-3 report: The page limit requirements (shown in parentheses) are based on 8.5" × 11" page size with minimum 1" margins. Use 12 point font for the text and minimum 10 point font for tables and figures. Present information in tabular and/or graphic format (charts, plots, flow diagrams, etc.) where possible. Maximum page limit on the entire report = 14 pages (excluding your cover page and table of contents). *Pages over the above page limit will not be graded.* Illegible text, figures, tables, plots, and illustrations will not be graded.

Appendix IV: Conceptual Design of the Proposed Vehicle and Technology Plan

OBJECTIVES

1. Search for additional information from benchmarking and trends in design and technologies and refine your vehicle definition
2. Illustrate vehicle configuration and preliminary packaging with key interior and exterior dimensions and details of the vehicle concept
3. Present a technology plan for selected technologies and features

PROCEDURE

Here, you will assume that the business plan that you submitted in Appendix III was accepted by the senior company management. Now, to kick off your concept design process, you will need to gather your team and provide them with information on vehicle details, for example, overall vehicle characteristics, program plan, timings, milestones, tasks of major teams, responsibilities, and key open issues.

In your next design team meeting, you should present an initial drawing of the vehicle concept. This drawing should help the team to visualize the vehicle in terms of its overall size, package, and engineering issues and to start work in the design studios—to create initial sketches and computer-aided design (CAD) models and to develop exterior and interior surfaces.

You also need to start a technology plan to define new features in the vehicle.

Thus, your assignments for this project are

1. Prepare an initial drawing containing the *side view and plan view (drawn to scale*, either hand drawn or using a CAD package) of the vehicle with the following details:
   a. Overall vehicle envelope showing overall length, overall width, and overall height.
   b. Show locations of the four wheels and specify dimensions of wheelbase, front and rear overhangs, and front and rear tread widths.
   c. Show locations of cowl and deck points, engine envelope, firewall and back of rear seatback, vehicle floor and headliner height, gas pedal location, front and rear seating reference points, and center of the steering wheel.
   d. Show locations and envelopes of major entities such as gas tank, batteries, and powertrain.

2. Prepare an initial technology plan *using a tabular format*. Your table should *include all major vehicle systems* (as rows). The major vehicle systems should be serially listed with the serial numbers in the first column of the table and the name of the systems in the second column. The third column should describe major changes planned (one-line bullet-point descriptions). The fourth column should briefly describe major technological challenges. The fifth column should provide comments on key open issues (where additional developmental and analysis work is needed to better understand the issues and associated problems and trade-offs), such as possible modifications of existing hardware/software, recommended technologies, risks associated with implementing the technology, alternative solutions, future actions that need to be taken, make versus buy decision recommendations, and potential suppliers.

   To develop your technology plan, search for information on the latest advances and developments in the following areas:
   a. Powertrain technologies to meet the upcoming Environmental Protection Agency (EPA) and National Highway Traffic Safety Administration (NHTSA) fuel economy and emissions requirements
   b. Other fuel-saving technologies, such as low-friction bearings, low–rolling resistance tires, power regenerative methods, and stop-start
   c. Applications of new lightweight and other recyclable automotive materials
   d. Safety technologies for active and passive safety devices (e.g., driver warning systems, collision avoidance systems, and driver assistance systems)
   e. Telematics devices (applications of information, communications, computers, and wireless and global positioning system [GPS] technologies)
   f. Automotive electronics (applications of microprocessors, sensors, actuators, and integrations of electronic control units (ECUs)
   g. Electrical systems architecture (configuration of the electrical system)
   h. Driver interface technologies (e.g., steering wheel–mounted controls, touch screens, Bluetooth, programmable/reconfigurable controls and displays, display technologies, voice controls, and gesture controls)
   i. Vehicle lighting technologies (light-emitting diode [LED] lamps, fiber optics, smart headlamps, etc.)

Briefly summarize your recommendations in the fifth column of the table prepared to present your technology plan in Step 2.

Appendix V: Systems Engineering Management Plan and Vehicle Brochure

OBJECTIVES

1. Develop a systems engineering management plan (SEMP) for the development of your target vehicle
2. Provide lists of important vehicle characteristics and features for inclusion in the brochure for your target vehicle
3. Prepare a brochure for the target vehicle

CONTENTS OF THE REPORT

1. *SEMP*: Assume that your business plan was approved by the senior management and alternate conceptual designs are being developed by your vehicle design and development teams. Now, your major challenge is to ensure that the right vehicle (with the right combinations of levels of attributes and trade-offs between attributes) is developed within the planned program schedule and budget. To meet the challenge, you must prepare a SEMP and present it to your teams so that they understand the process (i.e., the steps they need to undertake and the tools and techniques they must apply during the vehicle development process). Thus, your next assignment is to develop a SEMP and document it in an easy-to-read format.

   Make sure that your report begins with the inclusion of the following *information on your target vehicle*: (1) make, (2) model, (3) vehicle body-style, (4) market segment, and (5) two competitor vehicles.

   Your SEMP must include all *important steps, analyses and evaluations needed to implement the systems engineering process* shown in Figures 2.2 and 2.3. Each step must include specific design/development task(s) to be addressed in the target vehicle program.

   Present your SEMP using a tabular format as follows:

   Column 1: Step number

   Column 2: Description of the step (tasks/work that must be completed) and timings (i.e., step initiation and completion times in months before [] or after [+] Job #1)

   Column 3: Analyses, tests, and evaluations to be performed, including methods/tools to be used, and design reviews to be conducted

   Column 4: Disciplines/departments responsible for the step, and any comments and/or additional details on the step

2. *Provide three lists of engineering details to help the company's marketing department in preparing a vehicle brochure* for the prospective customers.

    The marketing department wants you to organize the relevant vehicle details (such as values of important vehicle dimensions and capabilities of vehicle features) by preparing *three separate lists* as follows:

    a. *Three* attributes of the vehicle that will be most desired by its customers

    b. *Three* new and unique features that will create a "Wow" impression among its prospective customers (i.e., the customers have not seen such features in vehicles of the market segment of your target vehicle)

    c. *Five* features that in your opinion would be most desired by the customers (i.e., "Must Have" for their buying decision)

3. *Prepare a brochure* for the vehicle for prospective customers. The brochure should include (1) vehicle exterior and interior dimensions, (2) key selling points, standard and optional features/contents of the vehicle, and technical superiority–related considerations (e.g., major engineering accomplishments, comparisons with leading competitors showing why your vehicle is better than some of its key competitors), and (3) sketches and drawings to illustrate the capabilities of the vehicle.

Index

Printed in the United States
by Baker & Taylor Publisher Services